Lecture Notes in Computer Science 975

Edited by G. Goos, J. Hartmanis and J. van Leeuwen

Springer

Berlin
Heidelberg
New York
Barcelona
Budapest
Hong Kong
London
Milan
Paris
Tokyo

Will Moore   Wayne Luk (Eds.)

Field-Programmable Logic and Applications

5th International Workshop, FPL '95
Oxford, United Kingdom
August 29 - September 1, 1995
Proceedings

Series Editors

Gerhard Goos, Karlsruhe University, Germany

Juris Hartmanis, Cornell University, NY, USA

Jan van Leeuwen, Utrecht University, The Netherlands

Volume Editors

Will Moore
Wayne Luk
Department of Engineering Science, University of Oxford
Parks Road, Oxford OX1 3PJ, United Kingdom

Cataloging-in-Publication data applied for

Die Deutsche Bibliothek - CIP-Einheitsaufnahme

Field programmable logic and applications : 5th international workshop ; proceedings / FPL '95, Oxford, United Kingdom, August/September 1995. Will Moore ; Wayne Luk (ed.). - Berlin ; Heidelberg ; New York ; Barcelona ; Budapest ; Hong Kong ; London ; Milan ; Paris ; Tokyo : Springer, 1995
(Lecture notes in computer science ; Vol. 975)
ISBN 3-540-60294-1
NE: Moore, Will [Hrsg.]; FPL <5, 1995, Oxford>; GT

CR Subject Classification (1991): B.6-7, J.6

ISBN 3-540-60294-1 Springer-Verlag Berlin Heidelberg New York

Typesetting: Camera-ready by author
SPIN 10485537 06/3142 – 5 4 3 2 1 0 Printed on acid-free paper

Preface

This book contains the papers presented at the 5th International Workshop on Field-Programmable Logic and Applications (FPL '95), held at Jesus College, Oxford University, August 30th-September 1st 1995. The previous workshops were held in Oxford in 1991 and 1993, in Vienna in 1992, and in Prague in 1994. The proceedings of the 1992 and 1994 workshop are also available as volumes in the Lecture Notes in Computer Science series (LNCS 705 and LNCS 849), and the proceedings of the 1991 and 1993 workshops are available from Abingdon EE&CS Books, 49 Five Mile Drive, Oxford OX2 8HR.

We were again pleased to be offered a large number and a wide range of papers from which, with the help of the programme committee, we hope we have selected the best. We extend our thanks to the programme committee for their efforts, to the other members of the steering committee for their advice and of course to all the authors.

We gratefully acknowledge the sponsorship and administrative support of the University of Oxford's Department for Continuing Education and particularly the dedicated and enthusiastic support of Maureen Doherty, Jane Heath and Laura Duffy. We are also grateful for the support of the Department of Engineering Science, Oxford University and the Department of Computing, Imperial College of Science, Technology and Medicine, London.

Our thanks go to Springer-Verlag for publishing these proceedings.

Oxford, July 1995

Wayne Luk
Will Moore
(Co-Organisers)

Program Committee

Doug Amos, Altera Corp., UK
Jeffrey Arnold, IDA, SRC, USA
Peter Athanas, Virginia Poly. Institute and State University, USA
Erik Brunvand, University of Utah, USA
Barry Fagin, USAF Academy, USA
Patrick Foulk, Heriot-Watt University, UK
John Gray, Xilinx Development Corp., UK
Herbert Grünbacher, Vienna Technical University, Austria
Reiner Hartenstein, University of Kaiserslautern, Germany
Brad Hutchings, Brigham Young University, USA
Sinan Kaptanoglu, Actel Corp., USA
Wayne Luk, Imperial College, UK
Will Moore, University of Oxford, UK
Peter Noakes, Essex University, UK
Ian Page, University of Oxford, UK
Franco Pirri, University of Firenze, Italy
Jonathan Rose, University of Toronto, Canada
Michal Servit, Czech Technical University, Czech Republic
Zoran Salcic, Auckland University, NZ
Steve Trimberger, Xilinx Inc., USA

Steering Committee

Herbert Grünbacher, Vienna Technical University, Austria
Reiner Hartenstein, University of Kaiserslautern, Germany
Wayne Luk, Imperial College, UK
Will Moore, University of Oxford, UK
Michal Servit, Czech Technical University, Czech Republic

Table of Contents

The Design of a New FPGA Architecture *

Anthony Stansfield[1] and Ian Page[2]

[1] SRF/PACT 10 Priory Road, Bristol, BS8 1TU.
(Formerly of Oxford Parallel, Parks Road, Oxford OX1 3QD.)
[2] Group Leader, Hardware Compilation Research Group, Oxford University Computing Laboratory, Parks Road, Oxford OX1 3QD.

Abstract. A new Field Programmable Gate Array (FPGA) architecture is described. This architecture includes a number of novel features not found in currently available FPGAs. It is believed to offer a significantly improved logic density in some common applications.
This paper describes the development of the FPGA and looks at the mapping onto it of some interesting application circuit elements. The design is discussed in approximately chronological order which allows us to explain other options which were considered and rejected during the development process.

1 Introduction

In late 1992 Oxford Parallel set out to develop a classification scheme for FPGA[3] architectures, with the ultimate aim of identifying the most appropriate architectures for use with the hardware synthesis tools previously developed in the Oxford University Computing Laboratory [1]. As a result of this work it became apparent that many FPGAs available at the time were broadly similar to each other from the point of view of suitability for use as a synthesis target, and in particular they all had similar weaknesses. Further work produced a way of modifying an FPGA to improve its usability and efficiency in this kind of application, and therefore in mid 1993 work started on designing an FPGA to include these improvements. The design was completed in early 1995, and the first silicon became available in late June 1995.

* This work was undertaken by Oxford Parallel, which is part of Oxford University Computing Laboratory. The work was initially funded by the Department of Trade and Industry under the Parallel Applications Programme, with the support of Inmos Limited (now part of SGS-Thomson Microelectronics Limited). Since May 1994 it has been supported by ESPRIT project P8202 (CoCo).

[3] Throughout this document the term FPGA is often used to refer to all types of field-programmable logic, both 'Traditional FPGAs' (of the Xilinx/Actel/Altera type) and also those more commonly referred to as PALs, PLAs and CPLDs (Complex Programmable Logic Devices)

2 FPGA classification

Oxford Parallel's primary interest was in FPGAs as a hardware synthesis target, especially for use with the methods of synthesis from high-level programming languages developed within the Computing Laboratory[1]. Manual programming via schematic capture and manual placement or routing was of secondary interest.

The first stage of this work was to identify the kinds of hardware structures generated by the synthesis tools, followed by a consideration of how these structures could be implemented in an FPGA. This resulted in a list of the features that would ideally be present in an FPGA to be used with the synthesis tools. Finally a number of commercially available FPGA designs were compared with this list.

2.1 Hardware Generated by the Synthesis Process

The Computing Laboratory's synthesis approach is based on a direct translation of statements in the source program to hardware elements. The individual elements of the program are converted into groups of logic gates, and then the overall circuit is assembled from these groups in a manner which directly reflects the structure of the original source program. An outline of the mapping between program statements and hardware elements is given in Table 1.

Table 1. Mapping of Program Elements to Hardware Elements

BOOL a:	(variable declaration)	Create a register holding value of 'a'.
INT [128]b:	(array declaration)	Create memory array.
c := d	(assignment)	Register update
b[3]:	(array reference)	Access memory array
x + y	(addition)	Create a row of full adder cells
	(similar rules for other arithmetic operators).	
c = d	(comparison)	Bitwise XOR; OR gate to merge results.
x >> 2	(right shift)	Reorder the wires within a bus.
SEQ	(sequential composition)	Connect up control circuits in series.
PAR	(parallel composition)	Connect up control circuits in parallel.
WHILE...	(loop)	More complex control, with looping.
IF...	(conditional branch)	Pass control to either substatement.
CASE...		Pass control to particular substatement.

2.2 Implementation on an FPGA

The circuits referred to above can be implemented on an FPGA as follows:

- Variables map directly to registers.
- Arrays can be made using dense memory elements within the FPGA rather than by using large numbers of registers.
- The arithmetic and logical operators (add, subtract, bitwise AND, OR etc.) require 2- or 3-input gates. Note that individual gates can become functionally complex - the sum and carry paths of a full adder cell for example.
- Shifts by constant amounts can be handled in the routing, they need no logic gates.
- Comparisons (equal, not equal) are a two stage process - firstly an exclusive OR of the individual bits in a word with the pattern, and then an AND or OR gate to combine the individual results. The number of inputs to this second stage gate can be quite large (up to the length of a word: 8, 16, 32... bits)
- The program constructors (SEQ, PAR, WHILE, etc.) determine the flow of control within the circuit. They typically use one-hot, decentralised, encoding to minimise routing requirements.
- Conditional branches (IF, CASE) can be implemented as a series of independent comparisons, one for each branch of the IF or CASE. Alternatively, CASEs (and also nested IFs with closely related conditions) can be implemented with an n-input, 1-of-2^n decoder.

From this, the following list of useful features for an FPGA can be developed:

1. An efficient implementation of small, complex, logic gates. ie: up to around 3 or 4 inputs, functions such as Exclusive OR, Sum, and Carry.
2. An efficient implementation of large, simple logic gates. ie: more than 10 inputs, but mainly AND and OR functions.
3. Registers, to be used for the registers to hold variables, and a good clock routing scheme, so that the registers can be run at high speed.
4. Space-efficient (ie: dense) memory within the FPGA.
5. Flexible routing structures, for wiring up the control and data paths, and able to handle shifts/rotates etc. Ideally, there should be sufficient routing resources to give a high utilisation of the logic resources.

Since the main interest was in FPGAs coupled with logic synthesis and automatic place & route, a further item was also added to the list:

6 The FPGA should have a regular structure, with as few special cases as possible.

This idea is borrowed from RISC microprocessor and compiler projects, which have concluded that having a very regular instruction set makes it easier to write and maintain the compilers and optimisers. There is no reason to suppose that the same is not true for FPGAs.

2.3 Comparison with Existing FPGAs

Table 2 is a summary of how a number of FPGAs available in 1992/3 compare with this list of features[4].

Table 2. Suitability of FPGAs as a compiler target.

	Xilinx 3000	Xilinx 4000	Actel 1280	Algotronix	Concurrent Logic	PAL PLA	CPLD (multi-PLA)
Small gates	√	√	√	√	√	×	×
Large gates	×	(1)	×	×	×	√	√
Registers and Clocks	(2)	√	√	√	√	√	√
Memory	×	(3)	×	×	(4)	×	×
Interconnect	√	√	√	×	√	×	√
Regular architecture	√	(5)	√	√	×	(6)	(7)

1. Up to 9-input AND/OR gate in 1 cell (ignoring special edge decoders).
2. Not an enormous number of registers available.
3. LUTs can be used as small RAMs (5 bit address). Bigger RAMs still hard.
4. Has a large number of registers, so RAMs could perhaps be emulated.
5. Regular arrangement of cells, but the large cell has many (non-orthogonal) options, which could complicate software support.
6. Locally regular. PLAs large enough for complete systems not available.
7. Individual blocks are very regular, but there would be a discontinuity in the implementation of a circuit as it grew to be just larger than one block.

A number of conclusions followed from the construction of Table 2:

- None of the available devices met all the requirements (although the Xilinx 4000 family came close).
- The distribution of ticks and crosses in the table seems to indicate that most of the devices have similar good and bad points. In particular:
  - They all have provision for registers and good clock distribution.

[4] Note that a number of the companies listed in the table now have different names. For example, Algotronix is now part of Xilinx and Concurrent Logic is part of Atmel.

- None (with the possible exception of the Xilinx 4000) has good support for embedded memories.
- All are good at either small or large gates. None is good at both. Traditional FPGAs are good for small gates, while the PLA-based devices are good for large gates but not the smaller ones, especially for more complex functions. The Xilinx 4000 series has limited support for PLA-style logic in special peripheral units.

3 The New FPGA design

After the foregoing analysis, our assessment was that available FPGAs had weaknesses in the areas of on-chip memory and the ability to combine support for large and small gates. We then tackled the question of whether there was an easy way to improve performance in these areas.

3.1 A Heterogeneous FPGA

One way would be to construct a hybrid system. Given that PLAs are good for large gates (and complex inter-related functions), while traditional FPGAs are better for small, complex gates, a device containing both PLA-like and FPGA-like elements might provide the advantages of both. Unfortunately, this approach has a major disadvantage, which can be summarised in the following questions:

1. What should be the ratio of PLA-like to FPGA-like elements?
2. Is it best to have just a few, large blocks of each type, or alternatively more, smaller, blocks?

It is easy to produce cases to justify answers to (1) ranging from 90%:10% one way to 10%:90% the other, and also to justify both the large and small blocks options. There appears to be no single, universally good, pair of answers[5]. Because of this problem, the hybrid approach was not followed any further.

3.2 CAM-based FPGA

Having failed to find a good basis for a heterogeneous array, we became convinced that only a homogeneous solution would meet all necessary criteria. Knowing that PLA-style logic simply had to be supported we began looking at dense array styles. The key insight was enabled because of the background of one of the authors[6] in the design of CAMs for cache memory applications.

Our concept of a CAM-based FPGA was based on the observation that a Content-Addressable Memory (CAM) has a similar structure to the AND-plane

[5] This is hardly surprising, as the extremes represent the pure PLA and pure FPGA cases, and there is sufficient commercial demand to justify the existence of both types of device.

[6] Anthony Stansfield

of a PLA. It is also compatible with Static RAM (SRAM) technology used to make the LookUp Tables (LUTs) in FPGAs such as the Xilinx 3000/4000 and Altera FLEX devices. The basic idea is to replace the LUT SRAM in such a cell with a CAM, in order to create an FPGA cell that can be used either as a LUT or as a PLA element.

A basic 10 transistor, 1-bit, CAM cell is shown in Fig. 1 and has 4 vertical and 5 horizontal bus lines. It can be seen that this is comparable in complexity to a static RAM cell, but the additional i/o lines allow the content addressability.

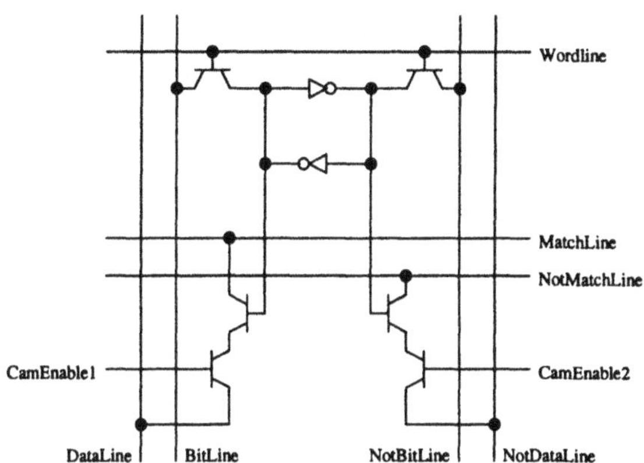

Fig. 1. The CAM cell

After some investigation, we determined that the best size of CAM array appears to be one with 16 bits, organised as a 4×4 array. This organisation is strongly indicated as preferable by a number of different factors. With this shape decided, we determined the following organisation:

- Used as a LUT, the array requires a 4-bit address, allowing it to generate any boolean function of up to 4 inputs[7].
- Used as a CAM, the array has 4 inputs and 4 outputs.
- Since both LUT and CAM (or PLA) operation use the same number of inputs they can share the same set of input buffers. The input options for the cell are therefore independent of its use as a LUT or a PLA.
- Since CAM operation has the same number of inputs and outputs it is easy to arrange to connect the outputs of one cell to the inputs of another in

[7] A number of studies([2][3]) have concluded that the optimal number of inputs for a LUT cell is around 3 or 4. These results might be slightly undermined since they assume that large gates are built from multiple LUTs. This is not the case with a CAM-based cell, where large AND/OR gates are made using the PLA-like features of the cell.

order to make the kind of dual-plane structures used in the classic AND-plane/OR-plane PLA.

Figure 2 is a simplified block diagram of the cell. It shows the following additional features of the cell:

- LUT and CAM operation share the same output buffers. Therefore the LUT can generate up to 4 outputs. This is used to tap some intermediate signals in the LUT column decoder, so that it can generate either 1 function of 4 inputs or 2 functions of the same 3 inputs[8].
- The CAM inputs and outputs are orthogonal - if the inputs are vertical then the outputs will be horizontal, and vice versa.
- Circuits are provided to link the matchlines (and/or datalines) in adjacent cells together. This makes it possible to create PLA-like arrays with more than 4 inputs or outputs.
- The links between adjacent cells are also available to be used as a fast local interconnect when the cell is used as a LUT (for carry paths etc.)
- The cells are also connected to a network of longer wires which link groups of cells in the same row or column.

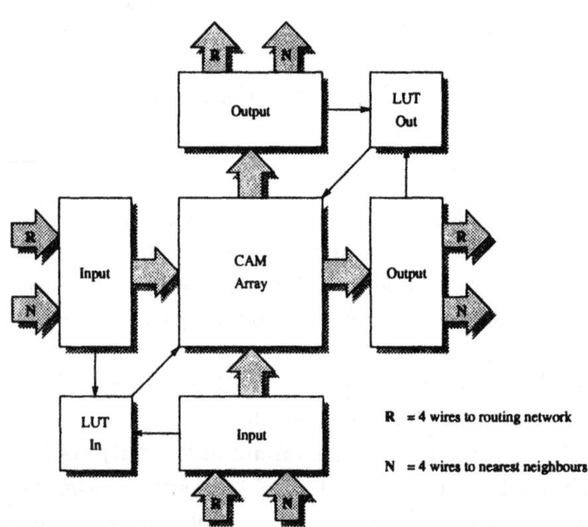

Fig. 2. Simplified block diagram of the new FPGA cell

This CAM-based FPGA cell appears to offer the advantages of the hybrid FPGA discussed earlier, but because it has only one type of cell it is not necessary

[8] for example, Sum and Carry, so that a full adder fits into one FPGA cell

to decide in advance on an appropriate mix of the two styles. It also has a number of other advantages:

- The CAM operation of the cell provides a fast, flexible, way of making connections between its horizontal and vertical inputs and outputs, and so of linking the horizontal and vertical routing wires. In other words, the 'logic' cell can also be used as a routing switch. This means that an appropriate mix of logic and routing can be chosen for each application, rather than having to be fixed in the architecture.
- The cell can easily be modified so that the CAM array be used as a small SRAM. This allows embedded memories be added to an application. Cells can easily be linked to create larger memory arrays. The row decoder needed for such an array is created using other FPGA cells operating in PLA mode.

The net result of all the above considerations is a single basic cell which can be used in any of the following modes (each of which are explored in more detail in Section 5:

1. a lookup table with up to 4 inputs,
2. a 4-input, 4-output expandable PLA element,
3. 16 bits of expandable SRAM,
4. 16 bits of expandable CAM,
5. a routing switch.

This kind of cell design seemed to be much better matched to the list of features required for our high-level synthesis approach than previous FPGAs and the other architectures considered. Consequently, in 1993 work began on a detailed design for an FPGA based on such a cell.

4 Lessons from the FPGA Hardware Design Process

4.1 Design Methodology

The major features of the design methodology which was followed for the FPGA chip were :

- Design entry using schematic capture, and all layout done manually (ie: no layout synthesis). All layout verified against the schematics.
- Analogue simulation models automatically derived from the schematics and used to verify the behaviour of the FPGA cell.
- A VHDL model (automatically derived from the schematics) used for simulation of both the individual FPGA cell and arrays of cells.
- An independent model of the cell written using a programming language, and used to generate test patterns for the VHDL and analogue simulations. The design only to be regarded as complete when all 3 models give the same results for the same set of inputs.
- The cell to use a mixture of CMOS and NMOS design styles, for compactness, but to be fully functional with a 3V supply.

Having two independent models of the cell was intended to provide a check on the interpretation of the cell specification. Also, writing a software model proved to be useful experience for the software development process.

4.2 Hardware/Software Development

In parallel with the hardware design another group began work on the support software (synthesis, optimisation, placement and routing). Before they started there was a worry that the cell was too flexible, and that it would be impossible to write software to fully exploit it[9]. This fear turned out to be unfounded, as the software group was able to find a viable solution to the problem. A detailed description of the software is outside the scope of this paper, but there are a number of places in which the hardware and software design processes have influenced each other:

- As mentioned above, the LUT column decoder is tapped in several places to give 4 possible output combinations. These were chosen based on the functions likely to be generated by the algorithm used for mapping onto LUT cells.
- All cell outputs are fully buffered (there are no pass transistors connected to the routing wires). This makes it easy to estimate delays, so that the software can easily identify (and hopefully fix) speed problems.

4.3 Registers

During the design process, a way of merging the input and output buffers of the cell was found. Furthermore, it was possible to arrange the compound circuit to contain a loop. This loop was used as the basis of a flexible register circuit, which makes it possible to register any output of any FPGA cell.

4.4 Problems

There was one area of the cell which caused major problems, both in its electrical behaviour and in the VHDL simulation. This was an attempt to create a bidirectional, analogue bus running through the middle of the cell, in the belief that it would be smaller than separate unidirectional buses. It took several weeks to get this to simulate reliably in VHDL, and the first identified hardware bug was due to this circuit. In the next revision it will be replaced with unidirectional busses.

[9] For example, with the same cell for both logic and routing, placement and routing are aspects of the same task, and so a simple 'place, then route' approach is inefficient

5 Applications

Here we examine some simple examples of kernel application circuits that can be mapped onto the CAM-based FPGA. Each of these applications is important in the context of high-level synthesis as determined by our initial analysis. In advance of the chip and its software tools being available we have only placed and routed relatively simple circuits by labourious manual methods. However these are some of the circuits that we wish to support well and it is crucial that they map densely onto the FPGA.

The cell uses an array of 4×4 one-bit CAM circuits, directly abutted to each other, to build the core of a CAM cell as shown in Figure 3. With the addition of the peripheral circuitry outlined in Figure 2 this forms the repeat unit of the CAM-based FPGA.

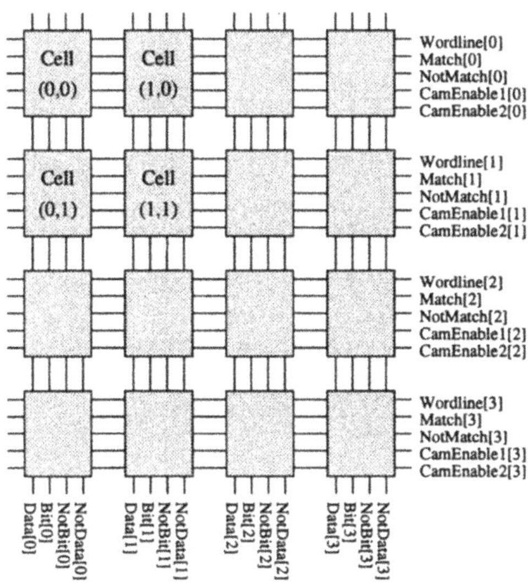

Fig. 3. The 4×4 CAM Cell Core

The fact that there are four output wires from each cell contrasts markedly with the single output of a conventional LUT. It allows much greater density of user circuitry in the array for those fragments of the circuit which can exploit it. These fragments are precisely the ones which are poorly supported by LUT-based architectures. In addition, the choice of a 4×4 cell gives a natural and highly useful input/output symmetry which results in simpler hardware which can be packed more densely and can be supported by simpler software.

5.1 The FPGA Block in LUT/RAM Modes.

Each cell can operate as a fully functional 16-bit LUT. The LUT mode is similar to conventional FPGA architectures except that it can be used to generate one function of four variables, or (the expected) two functions of three variables, four functions of two variables etc. Since the cell is designed to support four local links in each direction, the routing resources to support each of these different organisations is already in place.

The CAM-based FPGA also supports the construction of dense RAMs of arbitrary size in a way which we believe is superior to any competitive architecture. Fig. 4 shows how the CAM-based FPGA cells, operating in RAM mode, can be grouped to form a larger RAM structure. Again it is the dense local interconnection network that makes this organisation possible.

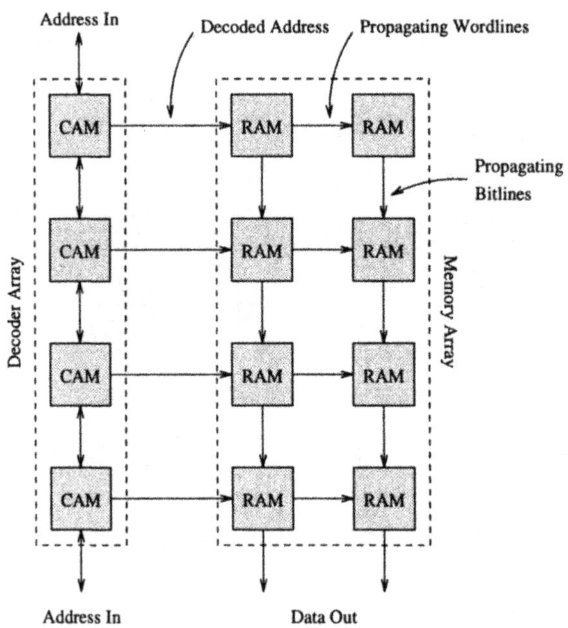

Fig. 4. A RAM structure in CAM FPGA

As can be seen in this figure, the CAM-based FPGA solves, rather neatly, the twin problems of building a dense RAM array and building the necessary decoder logic for the array. The decoder cells simply use the PLA mode with four horizontal outputs per cell, and the RAM cells use the RAM mode. Again there are no wasted cells in the core or periphery of this larger RAM. For larger RAM arrays the user can choose to use exhaustive decoding which results in the same rectangular structure, or can use a decoder tree which will release a few of the cells within the overall bounding box.

5.2 Using Cells in PLA Mode.

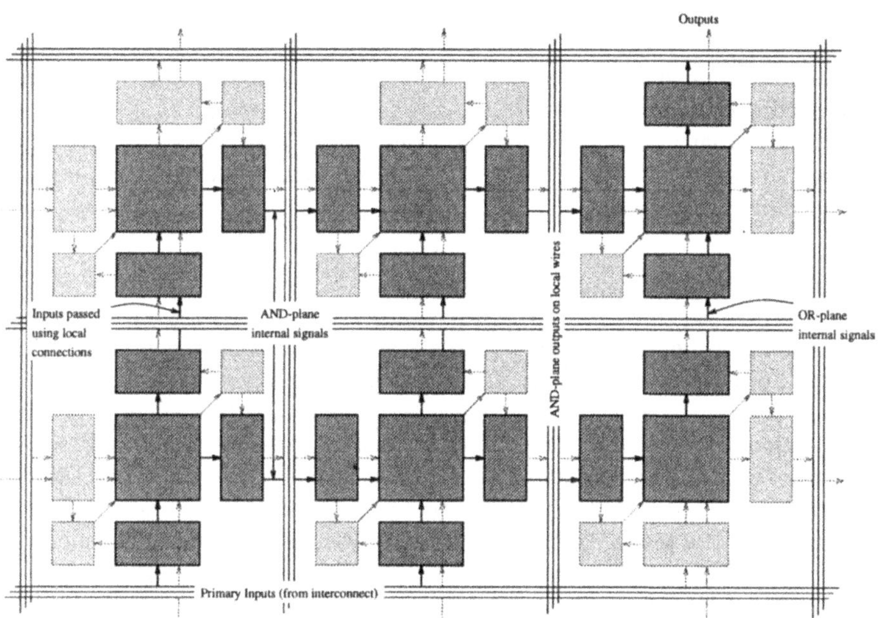

Fig. 5. A Larger PLA built from CAM-Mode cells

By design, the CAM cell is an ideal unit for building AND-OR planes. These cells, operating in PLA mode, can be grouped to form arbitrarily large PLA structures as shown in Fig. 5. When forming part of either the AND plane or the OR plane, each cell contributes sixteen programmable pulldown transistors at the intersection of the four input and four output wires.

For the AND plane, the four bit input pattern is propagated vertically (say) through the cell, with each bit in either normal or inverted form. The horizontal matchline will be high for any row which has a perfect match between the stored pattern and the input pattern.

Note that there are no wasted cells in the core of the PLA or around its periphery. Of course it is possible to apply standard PLA folding techniques to optimise the circuit, which might result in a non-rectangular PLA.

5.3 Using Cells to Build Larger CAMs.

The cells can also be used efficiently to build CAM arrays of arbitrary size. Although this was not an initial design requirement, it is nevertheless a natural consequence of using a CAM for the repeat cell. Having thought about the consequences of this, we are becoming convinced that CAM structures will themselves become an increasingly important building block in future applications.

As an example, we have recently implemented the Single Cube Containment (SCC) algorithm for combinational hardware optimisation using an actual CAM chip. The use of the CAM dramatically improves the performance of the algorithm and allows it to run in linear time. We have looked at a number of actual implementations of this algorithm in available CAD packages and they have all used a quadratic algorithm.

We believe that CAMs will find other uses in applications where previously they had not been considered as part of the solution space and we plan to continue studying this topic.

5.4 Using Cells as Routing Switches.

There is little to be said about this mode as the cell simply acts as a fully populated 16×16 crossbar switch. This is in fact not a new mode for the cell but rather a special case of its use as a PLA. This means that there is in fact more functionality available than a straightforward crossbar switch and some simple gates can be incorporated into the routing. These gates can be used to combine signals passing through a cell in the same way that signals are combined within a PLA. The major advantage of this functionality is that multiplexors can sometimes be implemented without using additional cells.

Naturally, these routing switches can be expanded to make larger routing structures merely by concatenating adjacent cells.

6 Combining Processor and FPGA on the Same Chip

All the evidence of the last four year's work at Oxford has indicated that there is a significant role to be played by FPGAs used as flexible co-processors interfaced to conventional microprocessors. We have implemented a number of applications on a microprocessor plus FPGA pair[4]. These have ranged from real-time video object tracking to spell checking and we have achieved speed-up factors of 6 to 30 times over a microprocessor alone. The novelty of our approach is that we take a purely programming approach to the construction of hardware/software systems. There was *no* user interaction with the design tools except via the high-level program. All optimisation, mapping, and place & route stages were completely automatic making the whole flow a 'single button-press' operation.

If our predictions about the relevance of FPGA co-processors prove correct, there will naturally follow a commodity market for an FPGA and processor on the same chip. This will obviously yield a reduction in the number of parts in a typical system implementation, and it will also speed up communication between the two processors, and may also simplify the communication and clocking arrangements. One of the authors (Ian Page) is actively engaged in designing FPGA co-processor systems and building the necessary software infra-structure so that the resulting hardware/software systems can be specified by a single, largely behavioural, specification. We expect that the CAM-based FPGA will have a significant role to play in these developments since it was designed from scratch with the needs of this style of high-level synthesis firmly in mind.

7 Conclusions

We have presented a novel FPGA architecture which we believe offers a significant advance on current FPGA implementations with regard to its ability to support the full range of sub-circuit organisations found in discrete logic applications. This is particularly true of those designs which are implemented by high-level synthesis, which we believe will become increasingly prevalent.

We have shown that FPGA co-processor combinations can usefully support applications and that entire hardware/software applications can be produced in actual practice via a purely programming approach. The development of the software to support this implementation route, the refinement of FPGA architectures, and the integration of conventional microprocessor cores and FPGAs are three of the most interesting areas for the future development of FPGAs.

8 Acknowledgements.

The authors would like to thank Jonathan Saul for his comments on this paper and for the implementation of the SCC problem referred to. In addition to the agencies acknowledged in the introduction to this paper, the work which underlies this paper has been supported over the last few years at Oxford by (in alphabetical order) Advanced Risc Machines Limited, Atmel Corporation, EPSRC (U.K. Engineering and Physical Sciences Research Council), European Union (Esprit/OMI programme), Hewlett Packard, Music Semiconductor, Oxford University Computing Laboratory, Sharp European Laboratories, and Xilinx Development Corporation, and others who wish to remain unacknowledged.

References

1. Ian Page and Wayne Luk. "Compiling occam into Field-Programmable Gate Arrays" in *"FPGAs" W.R. Moore, W. Luk (eds), Abingdon EE&CS Books, 1991*
2. Brown, Francis, Rose and Vranesic. *"Field Programmable Gate Arrays", Kluwer Academic Publishers*
3. Singh, Rose, Chow and Lewis. "The effects of Logic Block Architecture on FPGA Performance" in *IEEE Journal of Solid State Circuits, Vol.27 No.3 (March 1992)*
4. Lawrence et. al.. "Using Reconfigurable Hardware To Speed Up Product Development And Performance" in *this volume*

Migration of a Dual Granularity Globally Interconnected PLD Architecture to a 0.5μ TLM Process

J. Turner, R. Cliff, W. Leong, C. McClintock, N. Ngo, K. Nguyen, C.K. Sung, B. Wang, J. Watson
Altera Corporation, 2610 Orchard Parkway, San Jose, California, 95134

Abstract. A global interconnect architecture with dual granularity demonstrates considerable migration capability from the original product on 0.8μ two layer metal process by reducing die size to one third while nearly doubling system frequency when transferred to a 0.5μ three layer metal process.

Introduction

The FLEX8000 architecture, consisting of six family members, has a density range of 2,500 to 16,000 usable gates. It is a dual grained hierarchical architecture with Logic Array Blocks (LABs) of eight four-input logic elements connected internally using local interconnect resources. The LABs, arranged in rows, use global horizontal lines that span the entire row for interconnect. Connections between rows of LABs are made through vertical lines which may connect indirectly to LABs via the horizontal lines. Both horizontal and vertical lines span the entire distance of the LAB array making up a global interconnect matrix. These global lines are referred to as FastTrack interconnect.

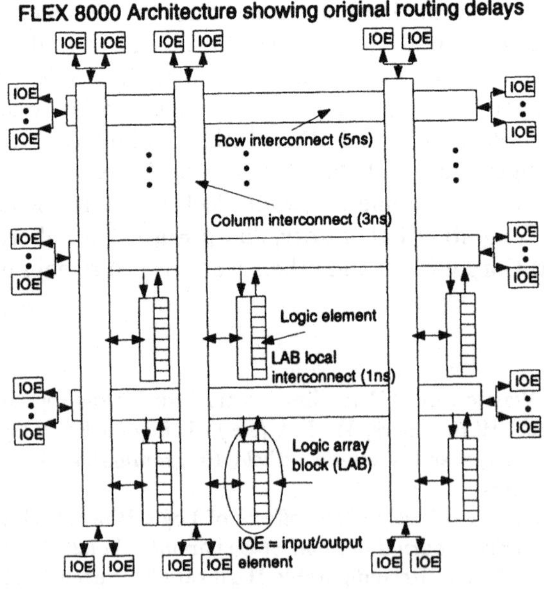

Figure 1

The logic elements can drive both vertical and horizontal FastTracks (see figures 1 and 2). This approach gives fast predictable interconnect. Because of the very high performance and small layout area of the local LAB interconnect, the architecture aims at maximizing the number of nets driven locally. The LAB contains cascade and carry chains to support high performance arithmetic, counter and wide input functions. Signals that route to other LABs or communicate off-chip are given maximum routing flexibility by using the global FastTrack lines. [1]

Logic Element Block Diagram

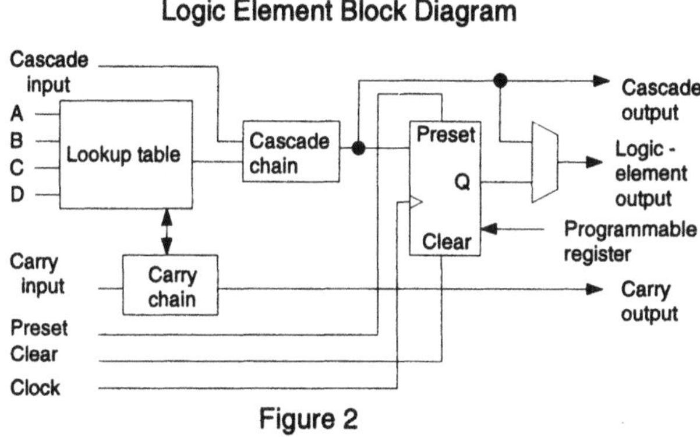

Figure 2

1 Design Considerations

1.1 Scaling of Design Rules

With advances in metalization technology through better planarization techniques and filled contact/via it is now practical to fabricate integrated circuits containing more layers of metal. These advances have allowed metal interconnect design rules to progress faster than those for base layer transistor formation. As a result interconnect rich structures can now be implemented with greater ease and efficiency.

1.2 Transistor Optimization

Transistor geometry's were optimized to account for changes in interconnect RC and gate loading due to diffusion and gate oxide. Since the relative parasitic loading differed in the new design reoptimization of transistor gate drive was done with Spice simulation. Each critical net was evaluated and adjusted individually, depending on the layout parasitic introduced with fine pitch geometry's and three layers of interconnect metal.

In addition to die area reduction, two other objectives were to be satisfied for the successful migration of the FLEX family to the 0.5μ process technology. These relate to the 0.5μ transistors electrical specifications. First, the transistors must provide reduced gate delays to support higher clock rate applications. Also, the 0.5μ family of devices must be a replacement for the earlier 0.8 micron products and serve as a substitute in existing customer applications. Power supply compatibility is essential for replacement purposes, placing some constraints on the 0.5μ transistor in order to tolerate the five volt power supply.

1.3 Change In Delay Model Parasitics

With all of these above changes, the factors affecting speed performance within the device have dramatically changed. For example, paths that may have been capacitively dominated by metal capacitance could now be dominated by diffusion capacitance, etc.

To take advantage of these technology trends, an architecture must be effective at utilizing the tight metal pitches on three layers of metal without sacrificing the W/L of transistors to keep pace with these metal area reductions. Or to put it another way, interconnect has become a lot cheaper with respect to transistors or muxes that connect it together.

2 FLEX on 0.8μ Technology

Since the FastTrack interconnect is global, it is very metal intensive, in fact, each row contains a cumulative wire length of 72 inches. This results in the row interconnect regions being dominated by metal 1 and metal 2. With metal real estate being at a premium, it was necessary to route the horizontal lines in metal 1 which has the most parasitic capacitance and greater resistance than the top metal. This was because the vertical FastTracks, LAB input lines and programming signals consumed the remaining interconnect real estate. To a lesser extent, the logic element and input muxes in the LAB were also metal bound. The configuration RAM address lines, power bussing and localized interconnect cause the transistors to be spaced further apart than otherwise. Internal nodes of the logic element, particularly the LUT are sensitive to the internal loading and so this had some impact on performance.

3 Migration To a Three Layer Metal 0.5μ Process

3.1 Logic Element and LABs

Migration allowed the logic element size to be reduced by freeing up area used for localized interconnect whilst still maintaining transistor strengths. Floor planning of the signal flow and power distribution was redefined. Signals that pass through the block, for instance, can be routed in higher levels of metal and are critical to skew clock signals.

Reoptimization of transistor drive strength resulted in halving the combinatorial logic element delay. This combined with reduced carry chain parasitics has improved the 16 bit counter frequency to 130MHz.

3.2 Global Interconnect

With the migration to 0.5μ, the size of a LAB and hence the internal loading of the LAB has been reduced. This means that smaller mux transistors are possible for signals entering the LAB. The FastTrack capacitance is a function of the metal area with fringing capacitance and the diffusion loading of all muxes (most of them turned off) on the line. Smaller mux transistor place less diffusion capacitance on the global lines.

The horizontal FastTracks are now implemented in metal 3 which have less parasitic capacitance per unit length (in only having capacitance to layers below) and in being twice the thickness of the metal 1 layer used in the prior design, have reduced resistance. With shorter, lower capacitance and less resistive metal lines, the horizontal FastTrack delay has been reduced from 5.0ns to 2.2ns. A further consequence of this is a 30% typical skew reduction. Global line skew is a function of both metal resistance and capacitance. With this improvement, the overall speed performance of the architecture is more predictable than before.

At the time of definition, the FLEX architecture provided a very competitive die area with the 0.8μ two layer metal process. It now has taken advantage of standard logic process technology prevalent today for use in gate array and microprocessor products to yield a higher performing low cost programmable logic device.

4 Device Performance

4.1 Performance Benchmarks

Industry standard PREP benchmarks were run analyzing the performance of the 0.8μ and 0.5μ devices. Average Benchmark Speed (ABS) is a measure of the average speed of the nine PREP benchmark circuits. A 75 percent improvement in benchmark performance was realized with the migration to 0.5μ technology. Comparison of the ABS is as follows:

	0.8μ	0.5μ
Average Benchmark Speed	43MHz	75 MHz

4.2 Design Examples

The speed up is also seen in real design examples. Listed below are performance values for several customer designs originally implemented on the 0.8μ and then on the 0.6μ device. [2]

	0.8μ	0.5μ
Design A	12.1MHz	21.5MHz
Design B	9.5MHz	18.9MHz
Design C	46.1MHz	72.5MHz

4.3 Delay Across Chip

One feature of this architecture is the predictability in timing, particularly of the interconnect. Reduction of interconnect parasitic delay has improved further the timing predictability. Placement and fanout timing dependency has been improved.

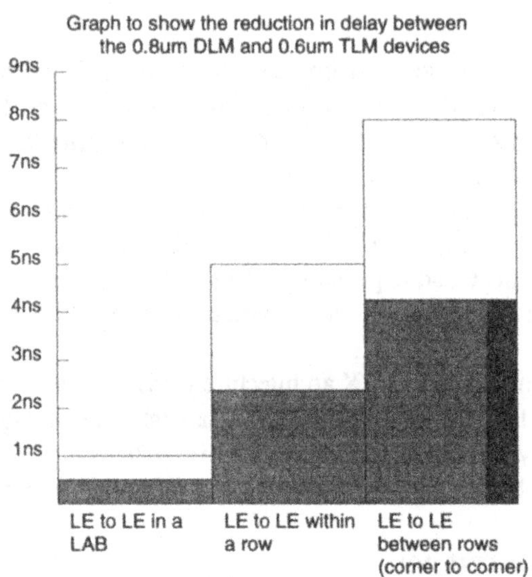

The above figure shows the comparison in delays, within a LAB, along a row and between rows such as from one corner of the chip to the opposite. The shaded area represents the 0.5μ TLM timings. The non-shaded shows the fastest 0.8μ DLM product speed grade timings.

4.4 Die Size

In migrating the EPF81188 from 0.8μ DLM to 0.5μ TLM the die size has been reduced to 33% of the original.

5. Conclusion

The FLEX architecture is well suited to take advantage of modern process capabilities. In migrating this product to 0.5μ process with three layers of metal, a replacement compatible family has been created with considerable cost reduction and speed up over the original family on 0.8μ technology.

The family is offered in a variety of PLCC, QFP, PGA and BGA packages. The largest part is the EPF81500 which has 1296 logic elements and 1500 registers totaling 16,000 usable gates. It has 156 user I/O, each with a peripheral register.

FLEX, FastTrack, EPF81188 are trademarks of Altera Corporation

References

[1] ALTERA FLEX 8000 Handbook 1994
[2] FLEX 8000 Programmable Logic Device Family Datasheet August 1994 v 4

Self-Timed FPGA Systems

Rob Payne
rep@dcs.ed.ac.uk

Department of Computer Science, the University of Edinburgh

Abstract. Recently, there has been a renewal of interest in self-timed systems, due to their modularity, robustness, low-power consumption and average-case performance. Additionally, this paper argues that there are specific benefits to adopting self-timed design for FPGAs. The mapping problems of placement, routing and partitioning are simplified by not having a global clock constraint to meet, so more mappings are available for mapping algorithms to choose from. Hence, there is greater potential for algorithms to improve utilisation and performance of a design, or instead, to increase design turn-around by taking less time to produce a mapping. Furthermore, the ability to perform mappings quickly enables new FPGA applications where the mapping to the FPGA is done on-the-fly. However, currently available FPGAs provide no support for self-timed design. The latter half of the paper describes the STACC architecture, an FPGA architecture targeted at the implementation of self-timed bundled-data systems.

1 Introduction

All current commercial FPGAs are designed for implementing systems synchronously, so have dedicated clock signals. However, the alternative approach of building systems in an asynchronous or self-timed fashion has many potential benefits. This paper examines the case for self-timed FPGA systems and introduces STACC, a dedicated self-timed FPGA architecture.

Section 2 introduces self-timed systems and examines the general advantages of building systems asynchronously. In section 3, the specific benefits of building self-timed FPGA-based systems are considered. Section 4 reviews the current work on self-timed FPGAs and argues that current FPGA architectures do not support self-timed designs. In section 5, the STACC architecture is introduced, concentrating on the development of a timing cell for the architecture, from an unreconfigurable structures to a fully reconfigurable timing cell. Finally, section 6 summarises the paper and looks at future directions for the work.

2 Self-Timed Systems

In *synchronous systems* (Fig.1, left), all modules are synchronised through a global clock signal. The clock places a global constraint on the system: all the modules within the system must have their data ready by the next tick of the clock. In *asynchronous systems* there is no global synchronisation of the modules within the system, but modules do synchronise locally through their communication protocols. Asynchronous communication protocols that use some form of handshake between sender and receiver (Fig.1, right) are known as *self-timed* (for a formal definition see [1]).

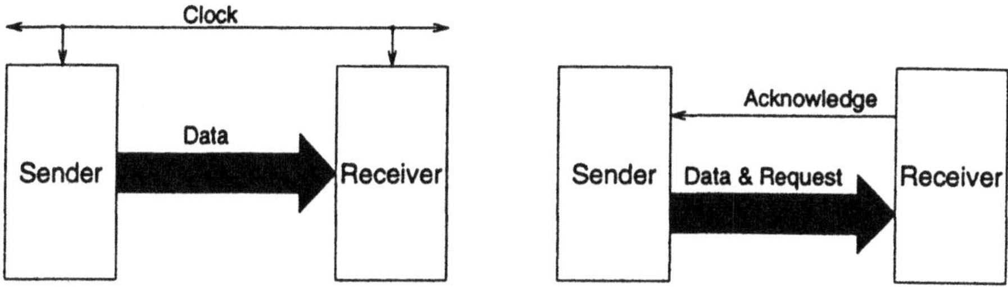

Fig. 1. Synchronous and self-timed protocols

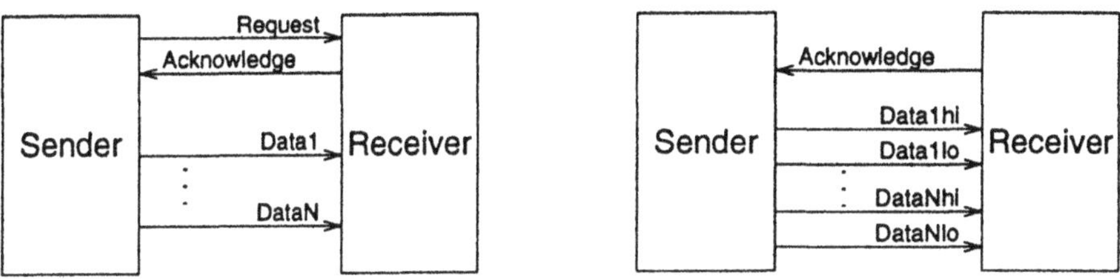

Fig. 2. Bundled-data and dual-rail encoded protocols.

Figure 2 shows two common self-timed communication protocols. Both protocols come in *two-phase* and *four-phase* variants that attach different significance to transitions within the request/acknowledge handshake. In two-phase or *event-based* signalling, all transitions are significant. In four-phase signalling, only positive-going transitions are significant, so an idle *return-to-zero* transition is required.
In the *bundled-data* protocol (Fig.2, left), a transition on the request signal indicates that the data is valid. The requirement that the request transition occurs after the data is stable is known as the *bundling con-*

straint. Sutherland's Micropipelines [2] popularised a 2-phase form of the bundled data protocol. Subsequently, Micropipelines were adopted by the AMULET group [3], at the University of Manchester, who built a self-timed version of the ARM processor.

In the *dual-rail* protocol (Fig.2, right), a pair of signals are used to send each data bit. One signal indicates the data bit is high, the other that the data bit is low. There is no explicit request signal, instead, a request is implied when a transition has occured on each pair of data signals. The dual-rail-code has been used by Martin [4] to build an experimental self-timed processor.

An important property of self-timed protocols are that they are *speed independent*: the protocols work regardless of the speed that the modules produce results. In addition, protocols such as the dual-rail code are *delay-insensitive*: arbitrary delays may occur on the wiring and the protocol will still work. The bundled-data protocol is not delay-insensitive due to the bundling constraint.

The removal of the global clock constraint in self-timed systems leads to several advantages over their synchronous counterparts. These are summarised below:

Modularity: In synchronous systems, all parts of the system are implicitly dependent on each other through the global clock signal. In self-timed systems, no global clock constraint has to be met, so any module with the required functionality can be used regardless of performance and the system will still work.

Robustness: Because self-timed systems are speed-independent, they are resilient to delays caused within modules by environmental conditions. Delay-insensitive systems are also resilient to arbitrary delays in wiring as well.

Average-Case Performance: The clock period in a synchronous system must be slower than the worst worst-case delay of all the modules within the system. In contrast, since self-timed systems are not limited by a global clock, they can go at their own speed, so tend to exhibit average-case performance rather than worse-case.

Low Power Consumption: For CMOS, the static power consumption is almost zero. However, in a synchronous system, transitions on the global clock are always causing transitions to be passing through the system causing power dissipation, even when the system is idle. Within a self-timed system, transitions and hence power dissipation only occur when data is passing through the system.

No Clock Distribution Problems: Problems such as clock skew are avoided since there is no global clock in a self-timed system.

The main drawback to self-timed systems is that extra circuitry is required to generate the local timing information. The overhead depends crucially on the protocol chosen. Delay-insensitive protocols, such as dual-rail code, generally need two wires per data bit, which represents a significant overhead. Bundled-data protocols only have the overhead of the acknowledge and request signals so the overhead is smaller and can be negligible for large data bundles. This is at the expense of requiring more careful design to ensure the bundling constraint is met. In both cases, the 4-phase variants of the protocols need less circuitry than the 2-phase equivalents.

3 Self-Timed FPGA-based Systems

In the previous section, the general advantages of self-timed systems were considered. However, there are also specific advantages to applying a self-timed approach to FPGA-based systems. By removing the global clock constraint, the mapping problems of partitioning, routing and placement are simplified. Since only the local constraints of the self-timed protocol have to be met, many more mappings of a circuit are possible. Hence, there is greater scope for mapping algorithms to search for mappings with improved utilisation or performance, or simply to find a mapping in less time. Below, the main implications of more flexible mapping to the FPGA are outlined:

Ease of Partitioning: The difference between off-chip and on-chip delays causes problems in any design but these problems are particularly acute in FPGAs where partitioning is more frequent as less functionality can fit on one chip. Using self-timed protocols, off-chip delays only limit performance when the signals between partitions are being used. A design can treat a array of FPGAs as a uniform array of cells since the self-timed protocols accommodate for the off-chip delays. Hence partitioning algorithms are only required to improve performance rather than to achieve a mapping that meets the global clock constraint.

Cope with Saturated Routing: In many designs with high utilisation, the interconnect can become saturated, causing many signals to be placed on long snaking paths through the FPGA. This can severely limit clock rate in synchronous designs. In a self-timed design, these paths can be allocated to infrequently used signals, which will only limit performance when the signal is being used.

Faster Mapping and Quicker Design Turn-around: In synchronous systems, detailed timing analysis of all signal paths is needed to ensure that the result of routing, placement and partitioning of a design meets the global clock constraint. Because of the insensitivity to delays of self-timed designs, any route, place and partition that implements the system net-list will produce a correctly functioning system, as long as it meets the local constraints of the self-timed protocol. Detailed timing analysis is only required to improve the performance of the mapping rather than to ensure a working system. Initial mappings may be done quickly, enabling faster design turn-around. Detailed timing analysis for improved performance can be reserved for the mapping of the final design.

Mapping on the Fly Enables New Applications: Fast mapping enables systems where the mapping to the FPGA is not fixed before run-time as in current FPGA systems. One way to exploit this advantage is to generate custom circuits for the problem in hand, rather than having a very general-purpose circuit. For example, custom pipelines for a specific processing task could be built from a selection of pre-designed generic modules and compiled together quickly to produce a circuit of the required data width and function.

The ability to map circuits quickly to the FPGA at run-time seems essential for proposed *virtual hardware systems* [5]. Virtual hardware systems are analogous to virtual memory: they try to emulate a larger circuits by swapping sub-circuits to and from configurable hardware. Unless the pattern of swapping is known before run-time, as in current virtual hardware applications (such as the neural-net simulations in [6]), the mapping has to be done on-the-fly, which requires mapping algorithms to the FPGA that are fast and robust.

4 Current Research on Self-Timed FPGAs

Most of the current work on self-timed FPGA systems has concentrated on building such systems using commercially available FPGAs. Brunvand [7] composed a library of self-timed data-bundled elements for the Actel FPGA, and subsequently built a processor with these elements [8]. Oldfield and Kappler [9] compared self-timed FIFOs implemented on CAL chips and on customised silicon. Shaw and Milne [10] implemented asynchronous circuits on the CAL-based SPACE machine. Other researchers have also built Micropipeline libraries [11, 12].

The advantage of using current FPGAs for implementing self-timed systems is that the chips are readily available standard parts, and can implement synchronous systems as well. However, whilst showing the potential of self-timed FPGA systems, these works have highlighted some limitations of current synchronously-oriented FPGAs for implementing self-timed systems. These limitations are listed below:

Arbitration: Arbitration and synchronisation are common functions in self-timed circuits. The arbiter elements used by Brunvand [7] in his self-timed library for Actel Chips have a small chance of failure. Arbiters are the only function that cannot be built using current FPGAs.

Local Register Clock Distribution: In bundled-data protocols, although there is no global clock signal, there are still local clock signals to control registers. For good performance, these local clock signals need to be distributed quickly to registers, but current FPGA architectures have little support for local clocking signals. One solution is to use delay-insensitive protocols: these do not have local clock signals.

Delaying Signals: Current FPGAs have no support for delaying signals. Delay elements are essential in bundled-data protocols, to delay the request signals to meet the bundling-constraint. Current architectures have little support for delays; Brunvand [7] has to waste cells by building inverter chains to build delay elements. A related problem is that routing architectures can re-order signals so that the bundling-constraint is not met. Even though delay-insensitive self-timed circuits do not have bundling-constraints to meet, Martin [13] shows that delay-insensitive circuits cannot be built without isochronic forks. Isochronic forks require one-sided delay bounds that can be hard to meet in current FPGA routing architectures.

The only FPGA architecture that has addressed some of these problems has been the MONTAGE architecture [14], which was designed for the implementation of both synchronous and asynchronous systems. MONTAGE solves the arbitration problem by sprinkling a small number of special arbitration cells amongst standard FPGA cells. Also, the routing architecture is designed to facilitate the construction of isochronic forks. However, no specific support is given for local clock distribution or for delaying signals so that they conform to bundling-constraints.

5 STACC: a Self-Timed FPGA architecture

Whilst the MONTAGE architecture includes many features to improve its implementation of self-timed circuits, it still retains a global clock. In contrast, the STACC (Self-Timed Array of Configurable Cells) is a dedicated self-timed FPGA architecture specifically for the implementation of bundled-data systems. The bundled-data protocol was chosen as it has a small area overhead compared to delay-insensitive protocols. Additionally, bundled-data systems use the same data-path as their synchronous counterparts, making it easy for designers to adapt to self-timed design. The four-phase variant of the protocol was chosen since it requires less circuitry to implement basic branching structures.

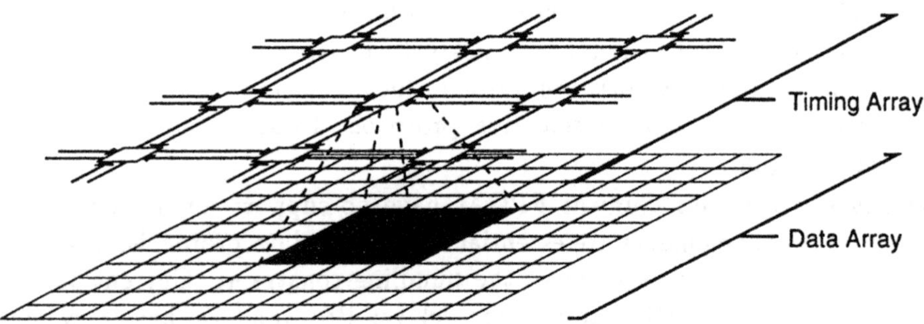

Fig. 3. Basic structure of the STACC architecture

In Fig.3, the basic concept behind the STACC architecture is shown. The *data array* consists of cells similar to current synchronous FPGAs, but the global clock has been replaced by an array of timing cells. Each timing cell provides local timing information to a region of FPGA. A timing cell and the data cells that it provides timing information for, are known as a *timed region* (illustrated by the shaded region in Fig.3).
Each timing cell is connected to its neighbours by two wires that are used to initiate request/acknowledge handshakes with neighbouring timing cells. Configuration data determines whether neighbouring timed regions communicate, and the direction of data flow between them. For a configuration to produce a correctly functioning circuit, the configuration of the data array must reflect the configuration of the timing array. No data should flow between timed regions that are not linked in the timing array.

Since the data array can use any standard style of FPGA cells, the rest of this section concentrates on the novel part of the STACC architecture, which is the timing cell array. The discussion starts by looking at unreconfigurable four-phase bundled-data pipelines, and then develops a timing-cell capable of implementing such pipelines with fan-in and fan-out. Successively more general versions of the timing cell are introduced that allow the implementation of branching and merging within the structures. The final timing cell is flexible enough to implement all the structures mention by Sutherland [2].

5.1 Four-Phase Pipelines

Figure 4 shows a four-phase bundled-data pipeline. As with Sutherland's Micropipelines [2], the basic timing element is the C-muller gate. The C-muller gate's behaviour is to cause a rendezvous between its inputs. In other words, the C-muller gate's output will not change until all of its inputs have changed. This behaviour is summarised in table 1. Bubbled inputs in Fig.4 indicate an inverted input. In effect, on initialisation, transitions are already assumed to have occured on bubbled inputs. At initialisation, the output of all the C-muller gates is logic zero (the reset signal has been omitted from the diagram).

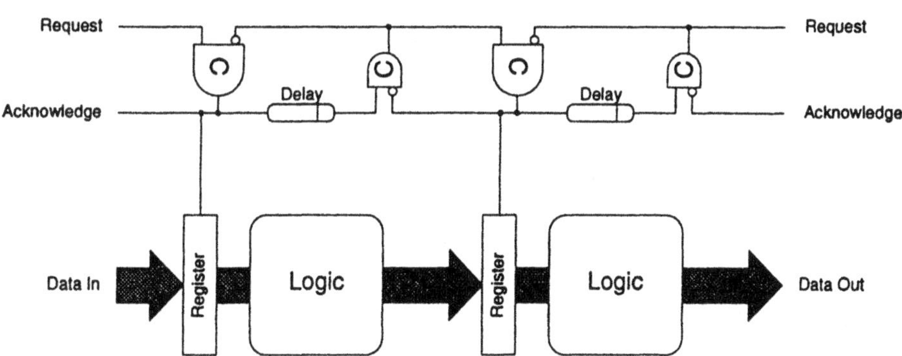

Fig. 4. A four-phase pipeline with two stages

Table 1. C-Muller gate behaviour

Inputs	Output
all logic 0	logic 0
dissimilar	previous value
all logic 1	logic 1

In Fig.4, the larger C-muller gates generate the local register clocking signal. The larger timing C-muller gates cause a rendezvous between the request signal from the previous stage and the acknowledge from the next stage in the pipeline. The rendezvous ensures correct operation of the pipeline, since the next processing cycle will not start until the previous stage has data ready (indicated by the request) and the next stage has got this stage's last result (indicated by the acknowledge). After the rendezvous, the C-muller's output latches the inputs into the register and sends an acknowledge back to the previous stage in the pipeline. The delay element delays the request to the next stage so that the logic block's outputs are valid.

Unlike Sutherland's two-phase Micropipelines, the four-phase protocol has an idle return-to-zero phase. Hence the delay elements and registers are only active on positive-going transitions. To allow the idle return-to-zero phase to occur concurrently with the processing phase, the smaller C-muller gates have been added between the stages of the pipeline in Fig.4. Their effect is to act as a holder for the return-to-zero transitions whilst the stages are processing.

The other main difference of the four-phase pipeline of Fig.4 from Micropipelines is the choice of register element. Instead of the capture-pass register used in Micropipelines, edge-triggered D-types are used. The advantages of using D-types are that only one control signal is required, and that the pipeline stages can retain state. Hence, by feeding their outputs back, the logic blocks can implement a finite state machine. The pipeline of Fig.4 can easily be generalised to deal with fan-in and fan-out in the pipe by having multiple request or acknowledge signals going into the timing C-muller gates.

5.2 A Basic Timing Cell

This section introduces a basic timing cell that allows the pipelines described above to be implemented. Figure 5 shows half of a reconfigurable connection between two such timing cells. The circuit is mirrored around the thick dotted line for the other side of the connection.

As in Fig.4, the timing and return-to-zero C-muller gates are retained. Various configuration bits controlling multiplexors are used to determine the nature of the connection. These configuration bits are described below.

DC: The DC (Don't Connect) configuration bit determines whether there is a connection between a timing cell and its neighbour. To connect with the neighbouring cells, the handshaking signals are

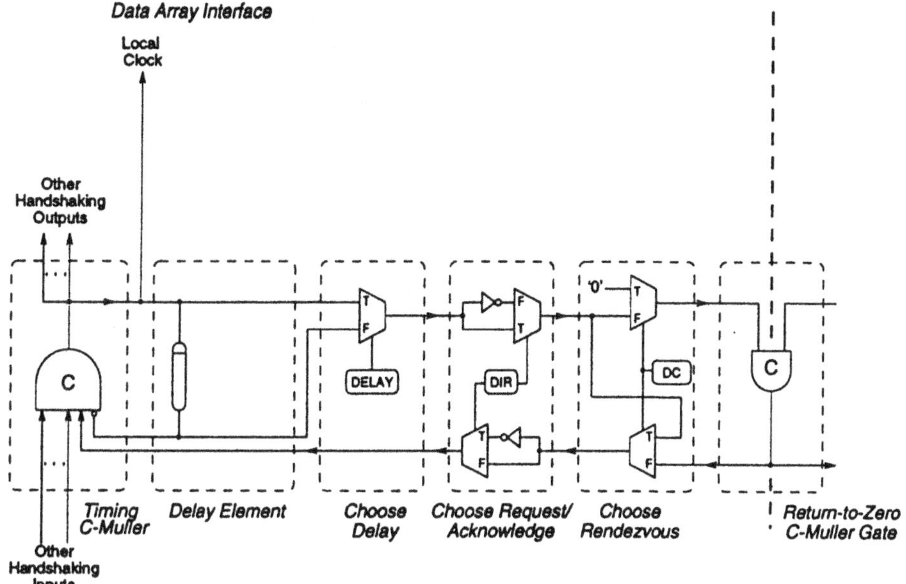

Fig. 5. A basic four-phase timing cell

passed to and from the return-to-zero C-muller gate. If there is no connection to the neighbouring cell, the out going handshaking signal is fed back to the timing cell, so that the C-muller gate effectively acknowledges itself.

DIR: The DIR bit determines which handshaking signal is the request and which is the acknowledge. Since the difference between a request and acknowledge is whether the signal is inverted (bubbled) or not, the DIR bit simply chooses which signal to invert.

DELAY: The DELAY configuration bit determines whether the out going handshaking signal is delayed. Normally, requests are delayed as they generally involve a data transfer that must meet the bundling-constraint, whilst acknowledges are not delayed since no data transfer is involved. However, by delaying acknowledges, data can also be passed on the acknowledge signal, allowing a two-way exchange of data. Alternatively, requests need not be delayed if no data transfer is involved with them.

An important part of the timing cell is the delay element. The delay must always be long enough so that the bundling constraint is met. However, for performance, it is important that the delay matches the logic block delay as closely as possible. One easy way to implement the delay element is to allow configuration data to choose from a range of fixed delays. An interesting unproven alternative is to use CSCD (Current Sensing

Completion Detection) [15]. CSCD utilises the fact that the static power and hence static current flow of a CMOS circuit is close to zero. By monitoring the current flow between the power rails, a CSCD circuit produces a variable delay that matches the time taken by the current computation.

5.3 A General Timing Cell

So far, a cell capable of implementing fan-in and fan-out pipelines has been described. However, when such pipelines branch, all of the fan-out pipes are followed. There is no way to choose that the computation should only continue along one branch. Conversely, there is no way for the cell to pick which of several fan-in pipes it wishes to accept data from. Sutherland's Micropipelines [2] use a variety of elements such as the Select, Toggle and Merge elements to implement branching and merging. In this section, the timing cell is developed to allow such branching and merging in four-phase pipelines.
Figure 6 shows an adapted timing cell that allows a connection to be chosen by an input from the data array. The select signal is latched after the delay, and determines whether communication takes place with the neighbouring cell in the current cycle. The RDZ configuration bits allows the choice of selected or inverted select signals so that the initial value

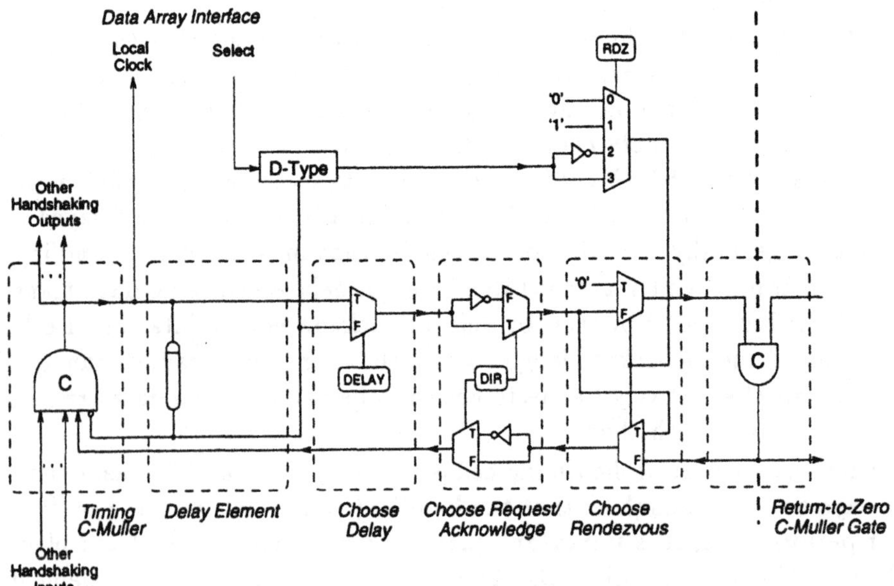

Fig. 6. Addition of select signals

of the select signal can be defined (assuming that the D-type is reset to a pre-defined value). The RDZ configuration bits can also choose the constants '0' or '1' so that the common functions of "never communicate" and "always communicate" can be implemented without using resources in the data array.

Another function of the select signal, not shown on the diagram, is to act as an enable signal for memory elements in the data array. Since neighbouring timed regions that are not selected for communication are not synchronised to the timing cell, sampling their data inputs can cause a meta-stable state in the data-array's memory elements. Hence, the select signals are also used to disenable memory elements not involved in the current communication.

There are still certain behaviours that the timing cell of Fig.6 cannot implement. Although, it can make decisions based on its own internal state, it cannot choose which connections to select on the basis of which neighbouring timed-regions are waiting to communicate. Figure 7 shows the development of Fig.6 that allows the state of neighbouring cells to be probed. The probe is fed in to the data array which can then make a decision on what connections to select. Since the probe signals are asynchronous to the cell, some form of synchroniser element must be used to sample the inputs. In this case, a Q-flop [16] is used. The Q-

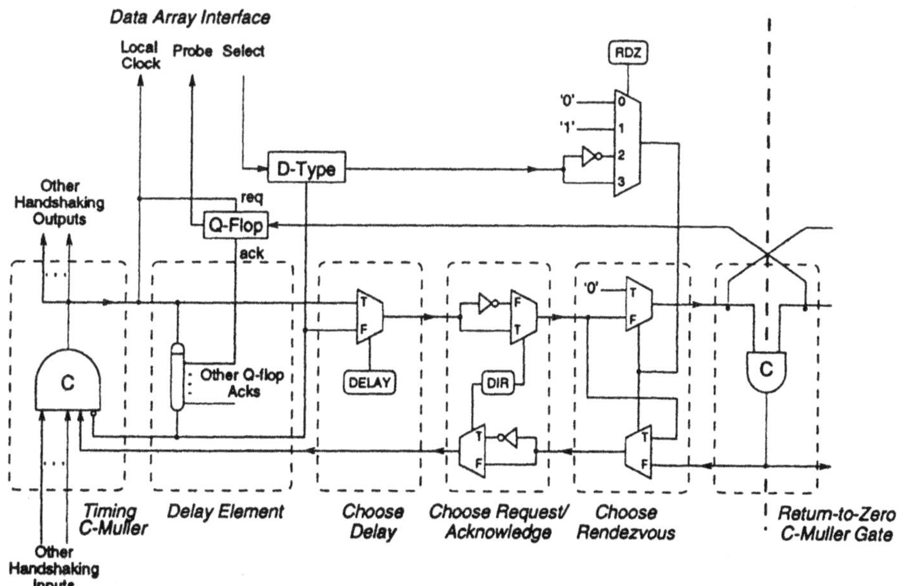

Fig. 7. Addition of probe signals

flop samples its input after a transition on its request signal. After any meta-stable state has been resolved, the Q-flop generates an acknowledge. This is fed to the delay element which delays the logic accordingly. Since Q-flops are relatively complicated elements to implement, alternative variants of the architecture are possible that just use a single arbiter element per timing cell.

6 Conclusions and Future Work

The first half of this paper set out to show the potential benefits of building a dedicated self-timed FPGA architecture. Many FPGA-specific benefits arise from the general robustness, modularity and average-case performance properties of self-timed systems. The mapping problems of routing, placement and partitioning are given a wider range of solutions to choose from, allowing a greater variety of trade-offs between performance, utilisation and time to find a solution.

The second half of the paper concentrated on the design of the STACC architecture, a dedicated self-timed FPGA using a four-phase bundled-data protocol. The discussion focussed on the design of the timing cells since these constitute the main novel part of the architecture. The final version of the timing cell could implement more than just a timing function since it was flexible enough to implement a wide range of control functions using a relatively small number of configuration bits. As presented, the timing cell only required four bits per connection. Furthermore, some configuration bits such as the DIR bit could be shared with neighbouring cells.

Many other aspects of the architecture were left undiscussed, and are part of on-going simulation studies. Such aspects include the style of configuration interface, non-local and off-chip communication links and the nature of routing, placement and partitioning algorithms. The results of the simulation work are intended to select an architecture variant for chip layout, so that comparison can be made with synchronous FPGA architectures.

Acknowledgements

My thanks to Gordon Brebner and Iain Lindsay for their advice and guidance. Also to Vinod Rebello and Rob Mullins for helpful discussions concerning self-timed systems and CAD tools.

References

1. C.L.Seitz. *System Timing*, chapter 7. Addison-Wesley, Mead and Conway Introduction to VLSI Systems edition, 1980.
2. I.E.Sutherland. Micropipelines. *Communications of the ACM*, 32(6):720–38, 1989.
3. S.B.Furber, P.Day, J.D.Garside, N.C.Paver, and J.V.Woods. A Micropipelined ARM. In T.Yanagawa and P.A.Ivey, editors, *Proceedings of VLSI 93*, pages 5.4.1–5.4.10, September 1993.
4. A.J.Martin, S.M.Burns, T.K.Lee, D.Borkovic, and P.J.Hazewindus. The Design of an Asynchronous Microprocessor. In C.LSeitz, editor, *Advanced Research in VLSI: Proceedings of the Decennial Caltech Conference on VLSI*, pages 351–373. MIT Press, 1989.
5. X.Ling and H.Amano. WASMII: a data driven computer on a virtual hardware. In *FCCM93: Proceedings of IEEE Workshop on FPGAs for Custom Computing Machines*, 1993.
6. P.Lysaght, J.Stockwood, J.Law, and D.Girma. Artificial Neural Network Implementation on a Fine-Grained FPGA. In *4th International Workshop on Field Programmable Logic and Applications*, 1994.
7. E.Brunvand. Using FPGAs to Implement Self-Timed Systems. *Journal of VLSI Signal Processing*, 6(2):173–190, August 1993.
8. E.Brunvand. Using FPGAs to Prototype a Self-Timed Computer. In *Workshop on Field Programmable Logic and Applications*, pages 192–198, 1992.
9. J.Oldfield and C.Kappler. Implementing Self-timed Systems: Comparision of Configurable Logic Arrays with Full Custom Circuits. In *FPGAs: International Workshop on Field Programmable Logic and Applications*, chapter 6.3. Abingdon EE&CS Books, 1991.
10. P.Shaw and G.Milne. A Highly Parallel FPGA-Based Machine and its Formal Verification. Technical Report HDV-28-93, U. of Strathclyde, 1993.
11. M.Gamble, B.Rahardjo, and R.D.Mcleod. Reconfigurable FPGA Micropipelines. Technical report, U. of Manitoba, 1994.
12. K. Maheswaran and V. Akella. Hazard-free Implementation of the Self-Timed Cell set for the Xilinx 4000 Series FPGA. Technical report, U.C.Davis, 1994.
13. A.J.Martin. The Limitations to Delay-Insensitivity in Asynchronous Circuits. In W.J.Dally, editor, *Sixth MIT Conference on Advanced Research in VLSI*, pages 263–278. MIT Press, 1990.

14. S.Hauck, G.Borriello, S.Burns, and C.Ebeling. MONTAGE: An FPGA for Synchronous and Asynchronous Circuits. In *Workshop on Field Programmable Logic and Applications*, 1992.
15. M.E.Dean, D.L.Dill, and M.Horowitz. Self-Timed Logic Using Current-Sensing Completion Detection (CSCD). In *Proc. International Conf. Computer Design (ICCD)*, pages 187–191. IEEE Computer Society Press, October 1991.
16. F.U.Rosenberger, C.E.Molnar, T.J.Chaney, and T.Fang. Q-Modules: Internally Clocked Delay-Insensitive Modules. *IEEE Transactions on Computers*, C-37(9):1005–1018, September 1988.

The XC6200 FastMap^{TM} Processor Interface

Stephen Churcher, Tom Kean, and Bill Wilkie
Xilinx Inc.

Abstract

The Xilinx XC6200 is the first commercially available FPGA to be specifically designed for use within microprocessor based systems. This paper discusses the architecture of the key element in this device : the *FastMap^{TM}* processor interface. Salient features of the user-programmable part of the XC6200 are also described.

1.0 Introduction

Almost all modern digital systems consist of three major functional components: microprocessors, memories and logic IC's. Logic IC's interface microprocessors to physical devices such as screens, keyboards and networks and perform computations which are unsuited to the microprocessor. Logic may be implemented using mask programmed devices or Field Programmable Gate Arrays (FPGA's) [1]. An interesting question is: what would the ideal logic part, from the point of view of a microprocessor, look like?

- Processors view the world as a sequence of memory locations, therefore the interface to the FPGA should be through memory mapped registers.
- Processors run different programs at different times. FPGA's should therefore be able to be reconfigured for different tasks at different times.
- Since processors are synchronous devices, the FPGA should be able to operate from the same clock as the processor, thereby allowing predictable interactions between them.
- The FPGA should be dense, fast and have good I/O capability.

2.0 Memory Mapped I/O

The concept of memory mapped I/O is simple: a particular register within the user's design selected via the processor address bus is connected to the processor data bus and a read or write operation is performed. The address selection and connection to the external bus is typically implemented using multiplexers and gates within the FPGA's user resources although this has several disadvantages.

- The interface to a modern processor is complex and high speed and can use up large quantities of user logic and IOB resources.

- If the registers which are to be accessed from the processor are placed near the centre of the device, long routing wires are required to reach user IOBs. If many registers are to be provided, complex selection circuitry is necessary.

When one considers that the control memory of an FPGA is itself a static RAM, with address and data busses available on the device, it becomes clear that these resources could be presented off the chip as the primary programming interface, rather than being hidden behind a serial channel. Naturally, this requires many more pins to support wide address and data busses, and so is not appropriate in many applications. However, where the device must work with a microprocessor it is advantageous.

Register resources on the FPGA can be addressed using bit and word lines within the RAM array in the same way as configuration memory cells, and their contents presented on the external data bus. This technique was first implemented in the Algotronix CAL1024 chip [2]. Algotronix technology was acquired by Xilinx in 1993, and has been continuously developed since then resulting in the XC6200 family. Figure 1 shows the first device in the family, the XC6216.

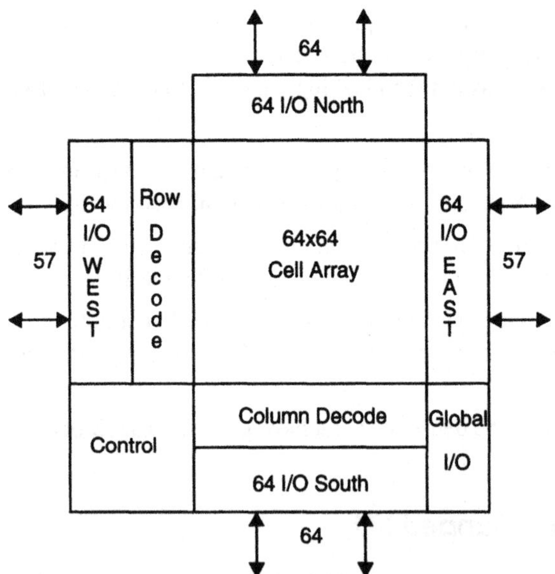

Figure 1 : XC6216 Architecture

Several additional steps are required to turn user registers, mapped into the device configuration memory, into a high bandwidth channel transferring 32 bit words between the processor and the user design on the FPGA. Firstly, the register bits, which will be physically dispersed among the configuration bits, must be collected together in the address space so that complete words of register memory are formed.

This is achieved in the memory row and column decoders. Secondly, some flexibility must be provided in the mapping of register bits which are to be accessed in this way onto device cells - ideally a set of user registers located in arbitrary parts of the device could be grouped into a word of memory accessible through the processor interface.

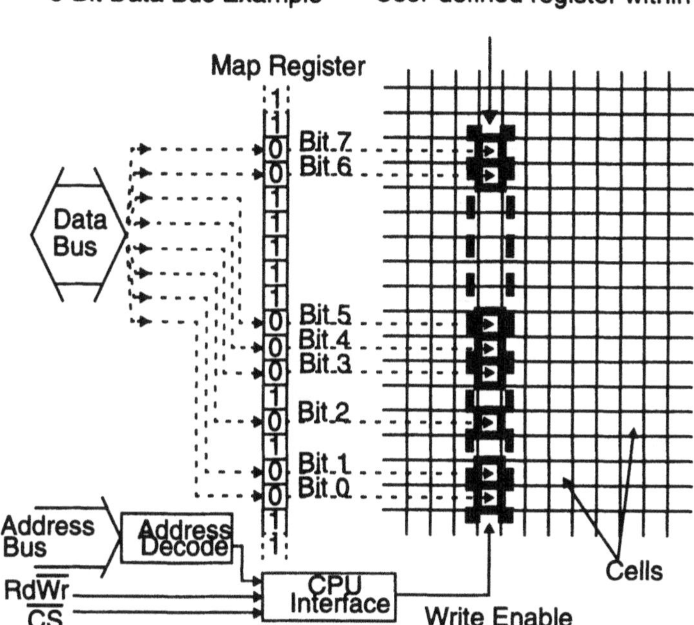

Figure 2 : Map Register - Principles of Operation

The row and column (bit line and word line) addressing scheme used by memories makes collecting arbitrary registers on the device into a single word difficult: a realistic constraint is that all the registers should be in the same column of cells. An issue also arises concerning the manner in which bits in the processor word are mapped into registers. If full flexibility was allowed, 6 bits (to select one of 64 cell positions in a column) would be required for each of the 32 data bus bits. This is too much information to change quickly, in order to allow selection between different registers. A realistic limitation is that the bits appear in order in the processor word with the register with the lowest cell y coordinate first. Given this constraint, the locations of the register bits can be specified with a 64 bit map register - a 0 in the map register (for example) indicates that the cell with the corresponding y coordinate will take part in the transfer. This technique allows the registers to be spread out over the column in

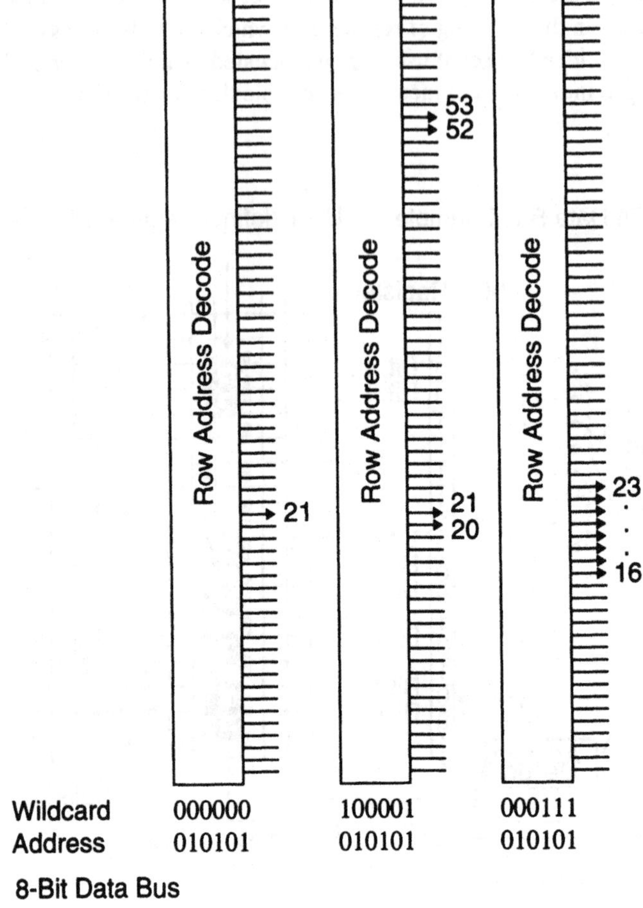

Figure 3 : Wild Card System

an arbitrary manner and can be implemented relatively efficiently in the RAM bit line logic. Figure 2 shows an example of accessing a register within user logic via an 8 bit external bus on the XC6216.

3.0 Dynamic Reconfiguration.

SRAM based FPGA's are inherently capable of dynamic and partial reconfiguration; all that is required is to bring the internal RAM data and address busses onto device pins, as was done on the Algotronix CAL1024 part. The XC6200 family is configurable to provide an 8, 16, or 32 bit external data bus, and a 16 bit address bus. Using these features, the entire configuration memory can be programmed in under 100 μs.

Normally, however, it is not necessary to program the entire device.In such cases, the random access feature allows arbitrary areas of the memory to be changed. In addition, a mask register is provided which allows a subset of bits within a word to be masked out of a transfer to the memory. This is useful because often a user will wish to reconfigure a particular logical resource (e.g. a routing multiplexer) which will represent only a small fraction of the bits within a word of program memory. This feature is particularly attractive in combination with the wildcard registers discussed below.

Another property of FPGA configurations, particularly datapath type designs, is that they are very regular i.e. the same pattern of bits may appear at many locations in the memory (e.g. each slice in a 32 bit datapath). The XC6200 supports writing the same configuration information to multiple locations in the control memory simultaneously, using so-called 'wildcard' registers. These registers modify the row and column addresses supplied to the chip, putting 'don't cares' on corresponding address bits. For example when one address bit is subject to a 'don't care' condition, two memory locations will be written simultaneously on every transfer (Figure 3). Using the wildcard addresses, the configuration memory for all the cells on the chip can be cleared in a single memory cycle. Most user designs will use a fraction of the chip's resources, and in such cases it will often be faster to clear the configuration memory before reconfiguring only those resources which are required,rather than configuring the whole chip.

4.0 Synchronous Access.

The communication between the processor and the FPGA can function most effectively when they are both running from a common clock. This synchronises input and output of data through the processor interface with computations running in the user logic and ensures that setup time requirements are met on write accesses to user registers and read accesses obtain valid data. For this reason the clock signal for the processor interface is also supplied as one of four low skew global signals to the user logic [3].

As well as synchronising FPGA and processor communication, it is useful to provide a mechanism for signalling to the user logic that a transfer to or from a register has occurred. This allows the user logic to begin processing an input value or to start computing a new output value. This function could be implemented by using a second user register as a flag to indicate transfers, but this would double the number of processor accesses required. Instead, bit and word lines used for transfers to user registers are made available as inputs to programmable routing switches within the array. By connecting these wires to logic gates within their design, users can monitor the transfers through the processor interface and take appropriate action.

5.0 Density and Performance.

The first member of the XC6200 family, the XC6216, contains a 64x64 array of fine grain programmable logic cells. Each of these can implement any logic function of two variables, or a 2:1 multiplexer. In addition to these combinatorial resources, each cell also contains a register, which may be used in consort with the logic function generator in a variety of ways, as shown in Figure 4. Finally, each cell in the array contains four interconnection multiplexers, which provide 'nearest neighbour' routing resources for up to four independent signals.

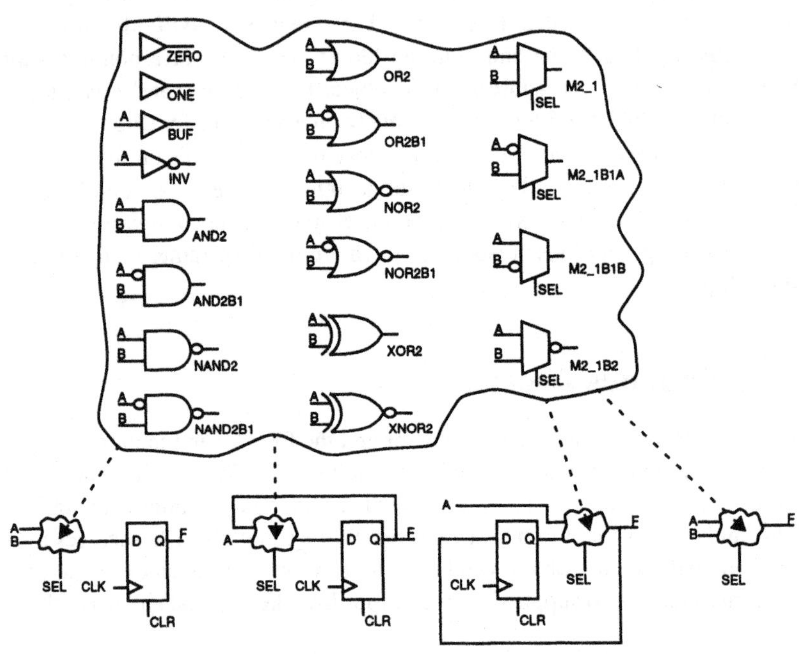

Figure 4 : XC6200 Function Unit Logic Configurations

The routing architecture of the XC6200 is depicted in Figure 5, and is a hierarchical structure with neighbour connections between cells, wires which span 4 cell blocks, wires which span 16 cell blocks and wires which cross the complete 64x64 cell array. There are also four global signals (clock, clear, and two user defined signals). These configurable connections, when combined with the 'wireless' I/O capability afforded by the *FastMap*™ interface, provide users of the XC6200 with a powerful, flexible combination of routing resources.

The XC6200 family is implemented in 0.6 μm triple metal CMOS technology, and is expected to meet state-of-the-art system performance criteria. At the time of writing, full performance figures were unavailable. The equivalent gate count for the XC6216 is 16k gates for typical applications, however this figure is potentially as high as 50k gates for designs which are particularly register intensive. Smaller and larger family members will also be made available.

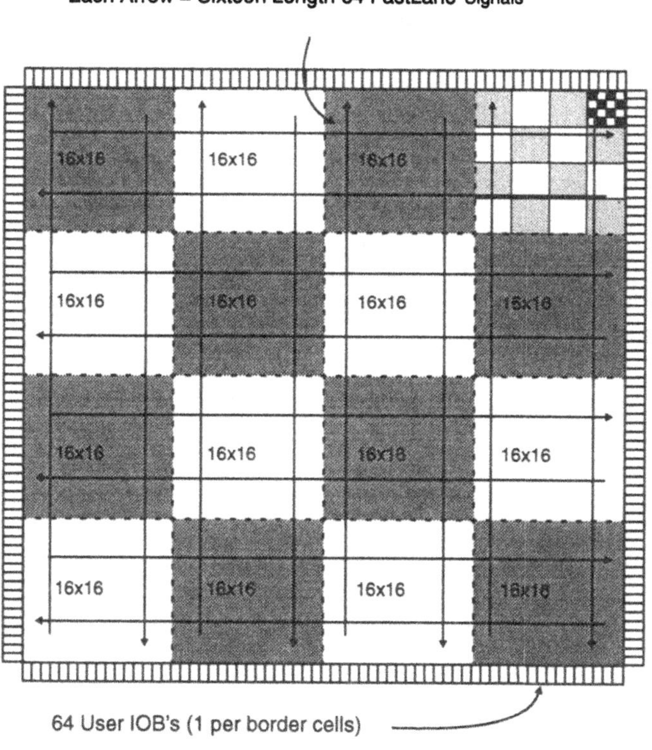

Figure 5 : XC6216 Hierarchical Routing

6.0 Summary

The XC6200 family [3] is the first commercial FPGA to address the requirements of interfacing programmable logic to microprocessors. With its combination of dedicated random access parallel interface, flexible hierarchical routing, and powerful fine grain logic function generator, the XC6200 is especially suited to 'embedded processing' applications where high bandwidth peripheral devices must be interfaced to a computer system, and processing of the incoming or outgoing data stream is required.

7.0 References

1. The Programmable Logic Data Book, Xilinx Inc, San Jose CA, 1994.

2. CAL1024 Data Sheet, Algotronix Ltd., Edinburgh UK 1990.

3. Xilinx XC6200 Family Preliminary Product Description, Xilinx Inc, San Jose CA 1995.

The Teramac Configurable Compute Engine

Greg Snider, Philip Kuekes, W. Bruce Culbertson,
Richard J. Carter, Arnold S. Berger[1], Rick Amerson

Hewlett-Packard Laboratories

1 Introduction

Research on special purpose parallel architectures and custom computing is very much an experimental science dependent on the existence of prototypes. We have built a custom FPGA-based configurable custom computing engine to enable experiments on an interesting scale. The system is being used to explore the potential of custom computing machinery.

We built Teramac as a testbed for computer architects who have an abundance of ideas for new and better parallel architectures and no hope of getting funding for silicon to try their ideas. The name Teramac is derived from *tera* (10^{12}) and *m*ultiple *a*rchitecture *c*omputer. With this tool, the architect can create a machine with a million 2-input Boolean functions being simultaneously evaluated at one megahertz—a trillion very small operations per second.

Like many others who have created a computer for people who could not make up their minds about what type of machine they wanted, we chose to use FPGAs with look-up-tables to perform the actual logic. The major innovations we brought to the task concerned wires and compiling.

1.1 Compiler-Driven Design

We realized at the beginning of the project that the problem of automatically placing and routing the equivalent of a million gates was the key problem to be solved. Before any hardware was designed we created an experimental compiler to place and route trial designs using a variety of interconnection schemes. Not surprisingly, the interconnect topologies which the compiler found easy to map to had a very large number of available wires.

1.2 Rent's Rule

Richard Rent [1] was the first to observe that as a logic design is partitioned, the number of signals crossing the boundaries of that partition is proportional to the number of

1. Currently at Advanced Micro Devices, Austin, TX

gates in the partition raised to the *r* power. The Rent's rule exponent *r* has since been found by other investigators to range from .5 to .7 for a variety of designs.

We carefully considered Rent's rule in designing Teramac. This paper details the design decisions which followed from this strategy: At the chip level we created our own custom FPGA, PLASMA, to ensure that we had enough wires on the chip and enough I/Os off the chip. At the system level, we balanced 300,000 inter-chip wires among MCM wiring, board wiring, and inter-board wiring using Rent's rule and careful attention to the state of the art in CAD and manufacturing technology.

1.3 Organization

The following sections show the decisions we made in FPGA design, MCM and system design, and compiler design. The early experiments we have begun with Teramac have been reported elsewhere [2].

2 PLASMA FPGA

PLASMA is a hierarchically structured FPGA (figure 1). At the top level of the hierarchy, 336 I/O signals connect through 336 I/O blocks (shown at the top and bottom of the figure) into a central crossbar (the vertical middle section of the chip). This crossbar, which is 1/4 populated with switches, is segmented by programmable buffers into four sections. This central crossbar connects to 16 peripheral crossbars (left and right sides of chip)—these crossbars are nearly fully populated. Finally, each peripheral crossbar connects to 16 logic units known as "PALEs." A peripheral crossbar, along with the PALEs connected to it, is referred to as a "hextant."

Each switch in the central and peripheral crossbars can be individually configured to be open or closed so that signals can be routed between the PALEs and the I/O signal pins. The PALEs are the heart of the chip: they contain configurable lookup tables which may be programmed by the user to implement small Boolean functions, and configurable latches for implementing storage elements. Certain PALEs may also be grouped together to form multiported register files, a feature we believe to be unique to PLASMA.

2.1 PALEs

A logic block in PLASMA is called a PALE (for *p*rogrammable *a*tomic *l*ogic *e*lement). It consists of a programmable 6-input, 2-output lookup table followed by a programmable latch module on each output. Each latch module can be configured as a D flip-flop, D latch, or as a pass-through gate which simply relays the lookup table output. Although research has suggested that lookup tables with 4 or 5 inputs and 1 output are

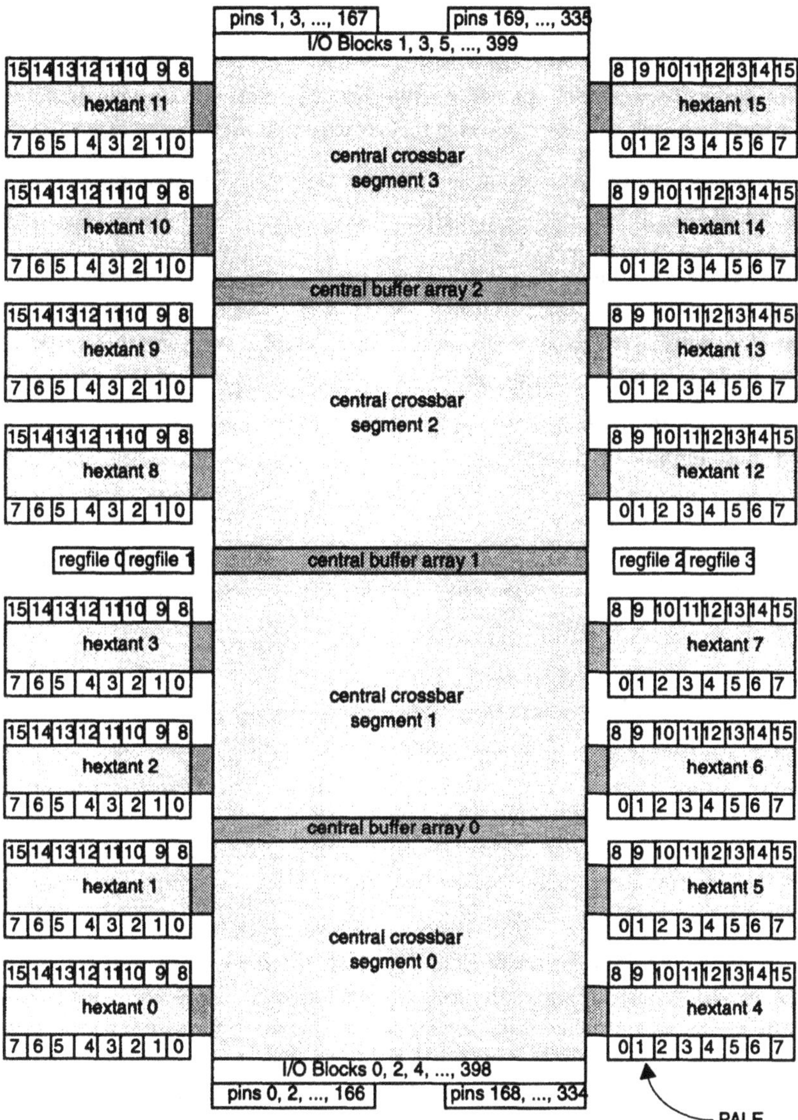

FIGURE 1. The PLASMA FPGA floorplan. Signal I/O pins (top and bottom) connect into the central crossbar. Sixteen "hextants," each containing 16 PALEs and a peripheral crossbar, also interface to the central crossbar.

more efficient, we chose the (6, 2) design since it allowed us to more efficiently support multiported register files, as will be described later.

2.2 Crossbars

Experiments with routing software indicated that the availability of routing channels would be the primary factor in getting the most logic into the system as well as getting the best performance from the placement and routing software.

Studies indicated a Rent's rule exponent in the range of .5 to .7 is appropriate for a general purpose logic system. Given that our PALE has 8 pins, Rent's Rule suggests that a group of 16 PALEs (a "hextant") requires 32 to 51 "pins" to interface the hextant to the central crossbar. Of course, Rent's rule only gives information about the mean I/O requirement, not the standard deviation. We wanted to encompass several sigma deviation and thus chose 100 signals for the hextant interface. Experimental routes show that 99% of designs will route fully in this configuration.

At the chip level, Rent's Rule predicted that only 128 pins would be needed on average, but to accommodate the outliers, we designed the chip with 336 signal I/O pins. Fortunately the standard deviation of the predication declines as more gates are incorporated into a partitions, so that the number of pins on larger and larger units, specifically the MCM and board, are closer to the actual Rent prediction. However, it is our experience that a Rent's Rule exponent closer to .7 than .5 is necessary for successful compilation of a wide variety of circuits.

2.3 Multiported Register Files

PLASMA directly supports multiported register files by "overlaying" the PALEs with additional circuitry to create four vertical *register file slices* two bits wide and 64 registers deep, with each slice having 8 read ports and 8 write ports. When a register file slice is configured, the PALEs associated with that slice no longer act as lookup tables, but instead act as read, write, and enable ports for that slice. A single slice consumes a total of 40 PALES for these ports.

To understand how this was accomplished, consider the model of a multiported register file shown in figure 2. Here a multiported register file is represented as an array of registers that can be simultaneously written by several write ports and read by several read ports. Each write port takes as inputs the address of the register in the file to be written, the data be written, and a write enable line that must be asserted to perform the write. Each read port takes as input the address of the register to be read, and supplies the contents of that register as an output.

A write port can be implemented with demultiplexers that: (1) select the desired register to be written, and (2) direct the data to be written to that register. A read port is simply a multiplexer that selects the contents of the register referenced by the input address and delivers it to the port output. When multiple write ports interface to the file, the write port outputs directed to a particular input of the file are ORed together.

A PALE lookup table uses a CMOS bidirectional switch tree for selecting the lookup table bit to be driven to the output; in implementing read and write ports, we steal this

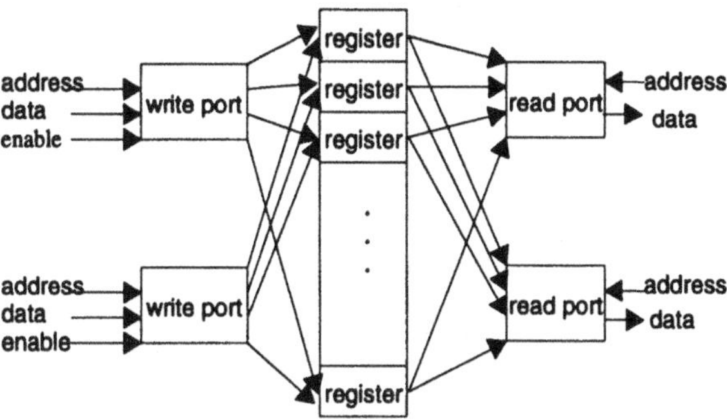

FIGURE 2. Multiported register file. Multiple read ports and write ports can simultaneously access an array of registers

tree for the multiplexer in a read port, and, by running it backwards, for the demuliti-plexer in the write port.

Figure 3 shows conceptually how columns of PALEs are grouped together to form multiported register files. The 6-input, 2-output lookup tables are shown here as two sets of 64 bit registers whose outputs are connected to two 64:1 multiplexers sharing address lines. When a PALE is configured as a write port, the lookup table bits are "disconnected" and the multiplexer switch trees are run "backwards" to generate the necessary signals for each of the 64 registers in the file. The corresponding demultiplexer outputs of each of the write ports are connected together on lines with programmable pull-up resistors to implement the ORing of the write port signals.

When configured as a read port, the PALE's lookup table bits are also "disconnected" and the PALE multiplexer instead accesses 64 lines which broadcast the state of each of the 64 registers to all of the read ports. The multiplexer then simply selects the desired register file bit and delivers it as the PALE output. The bits used for storing the actual register file data were implemented as separate structures that ran horizontally across the middle of the chip.

2.4 Scan Chain

The PLASMA chip is designed to make the debugging of a user's custom design easy and natural. All internal state in the chip is available on a scan chain which runs through the chip. This includes all flip-flops and register files as well as all the I/O pins of the chip. Both the D and the Q of every flip-flop are simultaneously observable. The outputs of all lookup tables are observable and the user-interface recreates signals that exist on the original schematic but were subsumed into lookup tables.

The clock is stopped before examining the scan chain and the user may change the state of the machine before resuming clocking. The PLASMA architecture allows the user full peek and poke capability while debugging. This scan capability uses different mechanisms than the read and write configuration mechanism. This allows fast debugging without risk of accidently modifying the configured design.

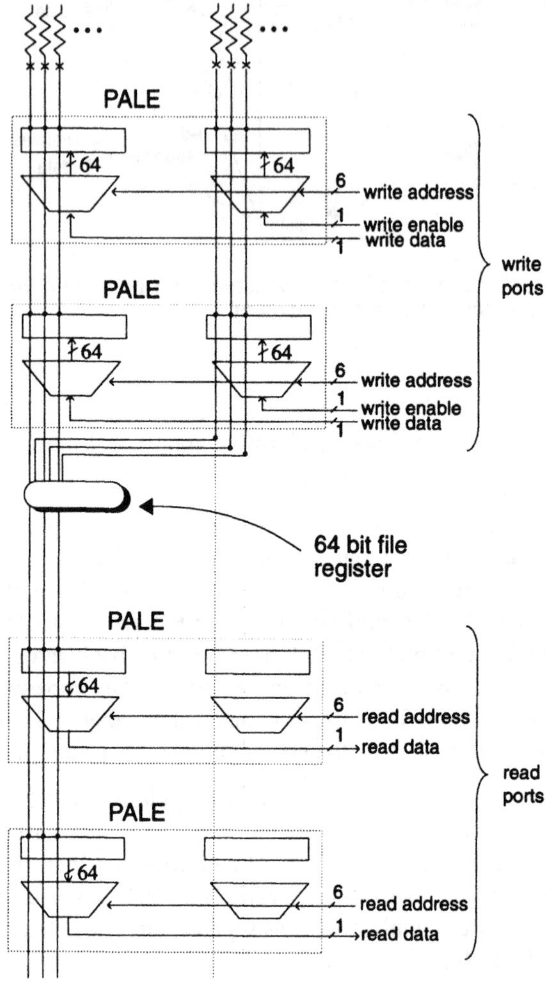

FIGURE 3. Multiported Register File implementatión with PALES.

3 System Design

The design of the system was dictated by the need to support an architecture comprised of 1728 individual PLASMA chips, each with 336 I/O pins. Much of what evolved to the final form of the Teramac system architecture and chassis design was dictated by the special requirements of this interconnect-rich system.

3.1 Mechanical Fixture

Bringing the PLASMA chips into close proximity was the first problem that we needed to solve. Our solution was to use three levels of interconnect. The first level mounts 27 chips on a multichip module (MCM) to provide a high level of local interconnectivity within the module [3]. The MCM, measuring 15.57 x 18.80 cm has a total of 39 layers: 27 signal and 6 power and ground pairs. The second level of interconnect connects four MCMs. This is achieved by mounting these four MCMs on a 44.45 x 73.66 cm, 12-layer printed circuit board, called a logic board. The final level of interconnect is between the PC boards.

Each board has approximately 4800 connections to other boards. Some of these are interspersed ground wires. Simple calculations showed that a design based upon a traditional model of one motherboard and multiple daughterboards would not work because the insertion force for such a system would be 300 lbs., assuming an insertion force of 1 oz. per pin. This led us to using standard, 100-pin insulation-displacement connectors to carry the signals between boards. The connectors are arranged in 8 groups of 6 connectors per board.

Each MCM has a total of 3264 I/O pins, arranged in a 48 x 68 pin grid array. The area consumed by the corresponding through-holes created a significant PC board routing problem. This problem was solved by a combination of hand routing to create the global routing architecture and then allowing the computer-based routing program to solve the individual interconnections within the global guidelines.

3.2 Control

Each Teramac logic board is mated with a dedicated control/memory board. Each control board has a SCSI control circuit which fans out to 16 synchronous serial channels. The serial channels have individual local memories which can be loaded under DMA control by the local processor. These channels can transmit or receive from a column of 7 PLASMA chips.

Each control board is managed by an MC68EC030 microprocessor. The microprocessor also performs all the housekeeping chores on the board as well as controlling the operation of the clocks.

Each board also contains a total of 32 MBytes of static RAM memory. Memory is dual ported to the microprocessor and the PLASMA chips. The local control board communicates with the logic board through twelve 80-pin connectors. Again, insulation displacement connectors are used because of their simplicity and low-cost.

Since all board pairs are identical, any group of board pairs can be combined to build a Teramac system. For example, sixteen Teramac board-pairs could be grouped into sixteen 1-board systems, four 4-board systems, two 8-board systems, one full system, or any combination. All these combinations could still be controlled by a single host computer, networked to multiple users at remote sites.

4 Compiler

Since fast compilation was a starting requirement for Teramac, the compiler was developed concurrently with the architecture and wound up being the principle driver of it—if we could not quickly compile to a candidate architecture or architectural feature, it was discarded. As implementation problems arose, the compiler became a tool for exploring alternatives and refinements.

The compiler consists of three major phases. The first phase maps the user circuit into PALEs and register file slices, and then partitions that representation (netlist) across the sea of PLASMA chips, assigning each PALE or register file slice to a specific chip, and each I/O pin on each chip to a specified signal. In the second phase, each subnetlist is mapped onto the PLASMA chip to which it has been assigned. The third phase analyzes the placed and routed circuit and, using precomputed SPICE simulation data of the PLASMA chip and system, determines the maximum clock frequency at which the configuration can run.

4.1 Global Placement and Routing

The first compiler phase consists of the following passes:

Netlist Filter: The first pass transforms the user's input netlist into an internal format used by the remaining passes of the compiler.

Merger: The second pass transforms the original circuit of logic gates into a functionally equivalent circuit of register file slices and lookup tables (PALEs); this is accomplished by "greedily" packing gates into groups respecting the 6-input, 2-output limitation of the lookup tables used in PLASMA, and then computing the truth table for the resulting function (additional heuristics in this pass also unpack and repack groups and replicate gates to further reduce the number of lookup tables).

Global Partitioning: The global partitioner takes the circuit of truth tables generated by the merger and partitions them into PLASMA chips in the Teramac system. Partitioning begins by constructing a tree to represent the user's circuit. Hierarchical information in the netlist, if present, is used to build the initial tree, which is then refined through the use of ratio-cut partitioning[4]. Applying the approach of Yeh and Cheng [5], the tree guides the recursive partitioning of the circuit onto the Teramac network using min-cut partitioning algorithms [6,7]. The majority of compiler processing time is spent in this pass.

Global Placement: Based on the results of global partitioning, subcircuits are assigned to specific PLASMA chips by a global placement pass.

Global Routing: The global router routes signals between PLASMA chips. The signals to be routed are sorted in order of fan-out, and routed one at a time (highest fan-out first) through the network using a "first-fit" algorithm. The output of this pass is an assignment of signals to all of the PLASMA pins in the network.

4.2 Placing and Routing within PLASMA

The PLASMA chip compiler accepts as input: (1) a netlist describing the circuit (in terms of PALEs and register file slices) to be mapped to the chip; and (2) an assignment of primary inputs and outputs to signal I/O pins.

The first step is placement of any register file slices. The heuristic for this is simple: grab any arbitrary slice and map it to the first available physical slice in the chip; then repeatedly grab an unplaced slice that is most strongly connected to the previously placed slice and place it physically adjacent to it. This simple approach works adequately since typically all of the slices placed within a chip comprise a larger register file structure and therefore share read and write port address lines. This placement of register file slices will consume the physical PALEs that comprise the physical slice.

Next, PALEs in the netlist are placed into the remaining physical PALE slots not allocated to register files. Recursive min-cut partitioning with terminal propagation is used to first divide the netlist PALEs between the top and bottom halves of the chip, then each half chip is similarly bisected, and so on. The min-cut algorithm used [6] has been modified so that PALE slots consumed by register file sliced are represented by cells that are "locked-down" in the appropriated partitions, thus guaranteeing that PALEs will not be placed into otherwise occupied PALE slots.

After placement, signal nets are ordered in terms of estimated difficulty of routing and routed one at a time. Our ordering heuristic is: (1) all nets driving primary output pins are routed first; (2) all nets driven by primary inputs are routed next; (3) nets spanning multiple hextants are routed next, with higher fanout nets routed before lower fanout nets; (4) nets confined to single hextants are routed last.

This placement and routing approach does not always work—some nets may be unroutable with the given placement. In that case, analysis code is invoked to determine the cause of the routing failure (e.g. too many signals within a hextant? incompatible I/O pin and pale placement?) and to specify a perturbation of the placement to correct it. The routing is then completely ripped up, and the chip is rerouted with the new placement. This process iterates until the chip has been successfully routed.

Because of the rip-up and retry, chip placement and routing time is not deterministic, but averages around two seconds on an HP 735 workstation.

5 Conclusions

The difficulty in creating a configurable machine lies in providing enough wires that placement and routing can be done with no human intervention. Several researchers have previously used tens of FPGAs to create configurable custom machines [8-11]; Teramac allows experiments using many hundreds of FPGAs by providing a routing-rich environment for implementing user designs by using custom FPGAs, MCM's and PC boards.

References

[1] B. Landman and R. Russo, "On a Pin vs. Block Relationship for Partitions of Logic Graphs," IEEE Transactions on Computers, December 1971, pages 1469-1479.

[2] R. Amerson, R. Carter, W. Culbertson, Phil Kuekes, Greg Snider, "Teramac—Configurable Custom Computing," IEEE Symposium on FPGAs for Custom Computing Machines, April 1995.

[3] R. Amerson and P. Kuekes, "The Design of an Extremely Large MCM-C—A Case Study," International Journal of Microcircuits and Electronic Packaging, Vol 17, No. 4, pages 377-382.

[4] Y. C. Wei and C. K. Cheng, "Toward Efficient Hierarchical Designs by Ratio Cut Partitioning," Proc. IEEE International Conference on Computer- Aided Design, 1989, pages 298-301.

[5] Ching-Wei Yeh and Chung-Kuan Cheng, "A General Purpose Multiple Way Partitioning Algorithm," Proc. 28th ACM/IEEE Design Automation Conference, 1991, pages 421-426

[6] Balakrishnan Krishnamurthy, "An Improved Min- Cut Algorithm For Partitioning VLSI Networks," IEEE Transactions on Computers, Vol C-33, No 5, May 1984, pages 438-446.

[7] Laura A. Sanchis, "Multiple-Way Network Partitioning," IEEE Transactions on Computers, Vol. 38, No. 1, January 1989, pages 62-81.

[8] Patrice Bertin, Didier Roncin, and Jean Vuillemin, "Introduction to programmable active memories," in Systolic Array Processors, Prentice-Hall, 1989, pages 301-309.

[9] J. M. Arnold, D. A. Buell, and E. G. Davis, "Splash 2," Proceedings of the 4th Annual ACM Symposium on Parallel Algorithms and Architectures, 1992, pages 316-322.

[10] J. Babb, R. Tessier, and A. Agarwal, "Virtual Wires: Overcoming Pin Limitations in FPGA-based Logic Emulators," Proceedings, IEEE Workshop on FPGA-based Custom Computing Machines, Napa, CA, April 1993, pages 142-151.

[11] S. Casselman, "Virtual Computing, "Proceedings of the IEEE Workshop on FPGAs for Custom Computing Machines, Napa, CA, April 1993, pages 43-48.

Telecommunication-oriented FPGA and Dedicated CAD System

Toshiaki Miyazaki, Kazuhisa Yamada, Akihiro Tsutsui
Hiroshi Nakada†, Naohisa Ohta

NTT Optical Network Systems Laboratories
Y-807C, 1-2356 Take, Yokosuka-shi, Kanagawa, 238-03 JAPAN
e-mail: miyazaki@exa.onlab.ntt.jp

† NTT Telecommunication Network Laboratory Group

Abstract. We developed a telecommunication-oriented LUT-based FPGA and its dedicated CAD system. The FPGA, called PROTEUS, has some unique features. For example, its logic block structure enables the user to easily realize the basic components used in telecommunication circuits such as binary counters and pattern matching circuits. In addition, PROTEUS has a lot of regularly-placed latches to accomplish pipelined data processing. With a logic synthesis system, the CAD system offers a top-down design methodology. The programming data downloaded into PROTEUS can be obtained directly from RTL language descriptions. Furthermore, the CAD system supports hardware-macro-based design to realize high performance circuits. In this paper, we introduce the PROTEUS FPGA architecture and CAD system. In addition, we show some experimental results proving that PROTEUS is applicable to real telecommunication circuits.

1 Introduction

The recent growth of multi-media services and progress in digital telecommunication networks toward B-ISDN (Broadband Integrated Services Digital Network) based on ATM (Asynchronous Transfer Mode)[1] require more functions and higher-speed data processing in every protocol layer. In general, data processing in the telecommunication systems can be divided into two parts; digital signal processing and data transmission. The former is mainly performed in the terminal equipment, and includes filtering and data coding. This type of processing often contains repetitive fixed bit-width operations such as "multiply-add", and they can be handled by DSPs (Digital Signal Processors) efficiently.

On the other hand, the data transmission part in telecommunication systems is often realized by dedicated hardware. This is because data transmission requires high throughput and various bit-level manipulations must be performed, which CPUs or DSPs cannot handle well. Thus, compared with digital signal processing, current data transmission circuits lack flexibility. To overcome this situation, we have presented an architecture for programmable hardware targeted at high-speed data transmission systems [2].

In this paper, we introduce a newly developed FPGA (Field-Programmable Gate Array) based on the architecture and its dedicated CAD environment.

2 Characterization of Data Transmission

Transmission processing consists of data synchronization and bit pattern manipulations. The former process synchronizes input data by matching bit pattern in the data stream. The latter consists of primitive operations such as bit insertion, multiplexing, demultiplexing, and latching.

We examined some parts of ATM-STM interface circuits before forming the PROTEUS architecture. Actual analysis was performed using a logic synthesis system: PARTHEON [3]. The circuits were divided into some pieces of sub-modules, and their behaviors were described hierarchically using SFL, which is an RTL language in PARTHENON. The result is shown in Table 1. This table shows which basic logic type, categorized from Type A to Type E, each sub-module has. As we can see, every circuit has Type A and Type B circuits, i.e. binary counters and pattern matching circuits. Thus, we considered easy implementation and high-speed execution of these kinds of circuits when designing PROTEUS. In Table 1, the numbers of gates, latches and *Wide Gates* (WGs) having more than three inputs are also shown.

3 PROTEUS FPGA

3.1 Chip Overview

PROTEUS is categorized Look-Up Table(LUT) based FPGA[4], and logic is realized using 3-input-1-output LUTs (3-LUTs) and 5-input NAND (5-NAND) gates. A chip overview of PROTEUS is shown in Fig. 1. This chip is made of a core and I/O interface part. In the core part, logic blocks called *Basic Cells* or *BCs* are placed regularly, and there are routing resources to connect the BCs. As shown in Fig. 1(B), a BC has four 3-LUTs and one 5-NAND gate. The I/O interface part has 192 I/O blocks called *Pad and Latch Blocks* or *PLBs*, 32 of which are located on the up and down sides and 64 on the left and right sides. They are used to connect the outside of the chip to the core part. In addition, there are unique routing resources called *direct lines* in the I/O interface part, and they can connect PLBs directly. The specification of PROTEUS and its picture are shown in Table 2 and Fig. 2, respectively.

The uniqueness of the PROTEUS architecture is summarized as follows:

- BC structure which can easily realize basic functions such as pattern matching circuits,
- A lot of regularly-placed latches which can perform pipelined data processing thus achieving high data throughput,
- Local lines which are routing resources to connect neighbor LUTs, and
- Direct lines which contribute inter-chip connections on a multi-chip board.

Table 1. Analysis results of transmission circuits

Sub module	*Type*	*Gate*	*Latch*	*WG*
S_VTG	A B	87	13	3
AU_STG	A B	88	13	3
VC4GEN	A B	164	13	4
AU4GEN	A B E	155	13	5
CELLGEN	A B C	429	54	1
CELLT	A B C D	410	63	4
SIGG	A B C E	1550	200	22
SIGT	A B C D E	2228	317	66
STM1GEN	A B C	647	93	18
STM1T	A B C D	1177	183	27
VC4T	A B	118	13	5
VC32DT	A B C D	2024	175	55

Gate = number of total gates excluding latches.
Latch = number of latches.
WG = number of wide gates whose inputs are more than 3.
Type = basic logic type.
A: small binary counter up to 16 bits.
B: pattern matching circuits which judge if the data match a given bit pattern.
C: bit-level computation circuits like scramble pattern generation, parity calculation and so on.
D: state machines. Most of them have several states and few transitions.
E: data stream selectors. Bit width is 8 at most.

3.2 Basic Cell

As shown in Fig. 1(B), there is one 5-NAND gate after four 3-LUTs in a BC. This structure achieves easy implementation of basic functions for telecommunication circuits such as pattern matching circuits. In addition, a latch is located at each output of the 3-LUTs and the 5-NAND gate, and it can be programmed to latch the output or not. The primitive components, i.e., 3-LUT and 5-NAND gate, were derived by an evaluation of ***supports***. Here, a ***support*** is defined as the total number of primary inputs and latch outputs reached when the paths are traced from a primary output or a latch input toward the signal inputs. If a primary output of a combinational circuit has support n, the circuit can be realized one n-input LUT. For example, the support of the circuit shown in Fig. 3(A) is 11, and this circuit is realized using an 11-input LUT.

Here, we assume covering a combinational circuit bounded latches with LUTs and WGs. First, as shown in Fig. 3(B), cut the nets at the inputs and outputs of WGs which have more than 3 fan-ins, and multi-fanout points. The former is considered to cover the portion of the circuit with a WG, and the latter contributes to easy covering with LUTs for the remaining circuits. A distribution of supports for a circuit is shown in Fig. 4 when the circuit is divided into small combinational circuits following the above rules. Compared to the original

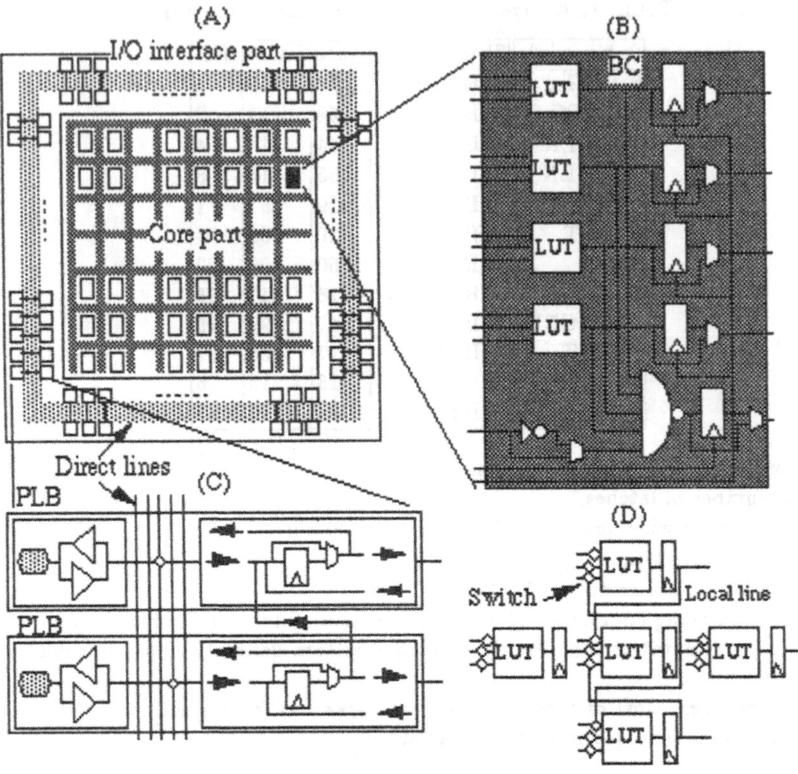

Fig. 1. PROTEUS Architecture (A)Overview (B)Basic Cell (C)PLB and Direct Lines (D)Local Lines

Table 2. Specification of PROTEUS Chip

Fabrication	0.5 μm CMOS
Die size	15 $mm^{\square}$
Package	372 pin PGA
Power supply	3.3 V
Usable I/O pins	192 (TTL-compatible)
Number of basic cells	16 x 32 = 512
Number of 3-LUTs	2048
Number of latches	2752 (including I/O latches)
Configuration	Bit-serial (18-bit address)
Scan mechanism	Full scan

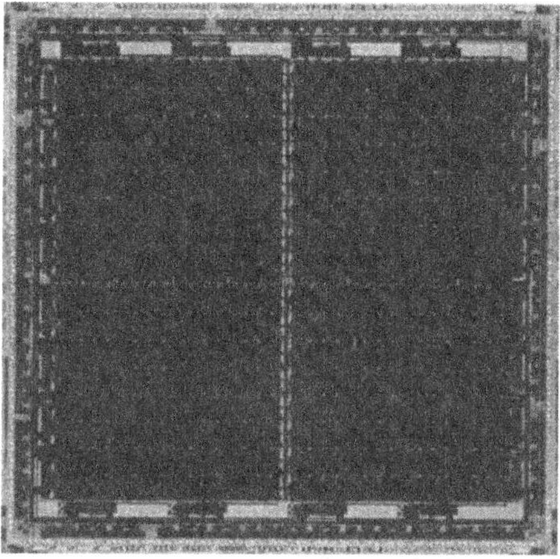

Fig. 2. PROTEUS Chip

circuit, the average support of the divided circuits was less than 3 as shown in Fig. 4. This trend was observed in all other circuits we examined. Thus, we chose the 3-LUT as a basic logic component.

On the other hand, the average ratio of the number of WGs, i.e. 5-NAND gates, in the proposed architecture is 6.7 percent if one 3-LUT realizes 3.5 logic gates in average. According to Table 1, it becomes 2.3 percent in actual circuits. Thus, we chose to introduce one 5-NAND gate for every four 3-LUTs. The triple number of WG ratio means that all WGs in an actual circuit can easily be covered with 5-NAND gates. In addition, it contributes to the easy development of the hardware macros such as pattern matching circuits by hand.

3.3 Data Latch and Clock Signal

As shown in Fig. 1(B), data latches are located at each output of the 3-LUTs and 5-NAND gates. The user can select which outputs should be latched when the PROTEUS chip is programmed. The total number of latches including latches in PLBs is 2752. Compared to commercial FPGAs, this number is rather big, and is enough to realize practical telecommunication circuits, considering the analysis results shown in Table 1.

A clock signal is delivered to all latches in a PROTEUS chip using dedicated clock lines. The clock lines were implemented in the fabrication process, and decreasing the clock-skew was, of course, considered. In addition, each latch has a clock enable pin. Thus, we can easily design the circuits that drive many sub-modules using a timing pulse. This kind of circuit is often used in data transmission circuits.

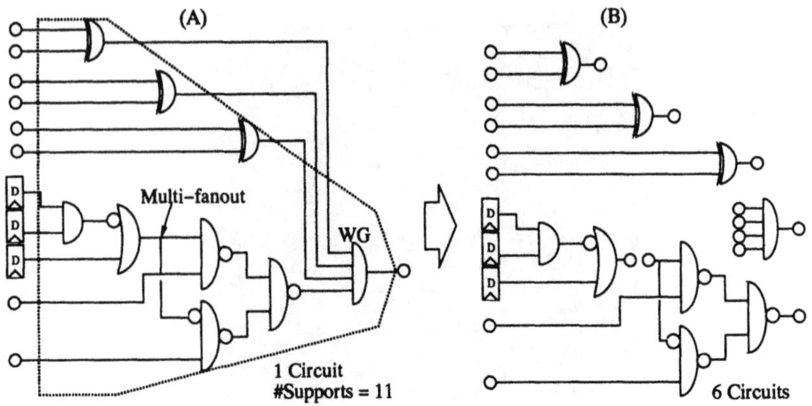

Fig. 3. An example of support and divided circuits.

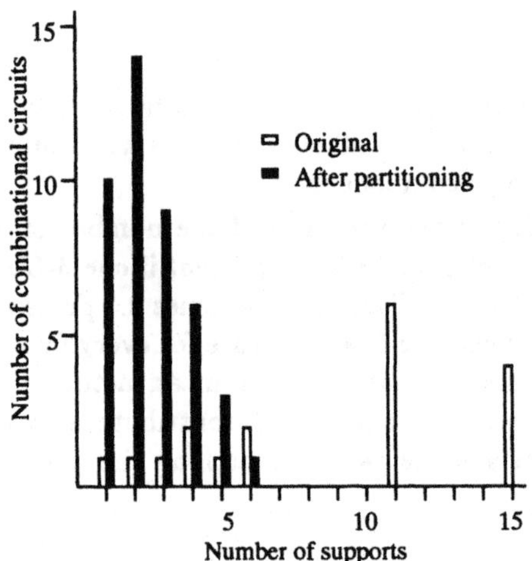

Fig. 4. Support distribution. (Circuit: AU_STG)

3.4 Routing Resources

PROTEUS has four kinds of routing resources; *local lines*, *middle lines*, *long lines*, and *direct lines*. Each routing resource has a different aim to achieve effective routing.

Local Line As shown in Fig. 1(D), local lines are used to connect neighbor LUTs directly. A mesh connection using all LUTs in the chip can be realized using the local lines without worrying about the boundary of each BC. Compared to the other routing resources, the local lines have the smallest propagation delays. Thus, they should be used if possible, to make high-performance circuits. Our preliminary experimental results showed that 20 to 50 percent of all nets in a circuit can be routed using only local line resources [2].

Middle Lines and Long Lines Details of the routing resource architecture are shown in Fig. 5. To connect BCs and PLBs, middle lines and long lines are provided. Long lines are used to establish long distance routes. A segment of the long lines has several times larger delay, compared to one of the middle lines. Thus, middle lines should be used for rather near connections. However, selecting long lines is better for long distance routes.

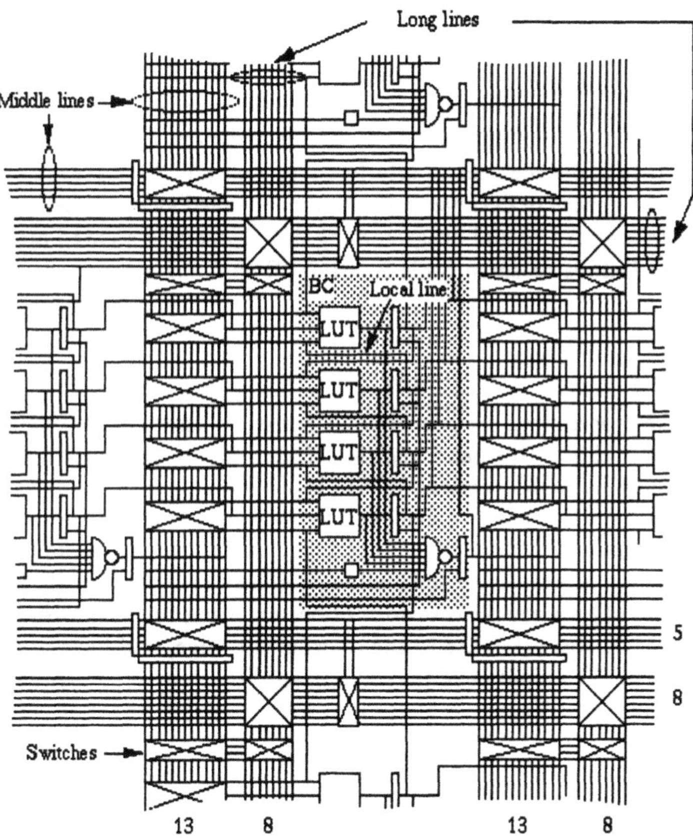

Fig. 5. Intra-chip Routing Resources in PROTEUS

Direct Lines One of the most unique features in PROTEUS is the fact that inter-chip connections are considered as well as intra-chip connections. To develop the huge circuits used in data transmission systems, we must use more than one PROTEUS chip. In addition, to ensure some board-level flexibility, we have to provide some mechanism on the boards. Our direct lines perform flexible board-level inter-chip connections without any inter-chip connection devices such as I-Cube FPID[5] and Aptix FPIC[6]. As shown in Fig. 1(A)(C), the user can bypass signals using the direct lines, without consuming any resource in the core part of the PROTEUS chip.

4 CAD System

4.1 System Overview

A dedicated CAD system called *PROTEUS-CAD* was developed for PROTEUS. An overview of the system is shown in Fig. 6. PROTEUS-CAD system supports manual design as well as automatic design.

The system provides a top-down design environment. With the PARTHENON logic synthesis system[3], the programming data downloaded into PROTEUS can be obtained directly from RTL descriptions. In addition, a visual design editor was developed to allow the user to interfere freely on any design level. Using the design editor, any piece of designed circuits can be saved at any time, and can be re-used as a macro. An actual programming data stream is created after design rule checking. Details of some tools in PROTEUS-CAD system are described after 4.2.

4.2 Mapping

Mapping is the process that assigns an input netlist to logic and routing resources in a PROTEUS chip. PLBs and registers (latches) are assigned one by one. Remaining combinational portions of the input circuit must be covered with 3-LUTs and 5-NAND gates. To achieve this, a heuristic algorithm is adopted. First, create a Boolean network from the input netlist. Next, the following two steps are executed alternatively until all logic gates are covered with 3-LUTs or 5-NAND gates; (1) the graph transformation including inverter elimination by adding a negative-edge attribute to the appropriate edge, and logic duplications if necessary, and (2) actual covering that is applied from the outputs using 5-NAND gates and 3-LUTs.

4.3 Placement

Our algorithm optimizes each LUT placement, considering the routing process. First, assume 34 x 66 slots which correspond to LUT and PLB locations, and formulate the placement problem as a slot allocation problem. A *simulated annealing*[7] method is used in the algorithm. The evaluation function considers the destination between current LUT location and the next location, usability of local lines, the number of feedback nets, etc.

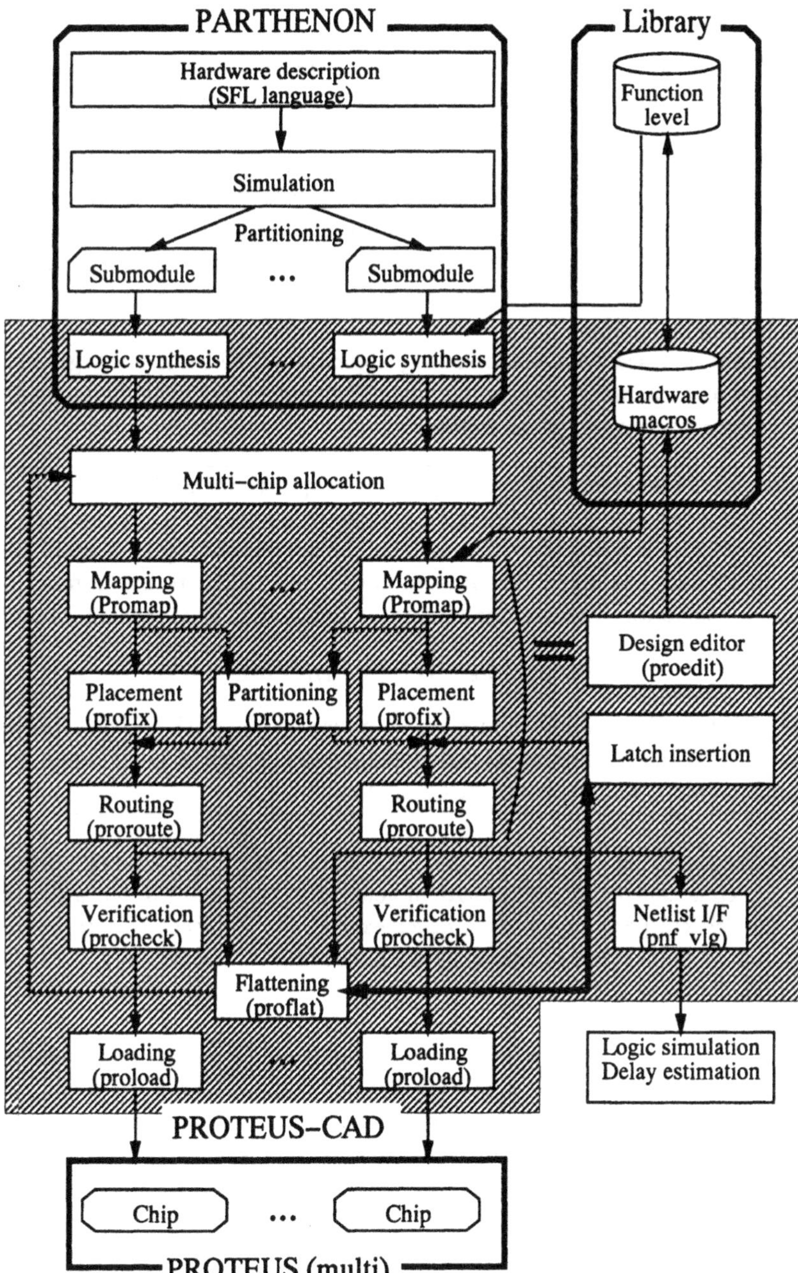

Fig. 6. PROTEUS-CAD system

4.4 Routing

Similar to other commercial FPGAs, routing is critical when implementing a circuit in PROTEUS. In this routing problem, it is very important to minimize propagation delays and to avoid unrouted nets. First, find the nets which can be routed by using just local lines, and route them. Local lines are dedicated resources to connect adjoining LUTs, and their propagation delays are the smallest among all routing resources in PROTEUS. Furthermore, local lines never disturb other routes. Next, remaining nets are routed using local, middle and long lines. Our algorithm uses unused 3-LUTs as extra routing resources by programming them as buffers if possible.

The router is based on a *maze routing* algorithm. The basic strategy of the routing algorithm is one by one net routing. The routing net order is decided by *averaged tolerated delay*, $D(e)$ as follows.

Averaged Tolerated Delay: Consider a graph whose nodes are BCs or PLBs, and edges are nets connecting among BCs and PLBs. In the graph, mark the nodes if the latches are used in the corresponding BCs or PLBs. Pick up all paths between two marked nodes, and calculate the number of nodes on each path. The averaged tolerated delay $D(e)$ is defined as follows.

$$D(e) = (D_c - W(e) * D_l)/W(e),$$

where $W(e)$ is the maximum number of nodes on the path which includes edge e, D_c is the maximum tolerated path delay, which is given when the algorithm starts, and D_l is the delay of one 3-LUT.

Clock frequency is decided by the maximum delay path among PLBs or latches, which is called the *critical path* in general. A constraint of our algorithm is that the critical path delay must never exceed D_c. To meet this constraint, nets are routed in increasing order of $D(e)$. According to this strategy, the delays of the nets containing critical path candidates are naturally minimized when routing is performed.

In addition, a mechanism to hide some routing resources temporarily is adopted in order to avoid areas of high density routed nets. Such areas often block other net routes, and cause unrouted nets. The hidden resources are released when the current net routing fails because of a lack of routing resources. Furthermore, our router has a "rip and reroute" algorithm to save finally unrouted nets.

4.5 Design Editor

It is effective to use hardware macros to program a large circuit in a PROTEUS chip. Here, hardware macros are the function parts in which BC locations are mutually fixed, and the routing of nets connecting the BCs has already been finished. For example, some function parts such as counters and pattern matching circuits, which are often used in telecommunication circuits, are suitable for preparation as hardware macros. An X-Window based design editor was developed to handle the hardware macros. In the editor, all resources, i.e. BCs, PLBs, and routing resources, are visible. The user can set or program the resources by mouse operations. One of the most unique mechanisms is that any BCs and routed nets can be grouped as in a conventional drawing tool, and the grouped part can be handled as one primitive such as a BC. A snapshot of the design editor is shown in Fig. 7. Main functions of the editor are as follows:

- setting up the function of the 3-LUT using a truth table or logic equation or Karnaugh map,
- setting up PLBs using visual sub-window,
- copying/moving/removing of any resources,
- grouping resources,
- creating and referring hardware macros,
- manual and automatic routing for any routing resources including the direct lines,
- user customization such as net ordering for automatic routing,
- resource searching by name, and
- unlimited undo.

When this design editor was developed, the MVC (Model-View-Controller) concept used in Smalltalk-80 was adopted for its maintenance and debugging. In addition, a PROTEUS chip architecture file was provided independently from the main program. This enabled us to handle the minor changes in the architecture without changing the main program.

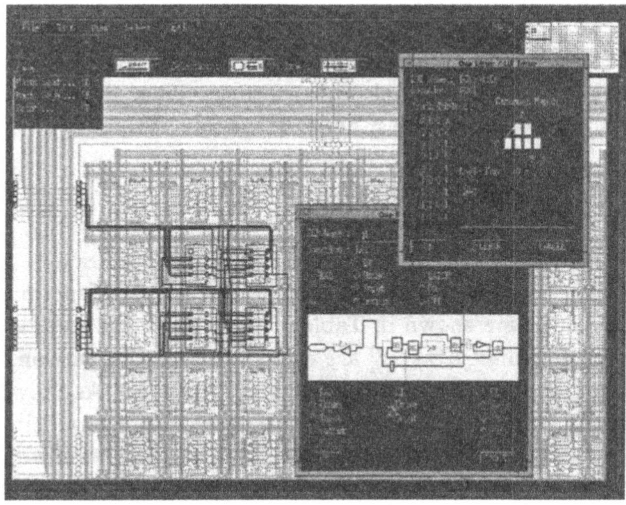

Fig. 7. PROTEUS Design Editor

4.6 Hierarchical Design Flattening

Logic level flattening can be done using the PARTHENON system. However, if there are hierarchically designed circuits using hardware macros, it is necessary to flatten the circuit hierarchy with partially routed network information. Thus, we also provided a flattening tool which can handle the hardware macros.

4.7 Partitioning

We feel that circuit partitioning for multi-chips should be performed in RTL or a much higher level to ensure high-performance circuits and effective design [8],[10]. However, considering emulator applications, a logic-level partitioning tool was also developed. Currently, we are evaluating the first version of the tool.

4.8 Throughput Improvement

Pipelining is often used to achieve high data throughput circuits in transmission circuit design. *Retiming*[9] and *latch insertion*[10],[11] methods are also known as performance improving techniques that do not change the original circuit except moving or adding latch locations. As mentioned before, PROTEUS has a lot of latches at the outputs of all BCs and 5-NAND gates, and the latches can be controlled without disturbing the routing. Thus, the presented speed-up techniques are easily applied in our PROTEUS chip.

4.9 External CAD Interface

PROTEUS-CAD system has a netlist converter. The converter produces Verilog-HDL descriptions from our original netlist format. Thus, delay evaluation and logic simulation are available with other CAD tools.

5 Experimental Results

Using the PROTEUS-CAD system, some ATM circuits and function parts were evaluated. The results are shown in Table 3. In Table 3, the results of realizing the same circuits as Xilinx LCA[12] are also shown for comparison. The upper table shows automatically designed cases using CAD tools. As shown in the table, it is better than Xilinx LCA case, and the performance of all circuits except one achieved better than 20MHz. This means that we can realize a transmission system which has a 156Mbps interface with a serial-parallel converter. On the other hand, the performance of the hardware macros programmed in a PROTEUS chip is noticeable as shown in the lower table. This indicates that hardware macro-based design would help developing very high-performance circuits.

Table 3. Experimental results

Design	*Gate*	*XFG*	*PLUT*	*XFQ*	*PFQ*
a_cell_c	231	46	87	21.0	20.0
a_tmg_t	157	24	46	22.5	23.0
count_21	362	45	59	9.2	26.0
inc_6	22	7	9	23.8	25.0
rts_mng	247	22	51	23.4	36.0
sc_mng	358	60	101	12.2	14.0
vpoaml	997	165	246	9.6	20.0
bip24	993	50	75	38.6	125.0
bip8	322	17	25	60.0	235.0
cnt_8	144	12	15	34.3	100.0
cnt_9	155	13	17	33.6	74.0
osync	413	58	84	28.7	113.6
scr	206	17	25	40.0	129.0

Gate = Number of gates
XFG = Number of used 4-LUTs in XC4000
PLUT = Number of used 3-LUTs in PROTEUS
XFQ = XC4000 clock frequency (MHz)
PFQ = PROTEUS clock frequency (MHz)
Upper table: automatically designed by CAD tools
Lower table: manually-designed hardware-macros

6 Conclusion

We introduced a high-speed telecommunication-based FPGA called PROTEUS and its CAD system. We also showed some experimental results that prove that PROTEUS is applicable to real telecommunication circuits. A programmable transmission system using multiple PROTEUS chips is under development. One of our future tasks is to evaluate PROTEUS FPGAs on boards when actual applications such as ATM-CLAD (Cell Assembly and Disassembly) circuits and protocol converters are programmed into them.

Acknowledgment

The authors would like to express their gratitude to Mr. Kazuhiro Shirakawa and Mr. Kazushige Higuchi for providing a part of the experimental results. They wish to thank all members in the Programmable Transport Research Group for their helpful suggestions.

References

1. Stallings W., "Advances in ISDN and Broadband ISDN," IEEE Computer Society Press, 1992.

2. Ohta N., Nakada H., Yamada K., Tsutsui A., and Miyazaki T., "PROTEUS: Programmable Hardware for Telecommunication Systems," IEEE Proc. ICCD'94, pp. 178-183, October 1994.
3. Nakamura Y., Oguri K., Nagoya A., and Nomura R., "A Hierarchical Behavioral Description Based CAD System," Proc. EURO ASIC, 1990.
4. Brown S.D., Francis R.J., Rose J. and Vranesic Z.G., "Field-Programmable Gate Arrays," Kluwer Academic Publishers, 1992.
5. I-Cube, Inc., "The FPID Family Data Sheet," April 1994.
6. Aptix Corporation, "Data Book," February 1993.
7. Kirkoatrick S., Gelatt J.C.D., and Vecchi M., " Optimization by simulated annealing," SCIENCE, Vol. 220, pp. 671-679, May 1983.
8. Ohta. N., Yamada K., Tsutsui A. and Nakada H., "New Application of FPGAs to Programmable Digital Communication Circuits," Proc. FPL'92, September 1992.
9. Leiserson C.E., Rose F.M. and Saxe J.B., "Retiming Synchronous Circuitry," Algorithmica, Vol.6, pp. 5-35, 1991.
10. Nakada H., Yamada K., Tsutsui A., and Ohta N., "A Design Method for Realising Real-time Circuits on Multiple-FPGA Systems," in *More FPGA,* Moore W. R. and Luk W. eds. (Proc. FPL'93), pp.129-137, 1994.
11. Miyazaki T., Nakada H., Tsutsui T., Yamada K., and Ohta N., "A Speed-Up Technique for Synchronous Circuits Realized as LUT-Based FPGAs," Proc. FPL'94, pp.89-98, September 1994.
12. Xilinx Inc., "The Field Programmable Gate Array Data Book," 1994.

A Configurable Logic Processor for Machine Vision

Paul Dunn

CSIRO Division of Manufacturing Technology, Locked Bag No. 9, Preston Vic., 3072 Australia

Abstract. This paper describes a digital VMEbus processing board which has been developed for real-time machine vision applications. The board uses FPGAs as programmable processing elements. The board is compatible with the Datacube video bus format and synchronisation. Programming is carried out at the circuit description level by means of a programming language developed for that purpose. Commercial applications in which this board has been used will be described. These include the location and tracking of road vehicles and high speed inspection of surfaces.

1 Introduction

Conventionally the processing required for video-rate machine vision has been carried out by circuit boards designed specifically for the operations required, for example histogramming or convolution. Such boards are relatively inflexible in their capabilities and are expensive. FPGAs offer the possibility of an inexpensive processor, capable of carrying out arithmetic and logical processing at video rate, which can be reconfigured to any processing architecture.

A VME processing board has been designed along these lines and has proved to be very effective. The board, called the CLP (Configurable Logic Processor), is compatible with the Datacube MAXBUS video bus and provides a parallel programming environment in which synchronisation and multiple frame storage are taken care of.

2 The Configurable Logic Processor

The **CLP - Configurable Logic Processor** is a VME circuit board containing several FPGAs which can be configured to carry out image processing operations or associated control tasks. The board includes digital input and output ports and associated timing compatible with the Datacube 10MHz MAXBUS format. A simplified block diagram of the CLP is shown in figure 1.

The VME FPGA normally configures itself from a PROM when power is applied. The other FPGAs are loaded over the VMEbus under program control from the host processor, the VME interface FPGA acting as loader.

The board contains four 512 × 512 8-bit frame stores which can be switched between various FPGAs. These use 25ns static RAM. and are connected via a

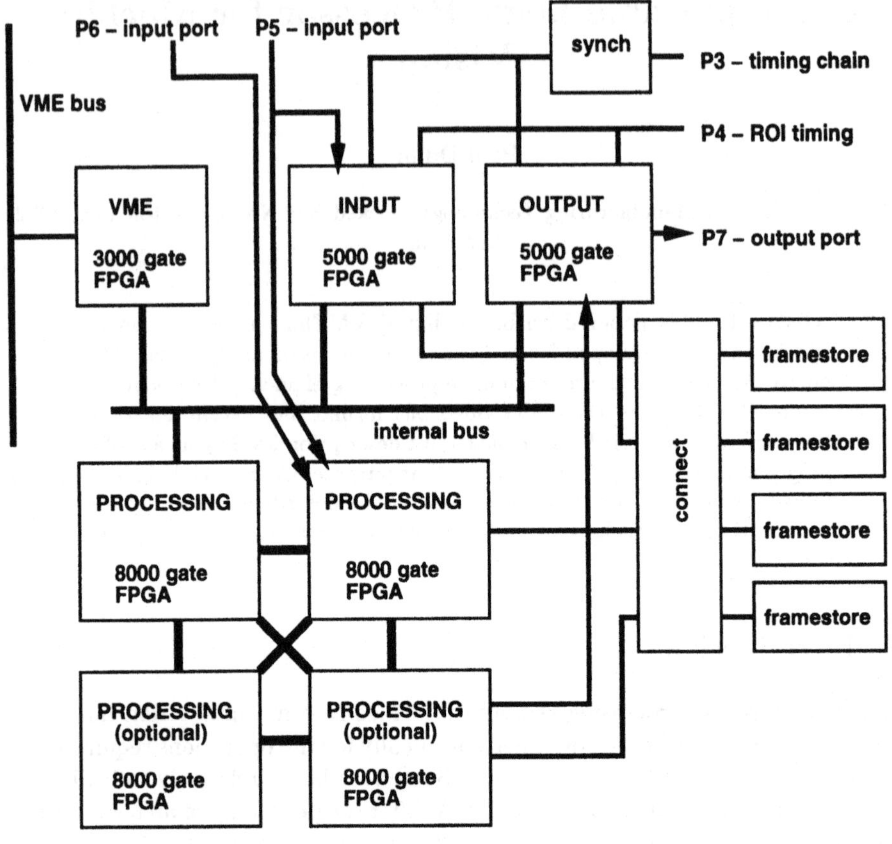

Fig. 1. CLP block diagram

shuffle network to four of the FPGAs (figure 2). All address, data and control signals are separate so as to allow independent access to each frame store.

The input and output FPGAs are normally responsible for de-interlacing and re-interlacing but can be used for other functions as they are configurable by the user.

The FPGAs are extensively interconnected to each other to aid logic partitioning and are supplied with pixel clocks and synchronising signals. Data transfer between the FPGAs and the VMEbus takes place over a synchronous bus controlled by the VME interface FPGA. 1 to 32 bit registers directly accessible through the VMEbus can be simply created in any of the FPGAs by software library modules. In addition it is usual to configure the input FPGA to directly address frame stores as part of VME address space.

Two of the processing FPGAs are socketed so that they can be omitted for simple applications or replaced by daughter boards where additional components are necessary. There are also connections to front panel sockets and to the second VMEbus connector for access to external circuits or other CLP boards.

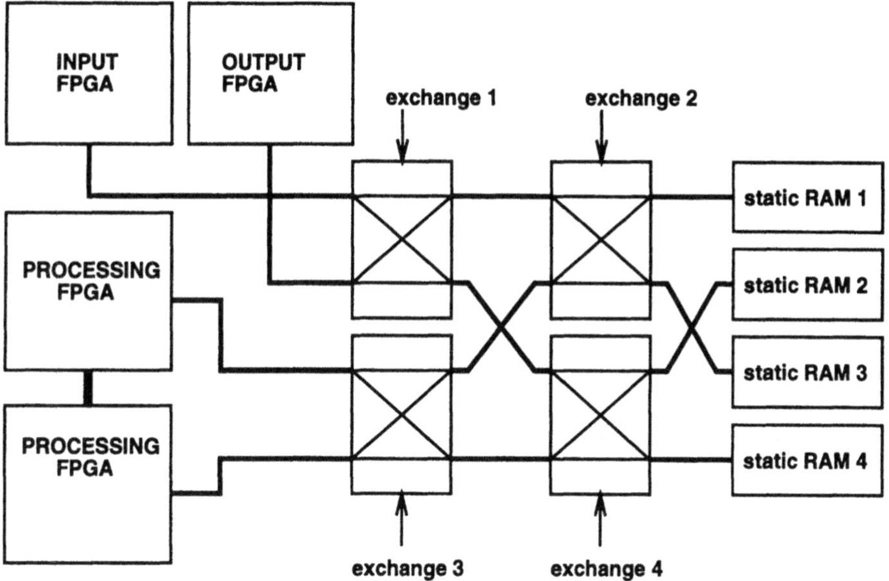

Fig. 2. Frame store shuffle network

3 Programming FPGAs

To enable programming of the CLP a circuit generation language based on C, called CCGL, was developed, along with a number of support tools such as a state machine generator, a back-annotated XNF [1] simulator and a netlist connectivity and delay analyser. The project was started at a time when VHDL was not readily available, but the adoption of our own hardware description language has conferred benefits in flexibility that otherwise we would not have had.

CCGL was intended as a step in the direction of a higher-level language in which algorithms, rather than circuit descriptions, could be encoded, however performance considerations have precluded this approach so far.

3.1 The CSIRO Circuit Generation Language

Overview. In addition to standard C data types, CCGL supports a signal data type. The assignment operator applied to signal variables implies connection, that is they now represent the same signal net. A signal variable argument to the netlist write function results in an appropriate signal name representing the net being written.

CCGL is not specifically designed for any particular netlist format. Netlist file code is created by a simple formatted write function. All netlist code creation

[1] Xilinx Netlist Form

statements would usually be contained in function libraries and source programs would not contain such low-level code.

Features. An example of CCGL code is given below. This creates a 4 bit loadable up-counter.

```
#include        "ttl_lib.cgl"
#include        "io_buffers.cgl"

        int     SIZE = 4;

#include        "connections.cgl"

        PART("4005APQ160-5");

        sig     input[SIZE], count[SIZE], clock, load;

        count = loadable_up_counter (SIZE) {input, clock, load};

        ibuf_con("clock");
        ibuf_con("load");
        ibuf_con("input", 0, SIZE-1);
        obuf_con("count", 0, SIZE-1);
```

The first `#include` statement inserts the standard library into the source code stream. This defines signal functions for gates, flipflops, I/O buffers and various other basic modules. It also defines some operators, such as `&`, `|` and `~` for logical AND, OR and NOT, so that infix logical expressions can be written inbstead of the more verbose functional notation. The second `#include` inserts the I/O library which contains functions with which to connect signals to package pins via pad buffers. The third `#include` inserts a file of declarations of package pins associated with signals to be externally connected, which might appear as;

```
        str     input_pin[SIZE] = {"12", "15", "17", "18"};
        str     count_pin[SIZE] = {"20", "22", "23", "25"};
        str     clock_pin = "37", load_pin = "39";
```

This file declares pin identifier strings which the I/O functions `ibuf_con()` and `obuf_con()` in the `io_buffers.cgl` library expect. The *str* type is a string (similarly a logical variable type *log* is available).

The syntax of the signal function call allows the passing of integers, logicals and strings in parentheses and output and input signals in braces. An output signal is returned as the value of the function (there may be more than one output signal and these may head the signal list with an intervening `<-` to separate them from the input signals).

The counter function would probably be defined in an advanced function library. The source code for such a function might appear as follows;

```
sig
loadable_up_counter (int n;) {Q[n] <- D[n], C, LOAD} {
        sig     in[n], carry[1..n-1];
        int     i;

        in[0] = mux2_n {LOAD, D[0], ~Q[0]};
        Q[0] = FD{D=in[0], C=C};
        for (i=1 ; i<n ; i++)  {
                if (i == 1)
                        carry[1] = Q[0];
                else
                        carry[i] = carry[i-1] & Q[i-1];
                in[i] = mux2_n {LOAD, D[i], Q[i] ^ carry[i]};
                Q[i] = FD{D=in[i], C=C};
        }
}
```

Note that the function is passed a size parameter. Signal functions can also determine array sizes by means of the *sizeof()* function applied to a signal array parameter. Some logical expressions have been written in infix notation to generate combinatorial logic. These could have been written using signal functions `AND{}`, `OR{}` etc.

The C-like code in these functions is interpreted sequentially and for this reason CCGL has been called a *circuit generation* language rather than a *hardware description* language.

Libraries. CCGL does not inherently generate netlist code in any specific format. Low level libraries must explicitly write strings in the expected format. For example, a low level library function for an inverter which produces Xilinx XNF netlist code might appear as;

```
sig
INV () {O <- I}  {
        net("SYM, %s, INV\n", block_name());
        net("PIN, O, O, %s\n", O);
        net("PIN, I, I, %s\n", I);
        net("END\n");
}
```

The *net()* function is a formatted write, similar to *printf()*, which writes to the netlist output file. Signal name arguments may be printed with a `%s` conversion which will produce a suitable hierarchical or flat signal name string.

Conditional Code. The usual C conditional operators are available for controlling code generation. These have been extended to the string type. The preprocessor conditional compilation operations available in C have not been implemented as a final CCGL program need only be run once to generate a netlist and hence there is really no advantage in conditional compilation over conditional code encountered during execution (in this case interpretation).

Constraint Generation. Library modules are usually written to provide include relative or absolute placement or timing constraint generation.

3.2 State Machine Generation

The CCGL compiler will execute source code pre-filtering processes named in an environment variable, but so far only one has been written. This filter parses state machine descriptions into CCGL code. A state machine description may thus appear within source code, the following being an example;

```
FSM (clock - idle, ready, run)  {
        idle -> ready = frame_flag & enable;
        ready -> idle = synch_error;
        ready -> run = start_flag;
        run -> idle = !frame_flag;
}
```

The state machine is implemented using the *one hot* approach, i.e. one edge-triggered flip-flop is used per state, hence states need no decoding. The first line declares the clock and allowed states, the body describing state transitions and the expressions which cause them.

3.3 Support Tools

xnfsim - An XNF simulator. *Xnfsim* will read a back-annotated XNF netlist file and carry out a logic simulation directed by a transaction file. The netlist file may be the result of merging several other XNF files, describing FPGAs and glue logic, using *xnfmerge*. The transaction file describes logic events, signals to print or plot and propagating logic transitions to label. It is pre-processed by the *M4* macro processor to provide functions and arithmetic expressions. The CCGL compiler produces a symbol table which lists all signal names used and the flat or hierarchical signal name used in the netlist code (flat signal names are usually generated to save file space and simplify Xilinx reporting files and displays). This symbol table is automatically sought by *xnfsim* and allows any of the names associated with a signal net to be used to reference it. If required, a history file of selected signals will be produced by *xnfsim* from which waveforms can be displayed by an interactive Xwindows program, *xsimgraph*.

xnfsurf - An XNF display program. *Xnfsurf* is an interactive Xwindows program which can be used to display connected modules in a back annotated XNF file. A signal name can be selected by clicking on its occurrence in the symbol table window or XNF text window, or by entering the name as text.

The module which is the source driving the signal will be displayed graphically. Signals can now be tracked back by clicking on displayed module inputs. The incremental or cumulative delays from the back annotated XNF file are displayed along with the signal name on each connecting wire. An example of part of such a display is shown in figure 3.

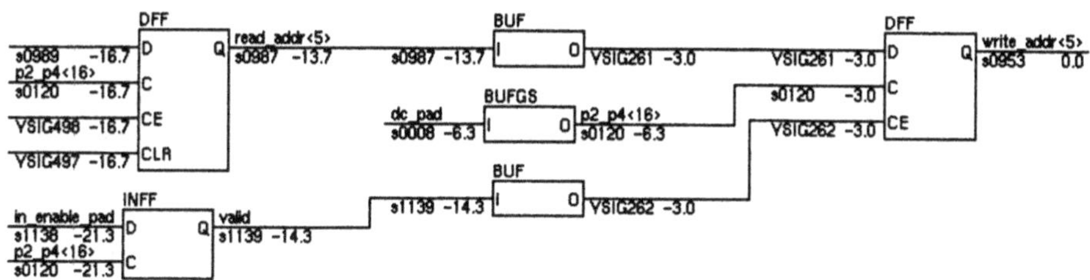

Fig. 3. Xnfsurf netlist display

3.4 Problems

Timing analysis. Path timing specifications can be used to constrain the Xilinx place and route software. The degree to which the constraints have been met can be determined by analysis of report files generated by the Xilinx software tools. Unfortunately neither the timing specification process nor the report analysis are easy. In particular the report files are voluminous, difficult to relate to the source files, and usually report many paths which turn out to be irrelevant.

It would be an advantage to formalise both the specification and the analysis of timing. At the moment the state machine generator produces netlist file timing specifications based on the clock. Generally it is easy to place global timing constraints on synchronous logic driven by one or more related clocks. The analysis of the resulting chip configuration is still difficult. A start has been made on a timing analysis program which will use the back-annotated netlist file as a source of timing information and verify that path delays do not exceed clock periods in synchronous logic.

Writing asynchronous static RAM. The use of on-chip static RAM in the FPGAs has proved to be of considerable advantage. The writing of asynchronous RAM has however been the cause of almost all timing problems [1].

The imminent availability of Xilinx FPGAs with synchronous static RAM access is expected to solve this problem.

Control of routing delays. Specification of logic module placement has resulted in a considerable improvement in path delays. For arithmetic modules using the Xilinx carry chains, relational placement must be used, however the extension of this to connected multiwire functions such as registers or multiplexers further improves layout and hence delays. Small critical timing generators can be placed in absolute locations selected to be close to associated pads or global clock buffers. Where systematic replication of modules, such as RAMs, is involved, absolute or relative placement can easily be generated in one or two dimensions by code loops.

Design level. The use of CCGL to program FPGAs requires that an algorithmic description be translated into a logical circuit. Automatic translation at the moment is challenged by the computational performance required in machine vision applications.

A compiler, using a small subset of parallel C, is being implemented based on the one-hot coding approach used by Ian Page and Wayne Luk [2] in their Occam compiler. It is hoped that the automated generation of timing constraints, the use of relative placement, perhaps some floor planning analysis and a final timing verification will augment the compiler to render it effective.

4 Applications using the Configurable Logic Processor

4.1 The Safe-T-Cam Project

The Safe-T-Cam project developed a system to automatically identify heavy vehicles using highways by means of automated number plate reading by machine vision. The project was successful and about 20 systems are to be installed this year as part of a final network of about 120 systems. The system has been described in [3]. Vehicle detection and size estimation was carried out by means of a normal interlaced camera. To avoid the problems of interlaced scans on moving objects, and to maximise the sampling rate, fields were processed individually at 50 fields per second. The pixel rate was 10MHz. The segmentation of moving vehicles in a natural outdoor scene was accomplished by subtracting a background image of the scene from the camera image. The background image was updated continuously from the camera image at a limited rate, controlled both by amplitude limiting the difference and by only updating every nth frame. The update rate was such that the background image converged on the camera image over 20 or 30 seconds. The difference image contained only moving objects. The modulus of this was taken, the result passed through a look-up table to eliminate shadow areas, thresholded and binary median filtered. Segmentation and feature extraction were then carried out by a real-time connectivity board which originated in this group and is now a commercial product [4] [2].

The image processing for the prototype was carried out by five Datacube VME boards. Production systems use the CLP board instead of those five boards as it was found that the image processing could be carried out in one 8000 gate FPGA.

[2] This board, the APA, was developed in conjunction with Vision Systems Ltd. and later Atlantek Microsystems and is now marketed by Datacube Inc.

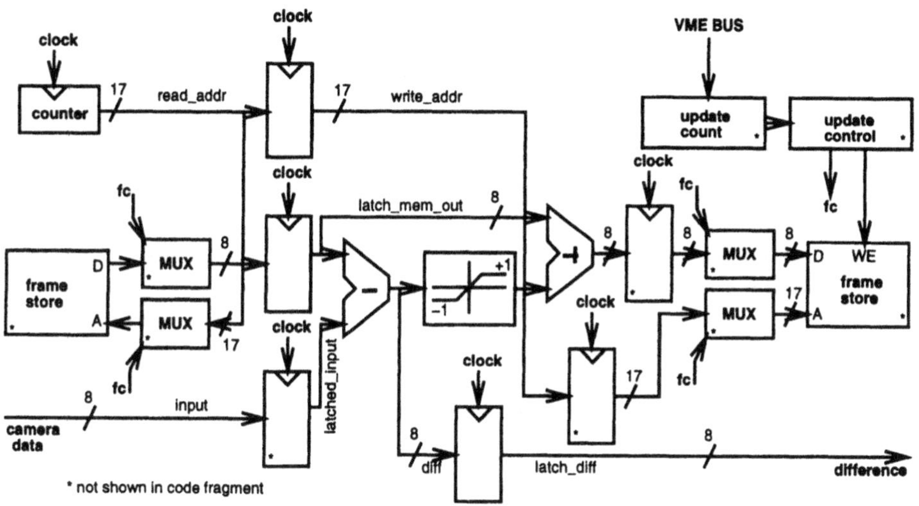

Fig. 4. Background subtraction processing

The core of the background subtraction logic is shown in figure 4. A part of the CCGL code for this is;

```
read_addr = cbce (17) {C=dc, CE=valid, CLR=~_in_vr};
write_addr = register(17){D=read_addr, C=dc, CE=valid};
latch_mem_out = register(8){D=mem_i_d, C=dc, CE=1};
diff = subs (8) {latched_input, latch_mem_out};
latch_diff = register(8){D=diff, C=dc, CE=1};
limited_diff = limit{diff};
mem_o_d = adds (8) {latch_mem_out, limited_diff};
```

While much of the detail of this code segment will be obscure, it is nevertheless interesting to compare it with the block diagram. The language allows a fairly cryptic description of the circuit. It would be preferable however to work with a description of an algorithm.

4.2 High Speed Surface Inspection

An inspection system is being designed which will digitise textured surfaces at about 25m/s and find 1mm defects. Strip width will be determined by the number of lateral channels, each 512 pixels wide. The pixel rate per channel will be approaching 20MHz. One of the filtering operations required is histogram equalisation.

The histogram code has already been developed and can carry out a 256 level 16 bit histogram at 10MHz using an 8000 gate 5ns FPGA at 93% CLB utilisation. The memory used is on-chip asynchronous static RAM with multiple adders local to smaller groups of RAM. In the first 50ns of the 100ns cycle the selected 16 bit RAM word is read and latched. In the following 50ns half cycle the value is incremented and written back. Both address and data are multiplexed to allow memory reading and clearing at the end of the frame and this adds significant further path delays. Absolute code positioning was generated systematically by loops in CCGL. The adoption of higher performance Xilinx FPGAs, which will be available shortly, is expected to double this speed and hence accommodate the 20MHz data rate.

More difficult is the computation of a cumulative histogram from the partial results and the scaling and the loading of this into a look-up table. This must be done in a frame gap time of a few lines, though the clock used does not have to run at pixel rate. The complexity of this sequential task has motivated an effort to develop a simple parallel C compiler which will generate CCGL. The algorithmic loops to carry out the function can be coded in a few lines.

5 Conclusions

The Xilinx FPGAs used in this project have two distinct disadvantages as processing elements. Firstly their internal routing contributes substantial delay between logic elements resulting in a significant limitation in performance. In video processing this has not proved insurmountable as parallelism and pipelining can still be employed successfully, even with these routing delays. The second disadvantage of configurable logic is that it is not possible to carry out very extensive arithmetic within the logic resources available. Added to this the programming of FPGAs to achieve maximum performance continues to prove difficult. Despite this we have found that an impressive amount of logical and arithmetic processing can still be carried out with FPGAs, and the CLP board has proved to be an extremely useful tool.

References

1. XILINX Inc.: The Programmable Logic Data Book. San Jose, 1994, pp8.127-8.147.
2. Page, I., Luk, W.: Compiling Occam into FPGAs. in FPGAs, W. Moore and W. Luk, Eds., Abingdon EE&CS Books, pp271-283, 1991.
3. Auty, G., Corke, P., Dunn, P., Jensen, M., Macintyre, I., Mills, D., Nguyen, H., Simons, B.: An Image Acquisition System for Traffic Monitoring Applications. SPIE proc. Vol. 2416, 1995, Cameras and Systems for Electronic Photography and Scientific Imaging, pp119-133.
4. Seitzler, T.: Area Parameters in Machine Vision: A Sampling of Multiple Applications. SPIE San Jose, Feb 1993.

Extending DSP-Boards with FPGA-based Structures of Interconnection

Peter Schulz
University of the Federal Armed Forces at Hamburg
Laboratory of Electrical Measurement Engineering

Abstract

A universal system of multiprocessors has been designed that can be used for real-time processing on measurement data. It is based on commercially available signal processor boards in a VMEbus environment. The functionality of these boards can be enhanced by freely configurable structures of interconnection. These structures are designed as reconfigurable hardware. Reconfigurability is achieved by using SRAM-based field programmable gate arrays (FPGA). Digital signal processors (DSPs) may be linked in a way that depends on the algorithmic structure of a task. Applications have to be implemented by combined hardware and software development cycles. The reprogrammability of the interconnecting structures allows recursive adjustment of hardware and software during design. One application has been implemented that performs short-time spectral analysis and computation of short-time correlation.

1 Introduction

At the Laboratory of Electrical Measurement Engineering we investigate structures of digital signal processors (DSPs) for real-time processing of measurement data. The main application is short-time analysis in both the time and frequency domains. The measurement data is sampled at rates from 100 kSample/s to 1 MSample/s on several channels in parallel. Real-time here means that data sampling takes place continuously and all data have to be processed without interruption. Processing this amount of data on-line calls for a lot of computing power. Therefore multiple DSPs in parallel are necessary.

The algorithms have to be partitioned on the processors and the communication between them has to be organized. Then the project started several VMEbus-based DSP-boards with multiple DSPs were available, but they had no special communications hardware. Structures of interconnection could be emulated by software via the VMEbus. Because of the real-time aspect the preferred hardware solution was one based on high speed serial links via the boards' front panels. Linking the processors directly is only possible by connecting cables to these high-speed serial ports. The resulting point-to-point connections are static and therefore do not satisfy the needs of the signal-processing algorithms. Separate communications hardware is necessary for more complex dataflow. This hardware should be very flexible so that interconnecting structures can be achieved for a wide range of applications.

The decision was therefore made to use the technology of reprogrammable FPGAs for designing a reconfigurable communications board. The board enables up to eight DSPs and two analog-to-digital converters (ADCs) to be embedded into a freely configurable structure of interconnection via the front panel. Multiple communications boards can be linked together across a local bus on the backplane. The board is based on a reprogrammable FPGA (XILINX XC 4013) [14] to achieve the desired flexibility. The interconnections can be configured in such a way that the dataflow is fully independent of the activities on the VMEbus. Other examples for using FPGAs in parallel computers can be found in [3] and [9].

2 Available Hardware Environment

The available VMEbus system is supplied with a PC-based master CPU and several DSP-boards.

2.1 The Host System

The VMEbus system has one master CPU. This CPU is based on an Intel 486 microprocessor and is compatible with PCs [8]. It is supplied with PC components such as VGA, floppy, harddisk etc. and is able to run DOS and Windows based programs. The DSP software development tools [2] are located on this machine. The CPU is connected to other hardware and software development platforms, which are also PCs, via a local peer-to-peer network. The master CPU has access to all VMEbus resources in a memory mapped way. The procedure is different between the real and the protected mode of the Intel processor.

2.2 The DSP-Boards

Each DSP-board holds two DSPs of the type AT&T DSP32C and has a parallel port (PIO) to the VMEbus [4]. Since the master CPU is not a real-time system, these parallel ports are only used for monitoring and development purposes. High-speed serial ports (SIO) are located on the front panel of each board (RS422 level).

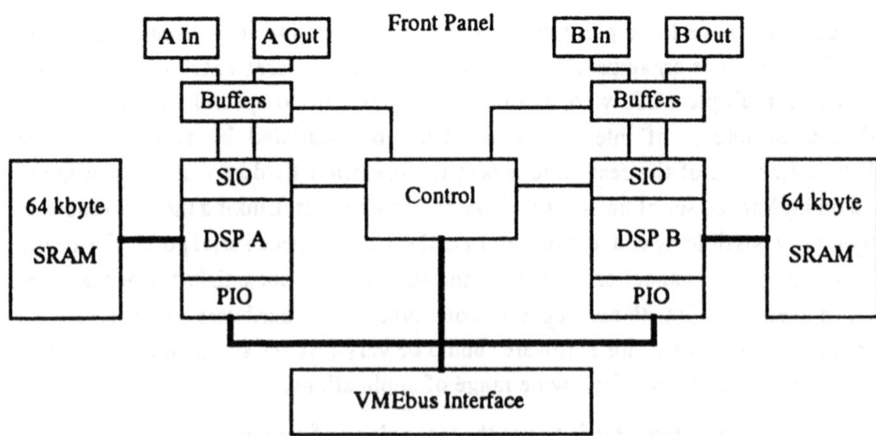

Fig. 1. Simplified diagram of the DSP board [4]

The DSPs have no possibilities to access the VMEbus as a master. They are only accessible as slaves via the parallel port. Each DSP has one port each for serial input and serial output. These are routed to the boards front panel, where they appear with RS422 level. The data transfer rate is guaranteed up to 10 Mbit/s but we have used bit clock frequencies of up to 16 MHz without any trouble.

3 The Communications Board

The purpose of the communications board is to provide dynamic structures of interconnection so that the complete system can be seen as a universal DSP-based parallel computer (see Fig. 2). This computer is a combined target and development machine.

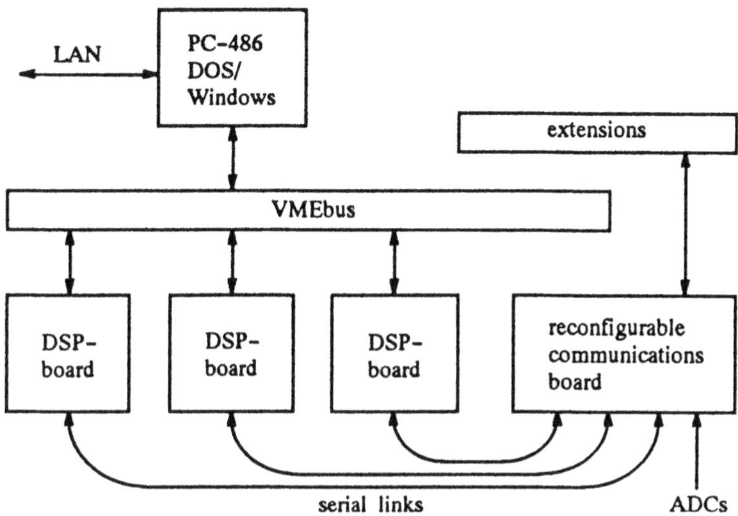

Fig. 2. Block diagram of the complete system

3.1 Concept of Dataflow-driven Communication

A multiprocessor topology is derived from the algorithmic structure of a given signal processing problem. This topology may be very unstructured. The topology then consists of data processing modules (e.g. FFT) and data transporting elements (e.g. multiplexer). The data processing modules correspond to the DSPs. The data transporting elements will become part of the communications hardware. If a data processing module has either two or more inputs or outputs, then it will be logically split into a new data processing module, which has exactly one input and one output, and hardware multiplexers or demultiplexers. Each data processing module can therefore be fitted into a DSP which also has only one serial input and one serial output port.

Because each DSP is embedded into the structure of interconnection through its serial ports its program execution can be controlled by this structure. Each DSP will not start executing its program unless its internal data buffers are completely filled with serial input data. Data from serial input is moved by direct memory access (DMA). The DSP makes no serial data transfer on its own initiative. Serial transfer can be fully controlled by external hardware. So the program execution is dataflow-driven and the communications structure is responsible for the systemwide synchronous timing.

3.2 Hardware of the Communications Board

The communications board has been designed as a VMEbus 6HE card (Fig. 3) [10]. A complex programmable logic device (FPGA) has been chosen as the boards central element to make dynamic point-to-point connections possible. A second device of this kind serves as a VMEbus interface. We use XILINX logic cell arrays (LCA) [14]. LCAs employ CMOS-SRAM technology for storing the logic configuration. It is therefore possible to reconfigure the device in system as often as required.

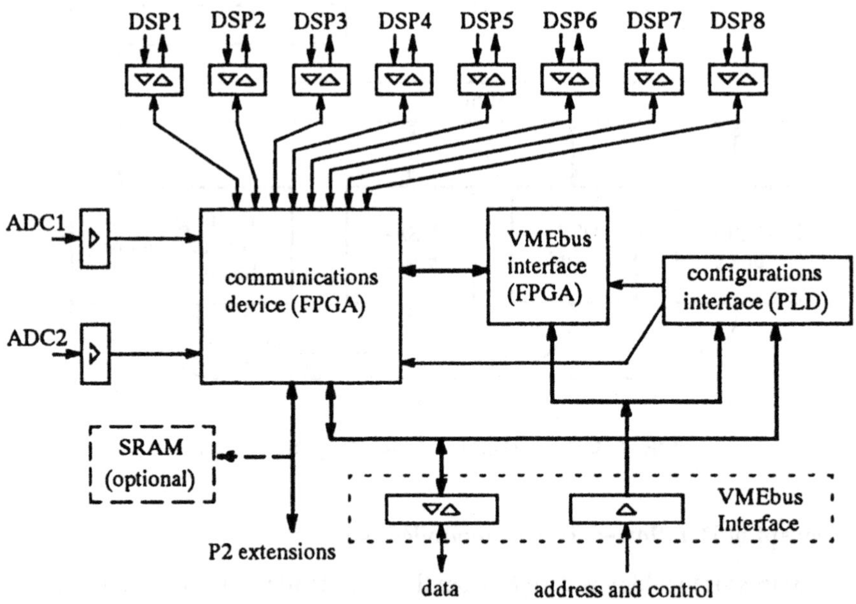

Fig. 3. Block diagram of the communications board

The main FPGA, called the "communications device", is connected to the front panel via RS422-drivers. Eight DSPs may each be connected with one input and one output port. Further on there are two serial inputs for ADCs. Hence 18 connectors are located on the front panel and the board only occupies the space of one slot! The communications device also has connections to the VMEbus data lines. So it is possible to control the structures of interconnection by the host CPU. The principal task of the VMEbus interface FPGA is to serve as an address decoder. Lastly, the communications device is connected to the 64 user-definable pins located on the VMEbus P2 connector. A local bus may be implemented there to link two or more communications boards together. Optionally some of the LCA pins used for the P2 connection can control two on-board SRAM chips (e.g. for data collection). These pins may also be used to connect a logic analyzer for debugging purposes.

The FPGA's configuration process may either be performed by a so-called download-cable (XCHECKER [15]) or by the on-board configurations PLD [5]. This device can be used to configure up to four FPGAs. It uses the so called slave serial mode [14] and the configuration process is software controlled by the master CPU.

4 Design Process

A special signal-processing application has to be implemented by hardware/software co-design [12]. The partitioning of the algorithm leads to a dataflow structure. The particular sub-algorithms are coded into programs for the DSPs. Adequate dataflow structure is achieved by the internal hardware design of the FPGA. In this way it is possible to create both static interconnections and more complex functions, such as multiplexing data streams or segmenting sets of data. The design has to be compiled into a configuration file (bitstream) for the FPGA on the hardware development system. The bitstream is transferred to the target VMEbus system via the local network and is downloaded to the communications board through the VMEbus, using the configurations PLD.

4.1 Tools

At present we work with four PCs arranged in a group network. One of these PCs is the VMEbus host. This and one other PC are used for DSP software development with the DSP manufacturer tools [2]. The two other PCs are dedicated to hardware design. There we use the FPGA manufacturer tools [15] and a schematic entry and simulation system [13].

4.2 Hardware/Software Co-Design

First the algorithm of a given problem has to be split into several sub-algorithms. Two aspects must be considered: meeting the real-time requirements and balancing the computing load. Each sub-algorithm has to be coded into a DSP program. DSP programs have to be verified by simulations and test runs. The design of the hardware/software interface is essential. For each DSP a protocol has to be defined for input and output that leads to a program frame. Using DMA for input and output means minimal overheads for the DSP software. Only changing buffers have to be managed. The design method gives enough freedom to put the main part of communications control into fast hardware functions. As far as the communication is concerned, the DSP is totally passive and therefore software development can concentrate on algorithmic tasks. The result of putting the sub-algorithm into the communications program frame is a so-called software module. Because of the changing buffers management a software module has a pipeline structure. This means that the data may be fed into it and retrieved while the module processes another set of data. Multiple instances of such a software module may be used in parallel. Each instance behaves in an identical way. The communications hardware has to ensure that each instance receives its individual sets of data.

Hardware design depends on the fact that the software modules wait for delivery or pick-up of data blocks. If the software module's processing time is longer than the communication time, then waiting states have to be inserted between two blocks of data and more instances of that module must be placed into separate DSPs, in order to guarantee processing of the continuous stream of data. These tasks require the design of state machines, counters, multiplexers and demultiplexers [6]. Multiplexers, for instance, must be switched when a certain number of data elements occur. The hardware has to be designed via schematic entry and simulation. The simulation stimuli must be derived from the DSPs hardware timing characteristics and from the input/output protocols. Last configuration bitstreams have to be generated. After configuration the complete system can be verified. The design cycles may be repeated recursively to achieve the desired functionality.

4.3 Sample Application

One sample application for the system is the short-time frequency analysis in real-time [7]. Two ADCs generate a continuous flow of data at rates of 200 kSample/s each. The communications network distributes the data to the DSPs, receives results from them and delivers intermediate results to other DSPs. The parallel working processors are synchronized by the communications board.

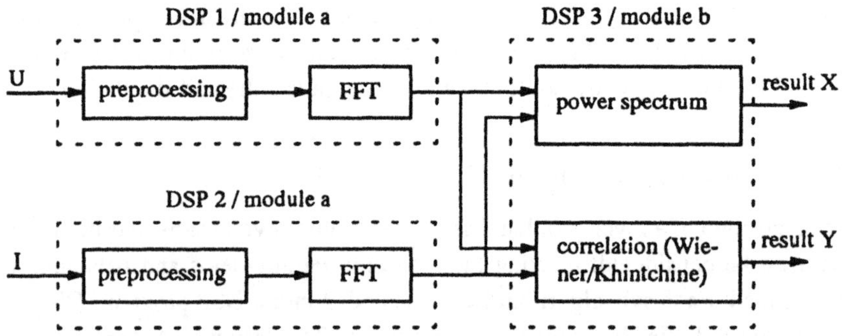

Fig. 4. Algorithmic structure and partitioning of a sample problem

Fig. 4 shows the algorithmic structure of the sample application. The measurement data are voltage and current samples taken out from electrical power supply system. Data processing takes place in data blocks of 64 Samples/segment. The main functions are separate fast Fourier transforms (FFT) on each channel and evaluation of power spectrum and correlation. Two calculated parameters, here called X and Y, will be passed on for further evaluation. The boxes with dotted lines in Fig. 4 show how the algorithm will be partitioned amongst three DSPs. The software module called "a" appears two times because the processing of U and I is the same at this stage. Thus two instances of one program can be used on two separate DSPs.

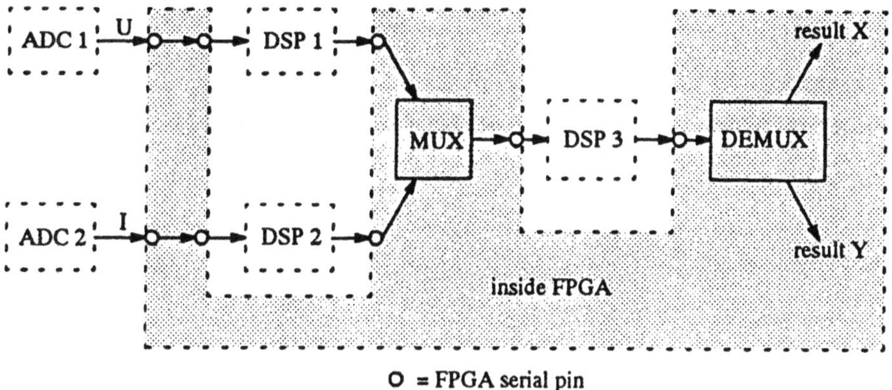

Fig. 5. Structure of interconnection for the sample problem

Fig. 5 shows how the DSPs from Fig. 4 are connected with the communications hardware. All functions built inside the FPGA are shown on a grey background. The sampled data from the ADCs are still passed through the network to DSPs "1" and "2". The other links have to be emulated by one multiplexer and one demultiplexer. Processing of all incoming data is guaranteed. If overlapped processing of the data segments is desired, then an additional group of three DSPs may be connected. The interconnection scheme is the same as above, with the exception that the second group has to be started with a delay of 32 samples. Additional multiplexers will also be needed to merge the interleaved results.

5 Conclusion

It has been shown how the facilities of commercial DSP boards can be extended by FPGAs. A multi-purpose parallel computer has been built by adding an FPGA-based communications board to available components. The system is extendable. The degrees of freedom in designing FPGA-based structures of interconnection allow realization of dataflow-driven processing schemes. The rudiments of a special design method, called hardware/software co-design, have been shown here. The hardware/software interface is still "hand-made".

The parallel computer may either be used as a target system for running varying signal processing tasks or for prototyping stand-alone applications. The concept can be ported to other DSP families or FPGA technologies. In future applications the FPGA will also be used for preprocessing the data, e.g. data decimation, moving average, offset suppression, scaling or digital filters.

Acknowledgements

The author would like to thank all colleagues at the Laboratory of Electrical Measurement Engineering, particularly Prof. Dr.-Ing. Trenkler for their assistance in carrying out this work.

References

[1] AT&T: "WE DSP32C Digital Signal Processor Information Manual" - 1990

[2] AT&T: "WE DSP32 and DSP32C Support Software Library - User Manual" - 1988

[3] J. Beck, N. Morgan, E. Allman, J. Beer: "A Multi-DSP Ring Array for Connectionist Simulations", 23. Annual Asilomar Conf. on Signals, Systems and Computers, Pacific Grove, 1989

[4] Burr-Brown: "ZPB 3200/3210 User's Manual" - 1990

[5] A. Erdbrügger: "Entwicklung eines Hard- und Softwareinterfaces zur Konfiguration SRAM-basierter FPGAs in VMEbus-Systemen" - internal report, Laboratory of Electrical Measurement Engineering, University of the Federal Armed Forces at Hamburg, Hamburg 1994

[6] V. Härtrich: "Entwicklung FPGA-basierter Kommunikationsstrukturen für einen DSP-Parallelrechner" - internal report, Laboratory of Electrical Measurement Engineering, University of the Federal Armed Forces at Hamburg, Hamburg 1995

[7] F. Hornung: "Entwicklung von Softwaremodulen zur Echtzeitsignalverarbeitung mit einem DSP-Parallelrechnersystem" - master thesis, Laboratory of Electrical Measurement Engineering, University of the Federal Armed Forces at Hamburg, Hamburg 1994

[8] Radisys Corporation: "EPC-5 Hardware Reference" - Beaverton, 1993

[9] G. Rosendahl, T. Paille, R. McLeod: "In-system Reprogrammable LCAs provide a Versatile Interface for a DSP-based Parallel Machine" - Oxford International Workshop on Field Programmable Logic and Applications, Oxford, 1991

[10] R. Schepull: "Entwicklung einer Kommunikationsbaugruppe zur Vernetzung digitaler Signalprozessoren" - master thesis, Laboratory of Electrical Measurement Engineering, University of the Federal Armed Forces at Hamburg, Hamburg 1994

[11] P. Schulz: "Einbettung von Signalprozessoren in ein flexibles Kommunikationssystem zur online-Messdatenverarbeitung"- accepted paper for ECHTZEIT'95, Karlsruhe 1995

[12] P. Schulz, F. Hornung: "Hardware/Software-Codesign mit Signalprozessoren in einer FPGA-basierten Kommunikationsstruktur zur Echtzeit-Meßdatenverarbeitung"- accepted paper for 40. IWK, Ilmenau 1995

[13] Viewlogic Systems Inc.: "Workview PLUS on Windows, Vol. 1-6" - Marlboro, 1993

[14] Xilinx: "The programmable Logic Data Book" - San Jose, 1994

[15] Xilinx: "XACT Reference Guide, Vol. I-III" - San Jose 1994

High-Speed Region Detection and Labeling Using an FPGA-based Custom Computing Platform

Ramana V. Rachakonda
Ross Technology Inc., Austin TX 78735

Peter M. Athanas and A. Lynn Abbott
The Bradley Department of Electrical Engineering
Virginia Polytechnic Institute and State University
Blacksburg, Virginia 24061-0111

Abstract. General purpose custom computing platforms, such as *Splash-2*, have demonstrated the ability to enter mainstream computing not only due to their near application-specific speeds but also because of their ability to run a wide variety of tasks. *Splash-2* is a second-generation FPGA-based system that can deliver processing performance rivaling application-specific systems, but is also reconfigurable. This paper describes a computationally intensive image processing task, known as region labeling, which demonstrates the effectiveness of such platforms. The design and implementation of a region labeling task on the *Splash-2* custom computing platform are described and the resulting performance is compared with that of other machines.

1. Introduction

Many image-processing tasks are inherently computationally intensive because of the large amount of data associated with each image. High speed image processing is further complicated by the fact that a video camera presents each two-dimensional image as a one-dimensional data stream. Fast image processing therefore requires specialized data paths, application-specific processing elements that can operate in parallel, and careful data sequencing. Most general-purpose systems do not provide these capabilities, and cannot perform image processing at near real-time rates. In essence, general-purpose systems are not capable of exploiting parallelism to the extent that is needed for high-speed image processing applications.

Reconfigurable custom computing platforms, such as *Splash-2*, are a compromise between application-specific and general-purpose platforms [1], [3], [4]. The typical design flow on such platforms starts with validating the algorithm to be implemented in a high level language. For *Splash-2*, synthesizable VHDL models of the algorithm are then simulated. The model is synthesized and then downloaded to the *Splash-2* custom computing platform. A detailed discussion of this design process is given in [4].

This paper describes the implementation of a computationally intensive image processing task, region labeling, on the *Splash-2* custom computing platform. *Splash-2* and the *VTSplash* image processing platform are briefly discussed in Section 2. The algorithm and the configuration of *Splash-2* for region labeling are presented in Section 3. Performance results are given in Section 4, and Section 5 contains a summary.

2. The Splash-2 Custom Computing Platform

Splash-2 is a second generation reconfigurable custom computing platform developed at the Center for Computing Sciences (formerly the Supercomputing Research Center) in Bowie, Maryland. The system contains a 1-D systolic array of Xilinx FPGAs and was designed as a proof of concept for reprogrammable platforms [7], [8]. The FPGAs may also be interconnected through a crossbar switch. For some applications, *Splash-2* performance has exceeded that of a Cray supercomputer. A more complete discussion of the features of the *Splash-2* attached processor is in [2], [5].

An image processing platform called *VTSplash*, which is based on *Splash-2*, was developed at Virginia Tech. The system consists of a video camera, a digitizer, a *Splash-2* attached processor, a frame buffer to grab data from *Splash-2* and send it to a monitor for display, a color video monitor for displaying the processed image from *Splash-2* and a Sun SPARCstation-2 to program and to send control signals to *Splash-2* attached processor. For a detailed discussion on the *VTSplash* environment refer to [4].

3. Region Detection and Labeling

3.1. Introduction

Region labeling is a computationally intensive image-processing task that has been successfully implemented on *Splash-2*. Briefly, a region in an image is a set of picture elements (or pixels) that form a connected set in $\mathbf{Z}^2$. A foreground pixel, P, is said to be *eight-connected* to another foreground pixel, Q, if there exists a path of adjacent foreground pixels between P and Q, traversing either horizontally, vertically, or diagonally. They are said to be *four-connected* if the path does not traverse diagonally. Three eight-connected foreground regions, represented by dark pixels, are shown in Figure 1(a). This separation of image pixels into foreground and background can be easily performed by thresholding in some situations. Many applications (such as motion estimation, tracking, or object recognition) require that a unique label be assigned to each region. This is illustrated in Figure 1(b), which contains integers have been selected arbitrarily as labels for foreground pixels. This labeling process is deceptively difficult because holes can lie within regions, and because equivalence lists must be maintained to handle cases in which different labels are tentatively assigned to a single region. Ultimately, such equivalencies must be detected and resolved to a single, unique label for a given region.

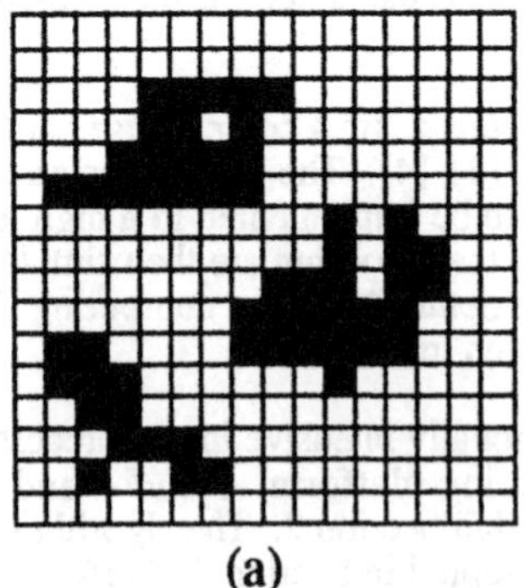

(a)

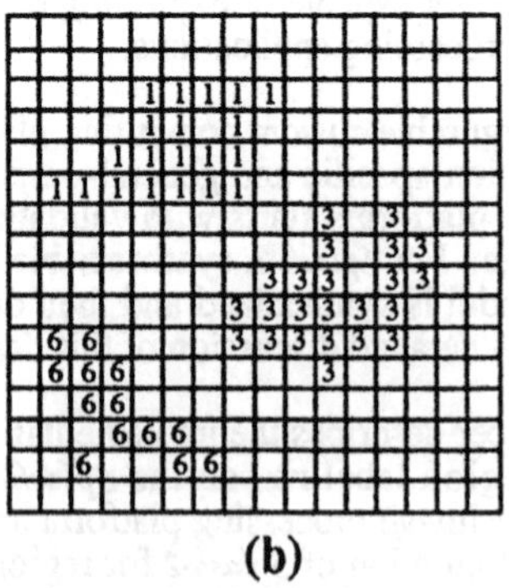

(b)

FIGURE 1. Illustration of region labeling.

The goal of the system described here is to assign a unique label to each 8-connected foreground region in an image. Labeling is performed in two passes over the image. During the first pass tentative labels are assigned as illustrated in Figure 2(a) for the image in Figure 1(a). A simple window-based operation results in connected regions, each of which may contain more than a single label. The second pass over the image resolves such equivalent labels, and assigns a single unique label to each region.

In this implementation a lookup table is maintained to store equivalences of labels. The lookup table is initialized at the beginning of the first pass for every frame to contain a value equal to its index ($M(i)=i$). The table is modified during the first pass to indicate equivalences of labels. The state of the lookup table after first pass processing for the above example is shown in Figure 2(b). Observe that the locations 1 and 2 have the label 1. Also, locations 3, 4 and 5 have the label 3.

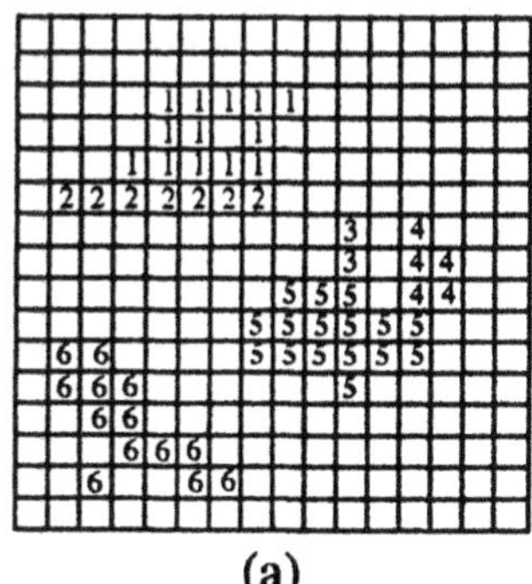

(a)

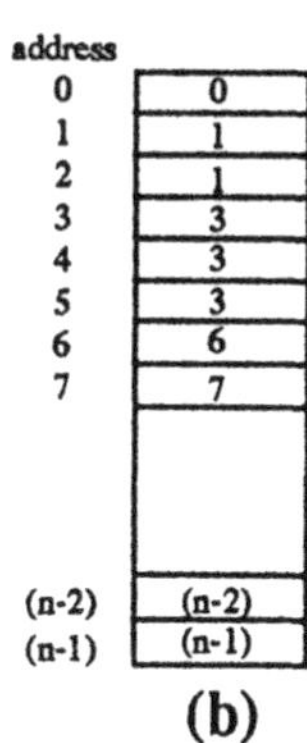

(b)

Fig. 2. Image and lookup table after first pass processing.

3.2. Algorithm

3.2.1 Region Detection by Image Thresholding

Thresholding reduces the number of quantized levels that a pixel of an image can assume. If the output image is to contain two quantized levels, foreground and background, one threshold, T, can be used as follows:

$$I_{out}(i,j) = \begin{cases} 1 & \forall (I_{in}(i,j) > T) \\ 0 & \forall (I_{in}(i,j) \le T) \end{cases}$$

For the implementation described here, 1 and 0 represent foreground and background, respectively. For simplicity, all pixels are represented using eight bits. The threshold, T, is selected before the design is synthesized.

3.2.2 Connected Component Labeling

The problem of region detection and labeling is the process of identifying connected components in an image [10]. Assuming a binary image in which each pixel is represented by background or foreground, a label R can be assigned to a foreground pixel, P, if and only if it meets the following conditions:

1. All the foreground pixels that are 8-connected by a path to the pixel in discussion are also labeled R.
2. All the foreground pixels that are not 8-connected by a path to the pixel in discussion have a label other than R.

For images processed in raster-scan order, in which the pixels are processed left to right and top to bottom, algorithms for connected component labeling consider labels of rows *Y* and *Y-1* only while labeling row *Y*. Figure 3 shows the four neighbors considered for labeling a current pixel C. The neighbors are named L, UL, UC, and UR to represent the left, upper-left, upper-center, and upper-right pixels with respect to the current pixel C, respectively. The current pixel C is given a label *R* if and only if one of its neighbors has the label *R* and C belongs to the foreground. If the current pixel C encounters two different labels, *P* and *Q*, given to its neighbors, it acquires one of the labels and an equivalent pair is generated indicating *P* and *Q* belong to the same region in the image. The equivalence pairs generated are stored in the equivalence table. The window then advances by one pixel and similar processing occurs. Conceptually, a moving window of 5 pixels passes over the entire image, and a label is produced for pixel C at each window location.

Row *Y-1*	UL	UC	UR	
Row *Y*	L	C		
Row *Y+1*				

Fig. 3. Window to be considered for labeling the current pixel, C.

An equivalence table can be implemented as a lookup table in memory. The table is initialized by writing the index *i* to a memory location *M(i)* (for $0 \le i < K$, where *K* is the maximum number of labels permitted by the algorithm). If an equivalence pair of (*P*, *Q*) is detected, the values *M(P)* and *M(Q)* are compared. If they differ, all the memory locations containing the value *M(Q)* are replaced with the value *M(P)*. The process is repeated for all equivalence pairs generated for that particular image. The equivalence table which results contains an identical value in the locations pointed by labels of the same equivalence class. Thus various labels given to the same object are mapped to one label. This processing is performed during the first pass over the image. During the second pass, the image is processed again to merge all the inconsistent labels and the output contains one consistent label per equivalence class. This is accomplished by replacing all first-pass labels, *f*, by the values *M(f)* from the lookup table. The result is that each 8-connected foreground region contains a unique single label.

The process described above takes more than 3*K* clock cycles for a lookup table in an external memory, where *K* is the maximum number of labels supported by the system. This is due to the fact that each memory location needs to be read and conditionally written to, and a read followed by a write to an external memory requires three clock cycles. The process becomes very inefficient for designs handling large number of labels, particularly for general-purpose processors.

3.3. Architectural Features of the Implementation on Splash-2

The *VTSplash* system accepts images of size 512 rows x 512 columns directly from a standard black and white video camera using a 10-MHz clock. Five pipeline stages are used to perform 1) image thresholding, 2) first-pass labeling and equivalence detection, 3) data storage and equivalence table management, 4) data retrieval for second-pass processing, and 5) merging. To provide near-real-time operation, the latter three stages are duplicated and used alternately. The complete implementation utilizes nine Xilinx XC4010 FPGAs [6], processes 30 complete images per second, and displays the results using pseudocolor on a video monitor. Figure 4 shows the five pipeline stages for two consecutive images. The pipeline stages shaded in stripes perform first pass processing on the incoming image, while the ones shaded in dots perform second pass processing on a previous image. The unshaded stages are idle. If the first pass processing of a previous image is not completed before the arrival of a new image, the new image is ignored. Hence, if the first pass processing of every image requires the algorithm to skip alternate images, the processing rate drops from 30 frames per sec to 15 frames per second.

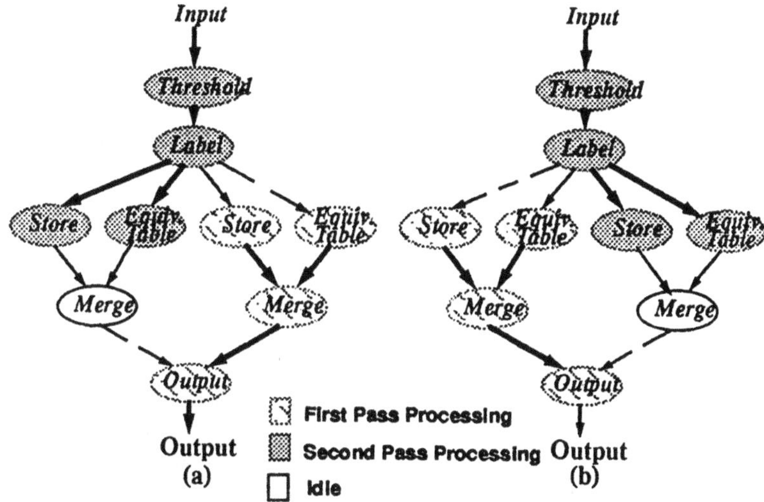

Fig. 4. Data flow through the algorithm.

3.4. Worst case analysis and limitations

If an equivalence pair *(P,Q)* has not been identified earlier in the image, the equivalence table needs updating and the time required depends on the size of the lookup table. For *Splash-2* this requires $3K+8$ clock cycles. Such a merger can be called a real merger. If the two labels from a merger already belong to the same equivalence class or are repeated, no label merging is required. This null operation is called a duplicate merge and eight clock cycles are needed for processing. A worst case analysis of the time taken to complete the processing was calculated. The example images shown below illustrate the process. Figure 5 exemplifies the process when the total number of real mergers are the maximum. For a 512x512 image, the maximum number of mergers is 65,536 and hence the number of clock cycles taken for completion is 64×10^5.

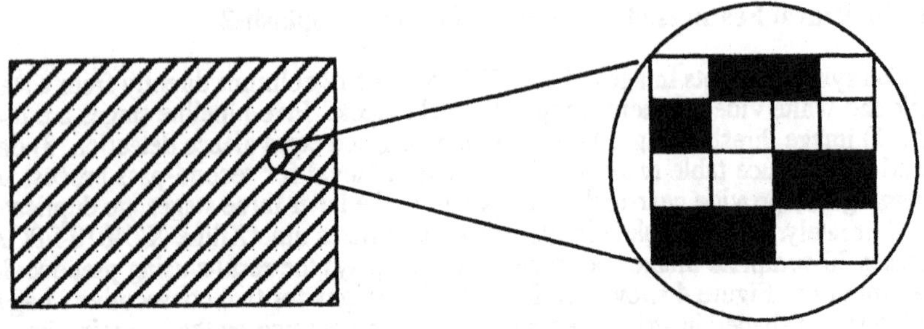

Fig. 5. Worst case analysis: Case I.

A second case is illustrated in Figure 6. In this case there are N real mergers (N is the number of rows), and $\left(4\sum_{n=1}^{\frac{N}{2}} N-n = \frac{3}{2}\times N^2 - N\right)$ duplicate mergers. Hence, this example considers the case when the total number of duplicate mergers are the maximum. Recall that a real merger updates the look-up table and takes $(2K+36)$ clock cycles to update the look-up table, which is typically 100 clock cycles for an equivalence table of size thirty two. The total processing time for such an image is given in the equation below. δ is the time taken to update the equivalence table.

$$T = 8\times\frac{3}{2}\times N^2 - N + N\times\delta = 12\times N^2 + (\delta - 1)\times N$$

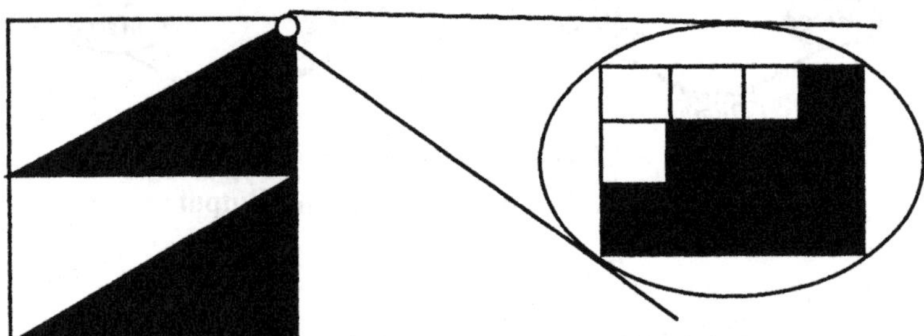

Fig. 6. Worst case analysis: Case II.

Since the equivalence table is implemented as a lookup table in external memory, the time taken to update the table is directly proportional to the size of the table. Hence for a huge lookup table and an image with a lot of real mergers, the algorithm trades off the ability to process the whole image with the real-time frame handling rate (typically 30 frames/sec).

4. Performance

For most images *Splash-2* executes this region detection and labeling task at the video rate of 0.033 seconds per image. A comparison with several general-purpose workstations, programmed using the C language, is given in Table 1. Initial memory load times and processed image store times have been ignored. As indicated in the table, *Splash-2* exhibits a speedup of approximately 17 over a SparcStation-10, an 11 fold improvement over Ross Technology's HyperSparc (SparcStation-20 HS11) and 9 fold over the HyperCache [9].

Platform	Time Taken (Normalized)
SparcStation-2	54
SparcStation-10	17
Ross HyperSparc	11
Ross HyperCache	9
Splash-2 (real-time mode)	1

TABLE 1. Normalized execution time required to process one 512x512 frame.

5. Summary

Custom computing platforms such as *Splash-2* can perform a wide range of tasks at speeds comparable to application specific hardware. Region detection and labeling is a challenging application because of the need to resolve equivalences, and because of the amount of data associated with each image. *Splash-2* performs this task at near video frame rate, which represents a 17-fold improvement over conventional architectures.

References

[1] A. L. Abbott, P. M. Athanas, L. Chen, and R. L. Elliott, "Finding Lines and Building Pyramids with Splash 2," *Proceedings: IEEE Workshop on FPGAs for Custom Computing Machines*, Napa, CA, April 1994, pp. 155-163.

[2] J. M. Arnold, D. A. Buell, and E. G. Davis, "Splash 2," *Proceedings: Fourth Annual ACM Symposium on Parallel Algorithms and Architectures*, pp. 316-322, 1992.

[3] P. M. Athanas and A. L. Abbott, "Real-Time Image Processing on a Custom Computing Platform," *IEEE Computer*, vol. 28, no. 2, Feb. 1995, pp. 16-24.

[4] P. M. Athanas and A. L. Abbott, "Image Processing on a Custom Computing Platform," *Fourth International Workshop on Field-Programmable Logic and Applications (FPL '94)*, Prague, Czech Republic, Sept. 1994. Published as *Lecture Notes in Computer Science 849*, R. W. Hartenstein and M. Z. Servit (Eds.), Berlin: Springer-Verlag, 1994, pp. 156-167.

[5] D. A. Buell, J. M. Arnold, and W. J. Kleinfelder, eds., *Splash 2: FPGAs in a Custom Computing Machine*, in press.

[6] *The Programmable Gate Array Data Book*, Xilinx Inc. San Jose, California., 1994.

[7] *Spectrum Reconfigurable Computing Platform Developers' Kit*, Giga Operations Inc., Berkeley, California, 1995.

[8] *Wildfire Programmers' Manual*, Annapolis Micro Systems, Annapolis, Maryland, 1995.

[9] *hyper ACTIVITY*, Ross Technology Inc., Austin, Texas, June 1995.

[10] B. K. P. Horn, *Robot Vision*, MIT Press, McGraw Hill, 1986.

Using FPGAs as control support in MIMD executions

Christophe Beaumont*

Laboratoire d'Informatique de Brest
Université de Bretagne Occidentale
BP 809, 29285 Brest, France
e-mail: beaumont@univ-brest.fr

Abstract. In the wide field of parallel architectures, machines involving FPGAs on each node have appeared during the last years. Connecting these reconfigurable components opens new research horizons. Expensive control tasks required by distributed applications may then be accelerated using application specific hard-wired elements.
A derivation from a standard algorithm for distributed simulation and its partial hardware implementation are presented in this paper, providing three magnitude orders speed-ups.

Introduction

In the wide field of parallel architectures, machines involving FPGAs on each node have appeared during the last years (a long list of machines using FPGAs is maintained on Internet[1]). Connecting these reconfigurable components opens new research horizons.

Only few work has been done on algorithms for architectures involving different communication mediums. When several distinct communication networks are used, the potentiality of each one may be difficult to express in terms of algorithm. One key point is the independence between the networks and thus, the possibility for messages to cross each other, or even to overtake one another, due to different transport delays. This is of particular importance when control and data messages do not use the same network: control mechanisms may not detect properties hold in the transmitted data.

These control mechanisms are crucial for a correct execution on MIMD machines where processors do not share variables. They require message exchanges and the contribution of the computing nodes, are always time-consuming and depend often on applications. The computation directly related to the application is stopped and the potential speed-up dramatically affected. Considering this last point, our major objective is to show that well-suited hardware may improve distributed execution by lowering control mechanism overhead, making

* This work is partially supported by the Doctoral-candidate Network for System and Machine Architecture of the DRED

[1] http://uts.cc.utexas.edu/~guccione/HW_list.html

computation and control executions overlap, decreasing the load of the main communication network.

The ArMen architecture has been presented in 1989 [5] and proposes the coupling of two distinct layers:

- a distributed MIMD machine communicating through message exchanges. The current prototype is based on a Transputer network;
- an FPGA ring with its hardware-like performance and software-like flexibility. An FPGA is connected to the Transputer of each node, sharing its system bus. The reconfigurability of the FPGAs allows circuit synthesis ranging over many areas: local coprocessors, global operators or global controllers. The two latter are shared by all nodes and the last one is used in this paper.

The use of these two layers offers the possibility of implementing all or the most part of control mechanisms in reconfigurable hardware while the calculation part of distributed execution remains on the MIMD level. This is a kind of hardware/software codesign. A compiler enables automatic generation of the whole hardware configuration from a unique source code written in a Unity-like language[6].

This paper presents latest development on the ArMen machine on speeding up the control tasks in MIMD execution. The whole development of a synthesized accelerator for a particular simulation protocol is described. We first describe more precisely the prototype machine we are working on. In the second section, we focus on the transformation of a standard algorithm[3] we propose for controlling distributed simulation. The way it is implemented in reconfigurable hardware is detailed in Sect. 3. We consider in this section the entire design process, including the decision of which part of the algorithm remain software and which will be synthesized. The next section gives first results considering both user and developer points of view. A short summary and further work on different exposed aspects end the paper.

1 The architectural support

The target architecture in this paper is a general-purpose machine called ArMen. This machine is based on a modular distributed memory architecture where processors are tightly connected to an FPGA ring (see Fig. 1). This section gives details of the implementation and possible applications.

Each ArMen node has a processor connected to a bank of memory and a large Xilinx FPGA[11]. The experimentation reported in this paper has been conducted on a prototype machine using Transputers and Xc3090 FPGAs on VME-like boards. The processor loads new configuration data from a bitstream into its FPGA in less than $100\,ms$. The topology of an FPGA is a CLB array (*Configurable Logic Block*) with four external 32-bit ports: its north port is connected to the local address/data multiplexed system bus, and its east and west ports to neighbours in a ring topology. The ArMen machine is a set of such

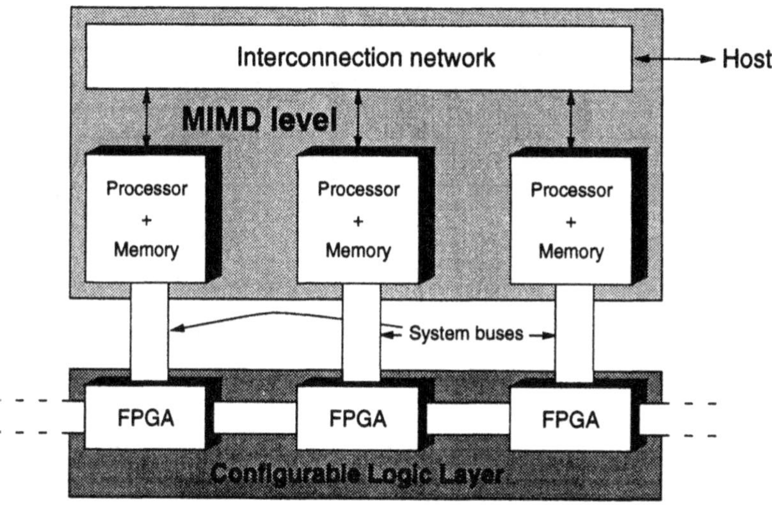

Fig. 1. Principle scheme of the ArMen machine

nodes interfaced to a workstation hosting a Transputer board. The Transputer nodes of the machine executes the TROLLIUS operating system[2] to ease the application developments. TROLLIUS provides concurrent input/output, routing facilities and support to load and observe processes on the nodes.

The ArMen FPGAs are local coprocessors for each node (such designs remain in the field of CAD approaches), or parts of a *global shared coprocessor* for the processor array. The logic layer (ring of interconnected FPGAs, later on called *configurable logic layer* or CLL) of the parallel machine provides support for a large diversity of global coprocessors. They combine operative and control functions for specific tasks.

Global operative units A global operative unit uses the set of local memories as a large contiguous data store, and combines the actions of the processor array with FPGA synchronous processing and micro-grain communications. Such units are currently produced from a high level language, called CCEL[10].

Global controllers An implicit model for global computation is based on pipelines with one stage within each node (see Fig. 2). In the pipeline mode, data is encapsulated into tokens to ensure the synchronism of their collection in the node. An automaton in the node 0 coordinates the pipelines[7].

The node processor interacts with its pipeline stage by reading or writing FPGA-based registers. Generally, local programs simply drop significant information into FPGA-based double buffered channels. Global predicates like termination detection or synchronization conditions can be computed with these global controllers. *Control circuits* are currently compiled from UCA[6], a lan-

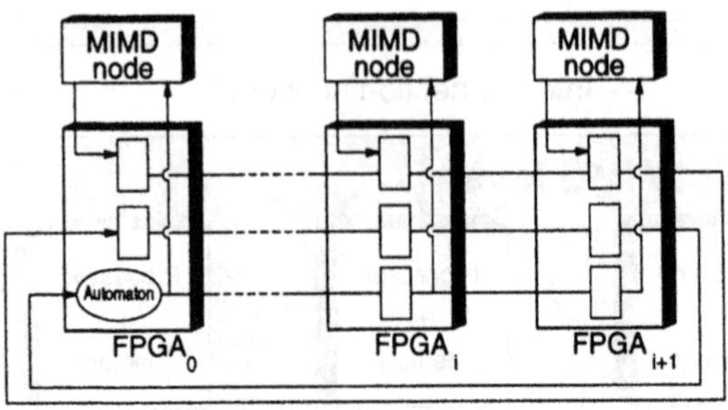

Fig. 2. FPGA nodes for pipeline implementation

guage based on the Unity formalism[4]. The rest of the paper will deal with this kind of controller.

2 Algorithm description

Parallel discrete event simulation (PDES), running on MIMD architectures, requires efficient control operators. The development reported in this paper aims this application field.

Although PDES sounds restrictive, research has shown that many different approaches reach good performance, depending on the simulated models. Interested readers may find further details on these approaches in [8].

Preliminary work on the ArMen machine has dealt with synchronous event driven simulation[1]. In this section, we describe an asynchronous approach based on a deadlock detection and recovery algorithm[3]. In brief, each simulation process simulates until it blocks. The control algorithm has to detect the deadlock (i.e., when every processor is blocked), and to provide a mechanism that breaks it. This is done by computing the global minimum of local time values owned by each site, taking into account messages in transit in the network (i.e., messages that are sent but not received yet).

Related work is an algorithm developed for an architecture using a so-called *Parallel Reduction Network* (PRN), built of general purpose processors associated with a MIMD machine[9]. This PRN has communication and computation capacities. Nevertheless, the algorithm requires serialization of the acknowledgement messages in the PRN. In contrast, our algorithm makes some kind of batch acknowledgement, at the cost of some simple memory management.

Let us first consider the control mechanism from a single site S_i. This site keeps track of everything it locally knows at a particular date about the global state of the machine: what it plans to do, what it has received from other sites, and what it has sent to other sites. S_i has a local counter (*lCount*) for each date (*lVal*). This counter is incremented when the site receives a message, decre-

On node 0	On node i, $\forall i \neq 0$
$nCount = lCount$ $nVal = lVal$ $GVT = \begin{cases} pVal & \text{if } pCount = 0 \\ pGVT & \text{if } pCount \neq 0 \end{cases}$ $pGVT = pVal \text{ if } pCount = 0$	$nCount = \begin{cases} pCount + lCount & \text{if } pVal = lVal \\ lCount & \text{if } \min(lVal, pVal) = lVal \\ pCount & \text{if } \min(lVal, pVal) = pVal \end{cases}$ $nVal = \min(lVal, pVal)$

Table 1. Operations performed by each node when owning the token

mented when it sends one, and left as it is when it decides to do something locally.

From a global point of view, the mechanism has to compute the minimum of the distributed dates, without forgetting in transit messages. Collecting these informations may be done using a token consisting of two values (Val and $Count$) and circulating along a logical ring in the MIMD machine. A particular node S_0 has the role of initiating the process. Table 1 gives detail on what happens when the token is on a node [2]. When S_0 receives a value with an associated counter equal to zero (meaning no message is in transit for the date Val), it broadcasts the received Val value to all sites (i.e., the current minimum).

For the deadlock detection and recovery protocol, such an algorithm requires when executing by software:

- memory management for the manipulation of the local counters;
- computation on the processors (for comparison and sum);
- overload on the primary communication networks.

Moreover, a token has to visit all sites, circulating on a ring of complexity $\mathcal{O}(n)$, at high cost as shown in Tab. 2.

3 Hardware implementation

In their paper, CHANDY and MISRA quote that any deadlock detection and recovery algorithm may be used with the simulation scheme, the only imperative being its correctness and safety.

The structure of the proposed machine is a good candidate for a hardware implementation of the previously described mechanism. At this point, it is the programmer responsibility to decide which part of the control algorithm will be done by processor or FPGA. Although possible, controlling memory with the FPGA is expensive in the sense that it will preempt the Transputer system bus. The processor is then stopped when the FPGA accesses the memory.

[2] when a variable is prefixed with n this means that it will be sent in the token, l means a local value, and p means the value received with the token

The whole part concerning the circulating token and the reduction operation, although not algorithmically optimal because performed on a ring, will take benefit of a hardware implementation on the ArMen machine. The algorithmical complexity will remain in $\mathcal{O}(n)$, but execution is done with hardware speed and do not require neither processor computation nor primary network overload.

Schematically, the final implementation of the whole application will look like Fig. 3.

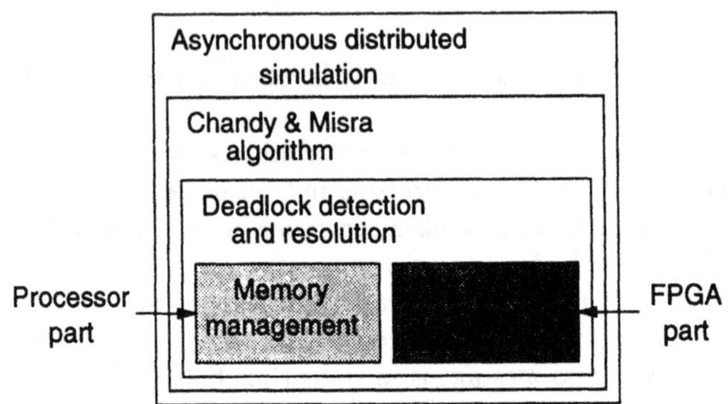

Fig. 3. Distribution of control execution

Table 1 represents all the operations that will be synthesized in the CLL. To pass from Tab. 1 algorithm to the final configuration bitstream, we first use the UCA compiler, and then the Xilinx design flow. This design flow is described in [6]. Writing a program with UCA is straightforward, and consists mainly of three steps:

- firstly, we must describe the global algorithm in an abstract program. Each node has a local clock, sequencing the logic element and implemented in the FPGA. An important hypothesis is that any logic element delay is smaller than the clock period. This is a key point for the final debugging phase. Two adjacent FPGAs communicate via an shared memory emulated between neighbour FPGAs. Communication between a processor and its FPGAis also done with two shared words;
- this first specification of the UCA program is validated by simulation and/or by generating its state graph. This permits the detection of inconsistencies and of possible deadlocks in the description;
- the validated abstract program is not directly synthesizable. Currently, manual transformations must be done using proved derivation rules. These solve communication handshakes and data consistency problems between adjacent FPGAs.

Hardware compilation of the final UCA program is done with a home built design flow. First, the UCA code is translated to an intermediate language, which is then optimized with a logic minimized. The obtained XNF file[11] is merged to the machine dependent interface: this interface provides primitives for interaction between the Transputer and its FPGA. After this merge, we continue with the standard Xilinx design flow. During the latter, tests are almost impossible. Once the circuit has been synthesized, the only possible way of getting execution information or detecting problems is to use hardware tools, such as logic analyzer. The signals to be observed are connected to pins on the south port of the FPGA and connected to probes of the analyzer.

The obtained circuit has to obey the hypothesis made in the first step of the UCA programming which imposes a delay smaller than the clock period. So, the longest combinatorial path should remain as short as possible. In our example, an integer comparator and a signed integer adder are synthesized. Their depth are 2, 4 or 6 CLBs for addition on 4, 8 or 16-bit operands, and $\lceil \log n \rceil$ CLBs for comparison on n-bit operands. With the technology of the machine, we assume $10\,ns$ delay per CLB level. Thus, the 20 Mhz clock is too restrictive for us.

Another important consideration in the synthesis process is the limited data path between adjacent FPGAs. As described in Sect. 1, the path width in the CLL between two node is 36 (32 on west or east port, plus 4 on the south port). Handshaking required for proper data exchange involves 2 bits. As the global operator works concurrently with the CLL, the token should be composed of:

- the currently computed minimum;
- the associated counter (as described in Tab. 1)
- the last computed minimum with associated counter equal to zero.

Thus, we have to maintain the following constraint on data sizes:

$$2 \times \text{sizeof}\ (lVal) + \text{sizeof}\ (lCount) \leq 36 - 2 = 34 \tag{1}$$

Considering the size/delay ratios for operators and (1), we have used 10 bits for *Val* and 8 bits for *Count* at a 10 Mhz clock frequency. Several solutions overcome this size limit, but they are not implemented: the simplest one consists in splitting the information in several tokens. The synthesized circuits consist of 102 CLBs and 87 IOBs on each node, except on node 0 where we have 73 CLBs and 77 IOBs. The placement of the operating elements in the FPGAs, checked with Xilinx CAD tool, is very similar to the scheme in Fig. 4.

4 Performances

In this section, we will focus not only on the performances from the point of view of a user, but also of a hardware developer.

The first expensive task in development is the understanding of the way the CLL is programmed. From UCA to the final raw-bit file, there are many steps one has to master. Once the developer has decided which parts are to be

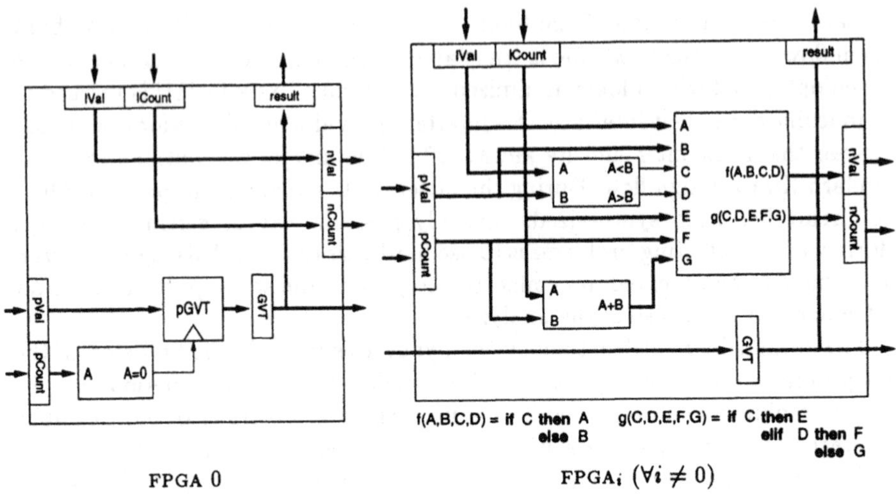

Fig. 4. Block diagram of the different CLL nodes

implemented in hardware and has learned the home designed compilation flow, the most expensive tasks are the minimization of the circuit description, and the routing and placing of the FPGA. These two tasks depend strongly on the circuits being synthesized, and more precisely on their size and wire requirement. The circuit built for our example, consisting of 87 IOBs and 102 CLBs, was minimized in about 20 minutes and placed and routed in one hour on a Sun SparcStation 5.

Due to the operator complexity (the adder requires at least 5 levels of CLBs), we had to slow down the FPGA clock on the Xc3090 chips. Therefore, the delay for a token to go through a FPGA node is about 200 ns with the local 10 Mhz clock and with inter-FPGAs asynchronous handshaking. Results for the complete mechanism are given in Tab. 2. One should note that the global minimum

	Software/Hardware solution	Pure software solution
Global minimum	3.2	10,250
Complete mechanism	6.2	19,000

Table 2. Execution delays (values are in μs)

computation could be computed with the CLL in about 3.2 μs, as its hardware complexity allows the use of the 20 Mhz clock. The timings are done with a logic analyzer plugged to the south port of the FPGAs. These measures are done with great accuracy, as the delays inherent to software monitoring are removed.

The gains on complete PDES applications are still to be evaluated. Nevertheless, the overlapping of the control mechanism with the processor computation and the lowered load on communication network will be of benefit to the distributed execution.

Summary and further works

We have presented latest work on the ArMen machine dealing with control acceleration in MIMD execution. First results show impressive results for the mechanism described. Timing problems encountered may be solved using latest technology. The biggest difficulty is the limited data path between neighbor FPGAs. A solution using several tokens solves this problem with decreased performances.

Some work is to be undertaken at different levels. Using the proposed algorithm, we have to quantify the impact hardware implementation has on an entire application. Deriving from this impact, we may then consider other algorithms where analytic complexity is too high for software implementations, but not too expensive with hardware considerations.

Having such speed-ups compared to software, one could consider the building of hard-wired cards for the control purpose. Compared to the benefits obtained, the cost in flexibility would be too high.

The ease of development offered by the description language and the flexibility of reconfigurable hardware gives the best opportunity for developing well-suited control circuit in short time.

The gain obtained for an entire distributed simulation is to be evaluated with the described hard-wired control mechanism. Considering the performances obtained for a complete application will give information on the extension of the number of control operation integrated in hardware.

References

1. C. Beaumont, P. Boronat, J. Champeau, J.-M. Filloque, and B. Pottier. Reconfigurable technology: An innovative solution for parallel discrete event simulation support. In *Proceedings of the workshop ACM/IEEE Parallel and Distributed Simulation PADS'94*, Edinburgh, UK, July 1994. SCS.
2. G. Burns, V. Radiya, R. Daoud, and R. Machiraju. All about TROLLIUS. *Occam User Group Newsletter*, pages 55–70, July 1990.
3. K. Chandy and J. Misra. Asynchronous distributed simulation via a sequence of parallel computations. *Communications of the ACM*, 24(11), Nov. 1981.
4. K. Chandy and J. Misra. *Parallel Program Design: a Foundation*. Addison-Wesley, 1988.
5. P. Dhaussy, J.-M. Filloque, B. Pottier, and S. Rubini. ArMen: an FPGA-based parallel architecture. In H. Siegel, editor, *8th International Parallel Processing Symposium (Parallel System Fair)*, Cancùn, Mx, Apr. 1994.
6. P. Dhaussy, J.-M. Filloque, B. Pottier, and S. Rubini. Global control synthesis for an MIMD/FPGA machine. In *Proceeding of IEEE Workshop FPGAs for custom computing machines*, pages 51–58, Napa, USA, Apr. 1994.

7. J.-M. Filloque, E. Gautrin, and B. Pottier. Efficient global computation on a processor network with programmable logic. In *Proceedings of PARLE'91*, number 505 in LNCS, pages 55–63, Eindhoven, Nl, June 1991. Springer-Verlag.
8. R. M. Fujimoto. Parallel discrete event simulation. *Communications of the ACM*, 33(10), Oct. 1990.
9. C. M. Pancerella. *Reduction operations in parallel discree event simulations*. PhD thesis, University of Virginia, May 1994.
10. F. Raimbault, D. Lavenier, S. Rubini, and B. Pottier. Fine grain parallelism on an MIMD machine using FPGAs. In *Proceeding of IEEE Workshop FPGAs for custom computing machines*, volume 3890-02, Napa, USA, Apr. 1993.
11. Xilinx. The programmable logic data book. San Jose, USA, 1994.

Customised Hardware Based on the REDOC III Algorithm for High Performance Data Ciphering

Fabio Guerrero and James M. Noras

Department of Electronic and
Electrical Engineering
The University
BRADFORD BD7 1DP
United Kingdom
telephone: 01274 384036
J.M.Noras@bradford.ac.uk

Abstract. The REDOC III algorithm for data ciphering is a potential replacement for DES. This paper looks at ways of customising the algorithm to increase security without reducing ciphering speed. Many valuable modifications are possible if reconfigurable hardware is used.

1 Introduction

Programmable chips have two characteristics that seem particularly suited to the construction of hardware for data ciphering: on a circuit board they are anonymous blocks with concealed functionality, and secondly they may be reprogrammed in situ and at will to change the nature or the details of their ciphering. Still, one must proceed with care. Secure ciphering generally relies on strong and tested algorithms that are in the public domain, so any temptation to use easily programmable hardware to implement homespun techniques must be resisted. For example, DES [1] requires nonlinear substitutions based on S-boxes [2]. It is easy to modify these to obtain new, secret algorithms that resemble DES in operation, but give completely different cipher output for the same data input. However, cryptological attacks do not require full knowledge of the algorithms concerned, and poorly designed choices may result in weaknesses [3]. Reprogrammability makes it possible to change the ciphering method frequently. This can offer greater security than frequent changes of key, but caution and balance is needed to ensure that compatible algorithms are used, lest management and control complications degrade system utility.

One reliable method of customising a system is to use a cascade of ciphers that are independent [4], or, as in triple-DES, to use the same algorithm repetitively in a carefully chosen manner, with unrelated keys and with encryption and decryption intermixed. In this paper we look at a third possibility, where different ciphering processes occur simultaneously in a synchronised way within a common algorithm structure, but differing in detail for every bit-slice and every session or block of data. With programmable logic it is straightforward to alter a suitable algorithm in this way, so that a given implementation is chosen from a family of ciphering techniques, which have the same basic structure, speed and security.

The work is based on the algorithm REDOC III, which was proposed by Wood [5] as a DES replacement in high speed data ciphering. A comparison made by Shepherd and Noras [6], shows REDOC III to be capable of much greater throughput, speeds of 2Gbps being attainable with a 64-bit data path in hardware. Here we look at some ways of customising hardware for data ciphering which increase security, but do not reduce ciphering rates or greatly complicate system management.

We envisage custom systems with two novel aspects:

> 1) The hardware blocks that operate on each bit-slice of data are distinct. In the original plan these differ only to the extent that they are loaded with different data-masking bits. We propose differences that are structural, with correspondingly more impact on the ciphering process.
>
> 2) In a system using programmable hardware that can be modified during and between sessions, not only can structurally different ciphering blocks be used for different sets of data, but their relative positions can be changed also, giving quite different ciphering results with each arrangement of resources.

Thus, to decrypt data, it will be necessary to know much more about the algorithm and system used than "just" the key and mask information.

Following sections first set out a description of the basic algorithm, and then hardware design is considered. The main section sets out various ways to customise hardware.

2 REDOC III: The Algorithm in Hardware

The algorithm consists of 16 rounds of XORing the data with parts of the key. In each round the data byte used to generate the address of the key is the only byte not altered in that stage. This Feistel structure allows the algorithm to be reversed by a similar sequence of operations, with the data being fed into the system in byte-reversed order. Secret key information consists of a table of 256 8-byte keys, and two 8-byte mask blocks, $Mask_1$ and $Mask_2$

To avoid shuffling key signals, which would require considerable hardware, two special registers are used. $Hold\_XOR_1$ is zeroed at the start of a new data block, and accumulates key information for the first 15 clock cycles. Addresses are generated using appropriate bits of this register XORed with data, without the stored data being altered. The accumulated key information transfers to $Hold\_XOR_2$ during cycle 16. Ciphertext is produced by XORing the data and $Hold\_XOR_2$ while new data are being loaded. The following description is directed towards hardware implementation.

```
rem:    transfer the contents of the first XOR register and then clear it
Hold_XOR2 [n] = Hold_XOR1 [n]  : n=1, 2, ... 8
Hold_XOR1 [n] = 0              : n=1, 2, ... 8

rem:    16 cycles for each block
for i  =  1, 2
for m =  1, 2, ... 8

rem:    calculate the address of the key
Address[m] = Mask_i[m] ⊕ Data[m] ⊕ Hold_XOR1[m]

rem:    store key in appropriate part of key register
for n = 1, 2, ... 8
Hold_XOR1[n] = Key_n[Address[m]] ⊕ Hold_XOR1[n]        : n ≠ m
Hold_XOR1[n] = Hold_XOR1[n]                            : n = m
endn
endm :  data loop

rem:    during cycle 16  Hold_XOR2 = Hold_XOR1 and Hold_XOR1 = 0
endi :  mask loop
```

During the next 8 cycles fresh data are loaded and the previous block is shifted out:
$Result[n] = Previous\_Data[n] \oplus Hold\_XOR_2[n]$

3 Implementation

Hardware has been designed and simulated, using Xilinx 4000 logic as the target technology. Partitioning the byte-wide design into parallel bit-slices allows the design to fit within four FPGA chips, each with two adjacent RAM chips holding relevant slices of the key. A suitable Xilinx part is the 4003APC84, using 29 pads and with 79 out of 100 CLBs occupied. Using a 30Mhz clock, data encryption with a 64-bit data path runs at 2Gbps. Results for Altera devices give the same kind of performance, because in both cases the critical delays are due to off-chip RAM access during key addressing and fetches. Higher rates could be obtained with full custom processes with on-chip RAM available.

4 Customising the Design

There are many feasible enhancements and alterations to the REDOC III algorithm. The following retain the framework and the speed of the method, but increase ciphering complexity with little hardware cost. The algorithm's structure fits well with a bit-slice architecture: only the key addressing is common, while all other operations are internal to each slice. This suggests that many different ciphering systems can be made by assigning different but compatible hardware units to each bit-slice: using programmable hardware means that this can be done easily, both during and between ciphering sessions. Each arrangement yields entirely different cipher output for the same plaintext input.

4.1 Extending the use of masks

The use of mask data and mask registers can be modified in many helpful ways. The additional design information associated with these custom changes increases security.

Loading mask data. With programmable logic, flip-flops can be loaded with "1" or "0" on a system reset. Each set of mask information is thus encoded within the binary design file for each particular chip, rather than loaded through the data channel.

Choice of masks. Although the original algorithm specifies masks related logically to the key, the key and mask data can be chosen independently.

An active data mask. At each stage the key address is formed by XORing particular bits of the data and the masks. These "active" data are not modified. However, at a cost of one flip-flop per bit and an XOR gate per bit-slice, an additional mask register can be set up which will cipher the active data and change it in store. This process, happening in parallel with the address generation, does not reduce speed.

Variable mask lengths. If the hardware is to be used to encrypt blocks of data that will be decrypted again as blocks, then the length of the mask registers may be a variable of the implementation, the lengths of mask registers varying between different bit slices. Moreover, these need not be arranged as linear shift registers. For example, recasting an 18-bit register as parallel 7-bit and 11-bit registers whose outputs are XORed together gives a cycle length of 77, with a marginal effect on resources. The size and configuration of mask registers can again differ from slice to slice, and different choices and positions of distinct blocks within the data-path produce quite different ciphering systems.

Clearly, control of block lengths of data, storage of pairs of initial states for encryption and decryption, timing of register resets, and configuration of the programmable hardware, all complicate the ciphering process and protocols; this has to be set against the corresponding gains in cipher complexity and security.

4.2 Further use of the Hold_XOR$_1$ register

First consider a simple model, ciphering only two bits of data, a and b, within a single bit-slice. Here f(a) represents the key obtained using a as the address. We ignore any masking and consider one round of XORing only.

Encryption:

[b]	[a]	: the raw data
$[b \oplus f(a)]$	[a]	: first stage
$[b \oplus f(a)]$	$[a \oplus f(b \oplus f(a))]$	: second stage, ciphered output

REDOC III is clearly seen to rely on generating a stream of bits that is XORed with the data to be encrypted. We can understand easily the inverse process.

Decryption:

$[a \oplus f(b \oplus f(a))]$	$[b \oplus f(a)]$	: the cipher data
[a]	$[b \oplus f(a)]$	: first stage
[a]	[b]	: the recovered data

These results follow from the XOR property: $f(b \oplus f(a)) \oplus f(b \oplus f(a)) = 1$.

Now consider two situations when the initial state of the Hold_XOR$_1$ register for a new set of input data is not all zeroes, first when a mask is preloaded at the start of the cycle, and secondly when the final state from the previous cycle is not cleared.

Non-zero initial state. An XOR mask forces "0" or "1" onto each flip-flop in the Hold_XOR$_1$ register at the start of the cycle. The above model becomes:

Encryption*:

[b]	[a]	: the raw data
[b']	[a']	: effect of the XOR mask
$[b' \oplus f(a')]$	[a']	: first stage
$[b' \oplus f(a')]$	$[a' \oplus f(b' \oplus f(a'))]$	: second stage
$[b \oplus f(a')]$	$[a \oplus f(b' \oplus f(a'))]$	: apply the XOR mask again

Note that the initial Hold_XOR$_1$ mask must be kept for use at the end of the ciphering cycle, which requires 8 additional flip-flops per bit-slice. To have the initial Hold_XOR$_1$ mask act on the incoming data needs no extra operations or hardware, as the required function is implicit in the hardware specification. The second application of the mask requires an extra XOR operation on the outgoing data, but this is not in the critical timing path. To recover the message from data ciphered in this way is equally straightforward:

Decryption*:

$[a \oplus f(b' \oplus f(a'))]$	$[b \oplus f(a')]$	: the cipher data
$[a' \oplus f(b' \oplus f(a'))]$	$[b' \oplus f(a')]$	: effect of the XOR mask
[a']	$[b' \oplus f(a')]$	: first stage
[a']	[b']	: second stage
[a]	[b]	: apply the XOR mask again

This shows how to introduce a mask that operates in a quite different way from the original masks, as it affects all the data each cycle, *except* the data being used in the address generation.

Ciphertext chaining mode. Here, the $Hold\_XOR_1$ register is not reset at the end of each cycle, but the value retained is allowed to interact with fresh data. The operation within each cycle is just as described in the modified sequences set out in the last section for the non-zero initialised state, but the starting point of each cycle is obtained by applying the blocks of data in the same order for decryption as for encryption. The hardware requirements are slight, with little effect on speed. However, this method gives no protection against error propagation. If there were incorrect cipher data, then until the $Hold\_XOR_1$ register was reset, all subsequent sets of data would be wrongly ciphered.

4.3 Expanding The Key Space

The more closely a modification of REDOC III could be made to resemble a one-time pad system, the greater would be its value. The cryptosystem has a large key space, and the sequence of key selections used for a given set of data is hard to reconstruct. In a custom installation it is feasible to expand the key space, by an arbitrary amount and by a variety of means. For example, the number of masks used in address generation could be doubled, giving 2 bits of address per bit-slice each clock cycle. Thus, instead of 8-bit addresses, a 16-bit address space could be used. Alternatively, some decoding of the control counter - perhaps also involving the data - could be used to generate additional address information.

This approach has real costs, as it demands additional memory resources, so that the benefits in a particular situation would need to be analysed. Note though, that not every bit-slice would need to generate more than one bit for the method to work, as the key returned to each bit-slice module is generated globally from all the active data each cycle. Thus, one or more bit-slices could be used to give 9-bit or more addressing, and the number and distribution of such enhanced modules would be part of the ciphering system and algorithm, the output of the system being radically different for each chosen configuration.

5 Conclusions

The use of programmable devices offers the possibility of making alterations to the basic REDOC III algorithm, while retaining its structure. Since the hardware uses bit-slice modules operating in parallel, different versions of the hardware can be used in different bit positions, and with reprogramming these can be changed at will. All the above modifications can be carried out with little effect on data rates, and most do not change the cipher chips' pinouts or system interactions. Also, they would have zero cost if, as intended, they make use of otherwise uncommitted logic.

The algorithm investigated is suitable for such an approach as it uses very simple operations, not requiring addition, multiplication or modular reduction, so that the design can be segmented and its various parts modified independently. The critical delays are largely due to memory access, which requires off-chip access, as the key space is at least 256 by 8-byte in size.

If it were really required to have significantly higher throughput, a custom process that offered both RAM and reprogrammable elements would provide a most interesting set of options. Such a system would open up investigation of other algorithms that do require appreciable arithmetic modules, but more theoretical work is needed to see how the approach of merging ciphering algorithms would apply in particular cases.

References

1. ANSI X3.92, "American National Standard for Data Encryption Algorithm (DEA)", American National Standards Institute, 1981.

2. Schneier, B.: "Applied Cryptography: protocols, algorithms, and source code in C", John Wiley & Sons, New York, USA, pp. 224-230, 1994.

3. Ibidem, pp. 238-239.

4. Maurer, U.M. and Massey, J.L.: "Cascade Ciphers: The Importance of Being First", Journal of Cryptology, vol. 6, pp. 55-61, 1993.

5. Wood, M.C., 1991, US Patent 5,003,596, 26 March 1991, "Method of Cryptographically Transforming Electronic Digital Data from One Form to Another."

6. Shepherd, S.J. and Noras, J.M., 1994, "A very fast hardware replacement for the DES", 3rd UK/Australian International Symposium of "DSP for Communication Systems", 12-14 December, University of Warwick.

Using Reconfigurable Hardware to Speed up Product Development and Performance

Adrian Lawrence[1], Andrew Kay[2], Wayne Luk[1], Toshio Nomura[2], Ian Page[1]

[1] Oxford Parallel, Oxford University Computing Laboratory, Oxford OX1 3QD, UK
[2] Sharp Laboratories of Europe Ltd., Edmund Halley Road, Oxford OX4 4GA, UK

Abstract. Harp1 is a circuit board designed to exploit the rigorous compilation of parallel algorithms directly into hardware. It includes a transputer closely-coupled to a Field-Programmable Gate Array (FPGA). The whole system can be regarded as an instance of a process in the theory of Communicating Sequential Processes (CSP). The major elements themselves can also be viewed in the same way: both the transputer and the FPGA can implement many parallel communicating sub-processes. The Harp1 design includes memory banks, a programmable frequency synthesizer and several communication ports. The latter supports the use of parallel arrays of Harp1 boards, as well as interfacing to external hardware. Harp1 is the target of mathematical tools based upon the Ruby and occam languages, which enable unusual and novel applications to be produced and demonstrated correctly and rapidly; the aim is to produce high quality designs at low costs and with reduced development time.

1 Introduction

The performance of many computationally demanding tasks can be improved by the use of matching parallel hardware. But the design of correct circuits is notoriously difficult and time-consuming. Harp1 is a circuit board designed to demonstrate that mathematical methods can overcome some of these problems.

Figure 1 gives a broad overview of Harp1. The architecture of Harp1 is inspired by the theory of Communicating Sequential Processes (CSP) [1]. The board contains two communicating elements: a transputer and a Field-Programmable Gate Array (FPGA) chip. Both elements are programmable and are capable of implementing many parallel processes internally; in particular, the FPGA hardware may be configured with almost complete freedom for major parallelism when appropriate. A board implements a CSP process, and may be used alone, in arrays, or as a component in a diverse system.

Harp1 provides external memories for the FPGA and the transputer. Each is also provided with external communication resources. The FPGA is normally clocked by a programmable frequency synthesizer.

Harp1 may be exploited using conventional methods, but it becomes far more powerful when coupled with mathematically based compilation tools. Both a variant of occam(TM) [3], Ruby [2] and a combination can be compiled directly

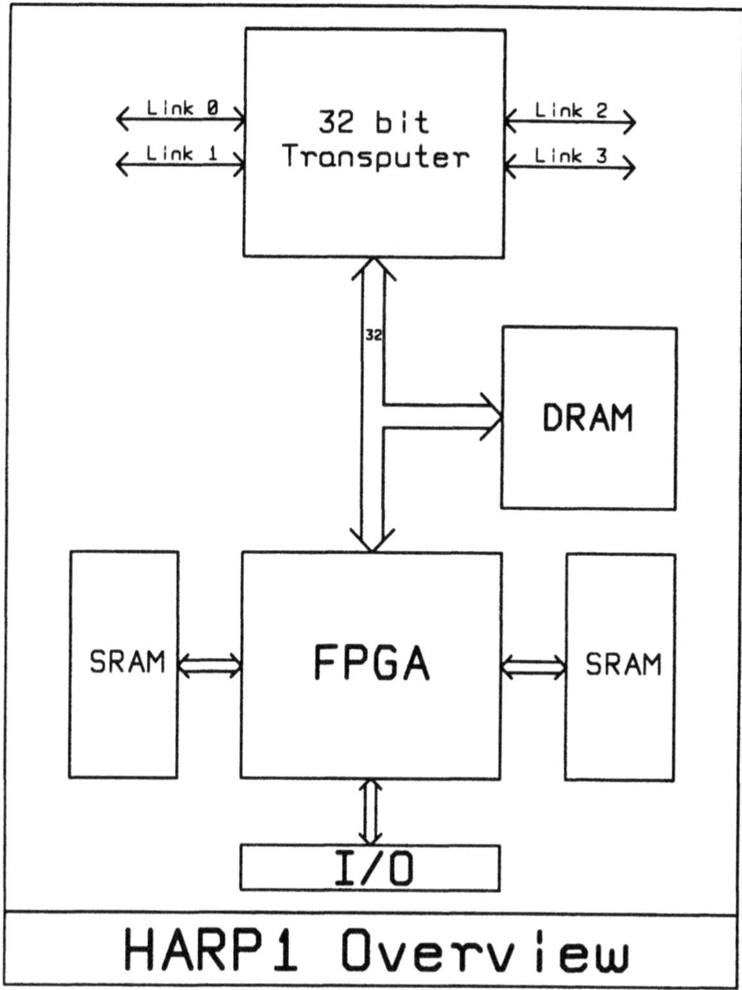

Fig. 1. Overview of Harp1

into the FPGA hardware. This allows correct hardware to be produced from high level descriptions very rapidly. Systolic arrays, for example, can be efficiently implemented on the FPGA.

The flexibility of Harp1 makes it useful in a variety of situations. It is a suitable development platform for FPGA designs or for prototyping application-specific integrated circuits. It may be used to accelerate a program: the speed-critical parts may be mapped onto the FPGA while the rest of the program runs on the transputer; different cost-performance trade-offs between hardware and software can be explored. Another use of the FPGA is to hold custom hardware for algorithms that require non-standard data size, or operations that are not included in the instruction set of microprocessors.

Here we give a brief overview of the hardware: aspects of the compilation tools have been described elsewhere (see [5], [6], [7], [8], [10], [9], [11] and [12]).

2 Field Programmable Gate Arrays

Field Programmable Gate Arrays are now widely used in electronic design. They provide directly programmable hardware and promise to change the way in which design is approached in a revolution similar to that engendered by the microprocessor. Programmable hardware has of course been available for many years, not least various forms of memory. Various PLDs (Programmable Logic Devices) have been used for several decades in implementing state machines and "glue" logic, among other things. But the available devices have tended to have restricted architecture, and to be rather small.

The last decade has seen a significant change with the introduction of a variety of field-programmable gate arrays, as well as an evolution of some PLDs into much larger devices with extended architectures. The term FPGA is normally understood to apply to programmable devices with a regular array of cells with distributed routing. But the term is sometimes also applied to "super-PALs" which may also be called CPLDs (Complex PLDs): these usually have what is essentially a large cross-bar switch as the major interconnection mechanism.

Some FPGAs may only be programmed once: these are often based on anti-fuse[1] technology. But others are based on SRAM, and may be programmed or *configured*[2] repeatedly.

There are a variety of FPGA architectures available, and no one style has emerged thus far as obviously superior for all purposes. Some have a hierarchical sub-structure for the cells and routing. However, there is one broad distinction that can be made: the architectures are either *coarse-* or *fine-grained*. This refers to the capability of a single cell. A fine-grained FPGA tends to have a very large array of relatively simple cells. An example is the AT6000 series [21]. But the FPGA used on Harp1 is coarse-grained. It is usually a Xilinx XC3195 [17].

The FPGA on Harp1 consists of an array of cells called CLBs (Configurable Logic Blocks). Each CLB contains two latches, and a function generator. The internal connections within the cell and the lookup table in the function generators are determined by configuration bits held in an integrated SRAM. This allows an individual cell to implement quite complex combinational and sequential elements. The routing resources allow the cells to be connected as required, at least in principle. In practice, the problem of routing a congested design is the major obstacle in obtaining the highest performance. The particular intercon-

[1] An *anti-fuse* makes a connection when it is "blown".

[2] Configuration here means the programming of the FPGA functions and interconnection.

nections required in a design are determined by further configuration bits held in the same integrated SRAM.

This ability to sculpt hardware to fit a computation is significantly different from the programmability of a microprocessor. In particular, the microprocessor has a fixed instruction set[3]. In contrast, an FPGA can be configured to fit a particular algorithm. We may even combine the two approaches, and compile a specialised microprocessor into the FPGA with a restricted instruction set chosen to suit the particular case. Often a computation falls naturally into two parts, one of which matches a microprocessor implementation, while the other can use special hardware. Harp1 provides just the right closely coupled environment for that situation.

Precisely because an FPGA is programmable in something like the same sense as a microprocessor, it is already becoming very widely used. Eventually it should become a commodity item, although prices are still very high. There is often a performance penalty: it is seldom possible to obtain the ultimate speed from a general purpose architecture as from a custom design. This is especially true of routing: a large number of programmable interconnections usually significantly slow signals. The anti-fuse technology has an advantage here, but is not reprogrammable.

The type of reconfigurable SRAM-based FPGA used in Harp1 provides an ideal testbed in which to explore compilation of algorithms directly into hardware. The methods apply to the design of any hardware, but the flexibility and reusability of the FPGA are exactly what is needed for a prototyping environment.

Although Harp1 was designed as a testbed for compilation techniques, it is useful in its own right. The performance of the FPGA is respectable, and is very adequate for many purposes.

3 Architecture

Harp1 is a size 6 TRAM [16] circuit board. A 32-bit transputer, usually a T805, controls the system, and is the relatively conventional processor. The processor has 4Mbytes of external DRAM memory which is also accessible to the FPGA via DMA. The transputer is responsible for configuring the FPGA, and controlling the FPGA clock. The four links are available for communication with other devices in the standard way.

Two banks of fast SRAM are associated with the FPGA. Communication resources beyond the 4 transputer links are provided: a connector provides a further 17 fully programmable signals. This permits high bandwidth interaction with external devices. It also provides a configurable port for connection to other hardware, particularly useful when prototyping applications that may need to control special hardware. Figure 2 gives a simplified description of the architecture of Harp1.

[3] Or micro instruction set.

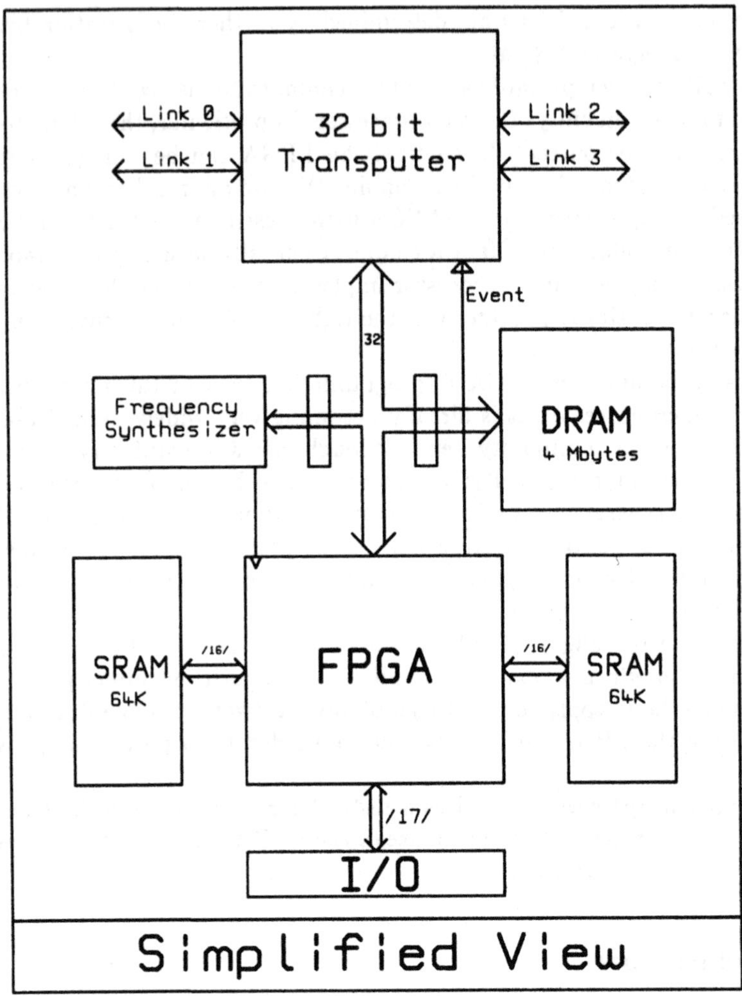

Fig. 2. Simplified view of Harp1

The configuration of the FPGA is carried out by the transputer whenever necessary. It is also possible to read back the loaded configuration when permitted: a security feature can be invoked to disable this option.

An important measure of the performance of an FPGA configuration is the maximum clock frequency. If a design has many layers of deep combinational logic, the propagation time, especially that involving inter-cell routing, is often the bottleneck that constrains maximum performance. Thus Harp1 has a high performance and very flexible clocking circuit, including a frequency sythesizer.

The FPGA can be clocked from various sources, including a single-step circuit. When the frequency synthesizer is selected, a very large range of frequencies with fine resolution is available. The range is normally of the order of 1 to 100MHz, with a resolution of better than 125kHz. This allows us to benchmark configurations, permitting investigation of compiler strategies, routing and placement algorithms and related matters. The frequency synthesizer is normally controlled by the transputer, and uses a Phase Locked Loop.

The FPGA may be used as a co-processor, dynamically reprogrammed when needed by the transputer. Or it may have a single application-specific configuration loaded at boot time.

In most parallel systems communication is paramount. Harp1 meets this need by connecting the FPGA directly to the transputer bus, and also by allowing it to activate the transputer *Event* channel. Thus there is a high bandwidth highway between the processors, and the ability for either end to signal the other at any time. In fact, we can implement many CSP channels between multiple processes spread across the transputer and the FPGA in this way.

The boards may be used alone, or as part of parallel arrays. Although designed primarily to support the compilation of Handel (a variant of occam) and of Ruby directly into hardware, the FPGA may also be programmed by standard commercial tools.

4 Applications

A few prototype applications have been developed to run on the Harp1 board. In most cases the applications were intended to exercise the compilation technology rather than the hardware itself. For this reason we have concentrated on functional complexity rather than speed or usefulness. The list is not complete.

The simplest application used the FPGA as a pattern matcher for a fixed string (with wildcard matches). It returns the set of addresses in RAM at which the pattern can be found.

A spelling checker was developed which checked the existence of words in a compressed, miniature dictionary (of 6000 English words) stored in SRAM. Here the FPGA is used as a "spelling coprocessor", looking up whatever word the transputer requires. An extension to the transputer part of the system allows close matches to be found for incorrectly spelt words.

By extending the data structures and improving the search strategy of the spelling checker, a morphological analyser was developed. This application is capable of splitting words into their constituent parts, and could be used as part of a natural language parsing system. For example, the word "uneatable" might be represented by three tokens denoting "un-(prefix)", "eat(verb)", "+able(suffix)".

An OCR kernel (optical character recognition) was implemented using the flexibility of the FPGA to provide the unusual binary operations used in the specific algorithm. This allows rapid searching to be done even at low clock speeds. The task of the kernel is to take a normalised input and compute a similarity measure with each entry in the database of characters (in this case, Japanese Kanji characters), keeping the best few for further analysis.

A merge sorter and a heap sorter were implemented to compare their relative efficiencies in space and time. The merge sorter works well, performing close to one memory access per clock cycle and achieving close to the theoretical minimum number of cycles for this algorithm. Sadly, the complexity of the heap sorter meant that it would not fit on the FPGA. This problem could be cured by improving the optimisation methods used by the compiler, by using a larger FPGA on Harp1 or by using an array of Harp1s.

Other applications of Harp1 include real-time digital filtering [22] and video motion tracking [23].

5 Summary

Because Harp1 is almost totally programmable, it is useful in a range of situations. The applications above demonstrate its utility as a development platform; it also has a place in education and familiarization with FPGA technology. The I/O connector means that it is easily used as a prototype for custom designs. It provides a powerful computational resource, particularly for algorithms which do not "fit" a microprocessor architecture. Systolic arrays, for example, can be efficiently implemented on an FPGA. More routine applications are as a "custom" interface to unusual hardware, or where a single piece of equipment needs to be reprogrammed to suit different hardware standards. By using the frequency synthesizer and programming the FPGA to pass the signals through to the external connector, Harp1 could even be used as a (exotic) signal generator.

Acknowledgements

Thanks to Bob McLatchie for suggesting the title of this paper. The support of a JFIT grant (IED/46/91/014) is gratefully acknowledged.

References

1. C A R Hoare. *Communicating Sequential Processes.* Prentice-Hall International, 1985.
2. G Jones and M Sheeran. *Circuit Design In Ruby.* In J Staunstrup (ed.), *Formal methods for VLSI design,* North–Holland, 1990.
3. *occam 2 Reference Manual.* Prentice-Hall International, 1988.
4. W R Moore and W Luk, Eds. *More FPGAs.* Abingdon EE&CS Books, 1994.
5. W R Moore and W Luk, Eds. *FPGAs.* Abingdon EE&CS Books, 1991.
6. I Page and W Luk. *Compiling occam into FPGAs,* In [5].
7. W Luk and I Page. *Parameterising Designs for FPGAs.* In [5].
8. I Page, W Luk and H Lau, *Hardware Compilation for FPGAs: Imperative and Declarative Approaches for a Robotics Interface. Proc. IEE Colloquium on Field Programmable Gate Arrays – Technology and Applications,* Ref. 1993/037, pp. 9.1-9.4, IEE, February 1993.

9. J He, I Page and J P Bowen. *Towards a Provably Correct Hardware Implementation of Occam.* In G J Milne and L Pierre (eds.), *Correct Hardware Design and Verification Methods*, Lecture Notes in Computer Science, 683, pp. 214–225, Springer-Verlag. 1993.
10. I Page. *Parametrised Processor Generation.* In [4].
11. W Luk, D Ferguson and I Page. *Structured Compilation of Parallel Programs into Hardware.* In [4].
12. W Luk and T Wu. *Towards a Declarative Framework for Hardware-Software Codesign.* In *Proc. Third International Workshop on Hardware/Software Codesign*, pp. 181–188, IEEE Computer Society Press, 1994.
13. A E Lawrence. *HARP1 User Manual.*
14. A E Lawrence. *HARP1 (TRAMple) manual, volume 2: Manufacturing Pack.* 1992. Confidential.
15. A E Lawrence. *Analysis of a Phase Locked Loop: Expanded issue for HARP1 documentation.* 1991. Confidential.
16. *Dual-in-Line Transputer Modules (TRAMs).* Inmos Technical Note 29, *Transputer Development and iQ Systems Databook.* 2nd Edition, 1991.
17. *The Programmable Logic Data Book.* Xilinx Inc., 1993.
18. *The Transputer Databook.* Inmos document 72 TRN 203 02, Third edition, Inmos, 1992.
19. *The T9000 Transputer Hardware Reference Manual.* Inmos document 72 TRN 238 01, First edition, Inmos, 1993.
20. M S Jhitta. *Introduction of a New FPGA Architecture.* In [4].
21. *Configurable Logic Design and Application Book.* Atmel Ltd., 1993.
22. M Aubury and W Luk. *Binomial Filters.* To appear in *Journal of VLSI Signal Processing.*
23. M Bowen. *Tracking Moving Objects in Video Images.* Project Report, Oxford University Computing Laboratory, 1994.

Creation of Hardware Objects in a Reconfigurable Computer

Steve Casselman, Michael Thornburg, John Schewel
Virtual Computer Corp. 6925 Canby Ave #103 Reseda CA 91335 USA
sc@vcc.com, jas@vcc.com

Abstract. We define reconfigurable computing systems as those machines that use the reconfigurable aspects of Field Programmable Gate Arrays (FPGA) to implement an algorithm. Researchers throughout the world have shown that computationally intensive software algorithms can be transposed directly into hardware design for extreme performance gain [1,2,3]. Hardware objects are algorithms implemented as dynamically downloadable hardware designs. Hardware objects execute on reconfigurable computing systems based on SRAM-style Field Programmable Gate Arrays (FPGA). A Hardware Object can be created via schematic and VHSIC Hardware Description Language (VHDL) or Verilog hardware description language. To use a hardware design in a software program, it must be converted into a Hardware Object. The Hardware Object can be used over and over or in combination with other Hardware Objects. This H.O.T. (Hardware Object Technology)™ method of programming reconfigurable computers is the subject of this paper [4].

1 Introduction

The advantage of reconfigurable computers is the hardwiring of software and the potential dramatic performance gains due to this technology. This paper demonstrates a method for programming and using a reconfigurable computing system. A Virtual Random Number Generator (VRNG, a Hardware Object) will be constructed and tested within the NAS Embarrassingly Parallel Benchmark Code[5]. We have included the performance results on a SPARC Workstation using the Engineer's Virtual Computer, a reconfigurable system [6].

2 Reconfigurable Platform

Hardware Objects are algorithms implemented as hardware designs. Hardware Objects execute on reconfigurable computing systems based on SRAM based Field Programmable Gate Arrays (FPGA). The Engineers' Virtual Computer™ (EVC1) is used for implementation of Hardware Objects in this paper. The EVC1 consists of a single FPGA (Xilinx® XC4010 or XC4013)[7] connected directly to the SBus of a

SPARC workstation. The EVC1 can be reconfigured in under 50 milliseconds. The EVC1 has been used as a logic emulator for rapid product development, as a evaluation/testing platform for new chips and designs or as an accelerator for repetitive computationally intensive algothrims.

This reconfigurable computer allows a user to implement the compute intensive parts of subroutines in hardware and swap them in and out on demand. Once a design is implemented as a Hardware Object, they can be placed in an ordinary library and used or reused repeatedly. The dynamic use of Hardware Objects offers the programmer the ability to maximize the performance of a given algothrim via a 'hardwired' design from within an application program. Hardware Object Technology (H.O.T.™) is a method for programming Hardware Objects into the application and using them much the same way a programmer uses software program routines.

3.1 Creating Hardware Objects -- Design Flow

The standard digital design flow for Hardware Object implementation.

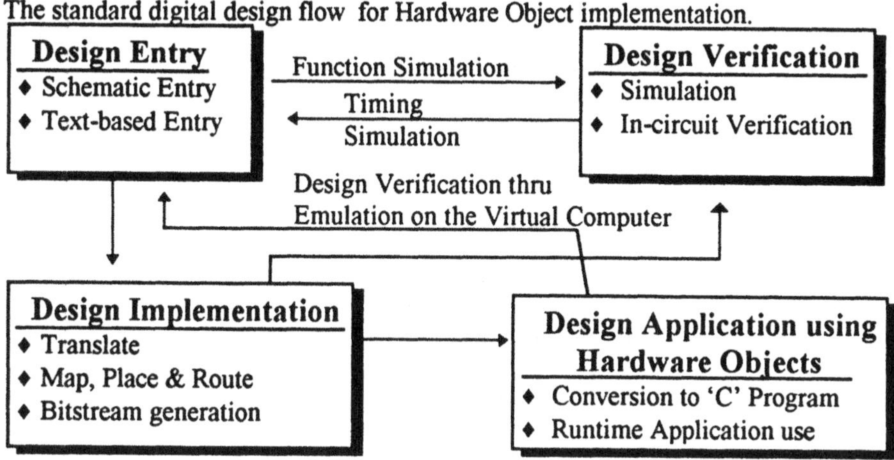

Fig. 1 Design Flow Diagram for Creating Hardware Objects

Except for the design entry, the following processes are handled automatically by the Xmake program included with XACT™ Develpoment Program by Xilinx.

- Create a schematic with the Viewdraw editor [8], using symbols from the Xilinx libraries.
- Translate the schematic into a Xilinx netlist format (XNF) file by the Xilinx WIR2XNF program.

- The design is partitioned into configurable logic blocks (CLBs) and input/output (IOBs), and the routing between them is established, the Xilinx Partition Place and Route program (PPR).

The partitioned, placed and routed design is converted to a configuration bitstream by the Xilinx MakeBits program. The result is a rawbits file in the name of your design (<your_design>.RBT). The rawbits file is converted into a **.h** header file by a program called **r2h**. This header file (the Hardware Object) can be inserted into 'C' language application code and used as anyother software function or routine. All handshaking between the host processor, host bus, host resources and the Hardware Object loaded into the EVC1 is handled by the bus interface macro included in the design. The application can now run the 'hardwiried' version of the algorthim instead of the software version.

3.2 Bus Interface Macro for Viewlogic™ schematic design entry

The Bus Interface Macro contains all connections and logic necessary for interfacing a design to the EVC1 platform. To port an existing ViewLogic® design into the EVC1, simply place the appropriate SBus Interface Macro Symbol on the left hand side of the schematic and connect the necessary pins [9].

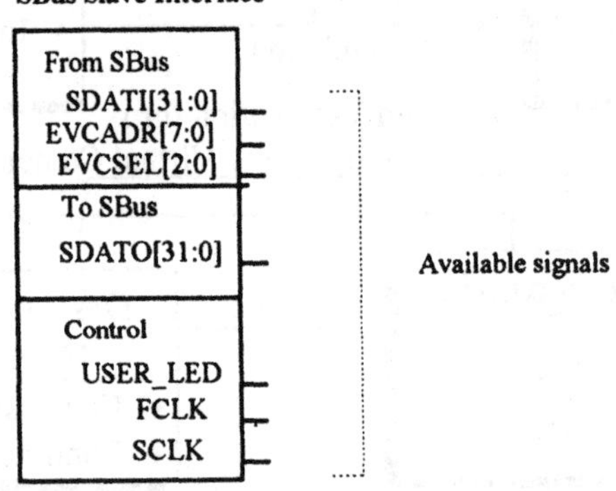

Fig 2 The symbol for the SBus Slave Interface Macro (**EVCSLVIF**).

- SDATI[31:0] --- The SDATI[31:0] is a 32 bit data bus that is used for writing data to the EVC1s. These are unidirectional signals from the host system.

- EVCADR[7:0] --- The EVCADR[7:0] signals are used for control signals and writing data to the EVC. When referenced by a "C" program these signals go to a logic "1", or true, for exactly one clock cycle. When used as a write signals, they can be used to latch data into the design from a user program. When used as a control signal they can be used to start a function or control some part of the design. These are unidirectional signal from the host system.

- EVCSEL[2:0] --- The EVCSEL[2:0] signals are used for read ports. By connecting these signals up to the selects of a mux, data can be read from up to eight different read ports. These are unidirectional signals from the host system.

- SDATAO[31:0] The SDATO[31:0] is a 32 bit data bus that is used for reading data from the EVC1s. These are unidirectional signals to the host system.

- USER_LED --- access to EVC1's LED.

- SCLK --- The SCLK clock signal is the clock provided by the host over the SBus. This clock is used in the design and programming examples presented in this guide. The SBus clock frequency maximum is 25Mhz. In some systems this clock frequency is usually ½ the CPU clock, i.e. SPARCstation 2 has a 40Mhz CPU, and a 20Mhz SBus. This clock signal can be replaced by the "FCLK" ('Fastclock') signal as required.

- FCLK --- The FCLK clock signal is connected to the oscillator socket on the EVC1s PCBA and can be used when a clock frequency different from the SBus clock "SCLK" is required.

Fig. 3 The relationship between the EVC1, Bus Interface Macro and SRAM Macro

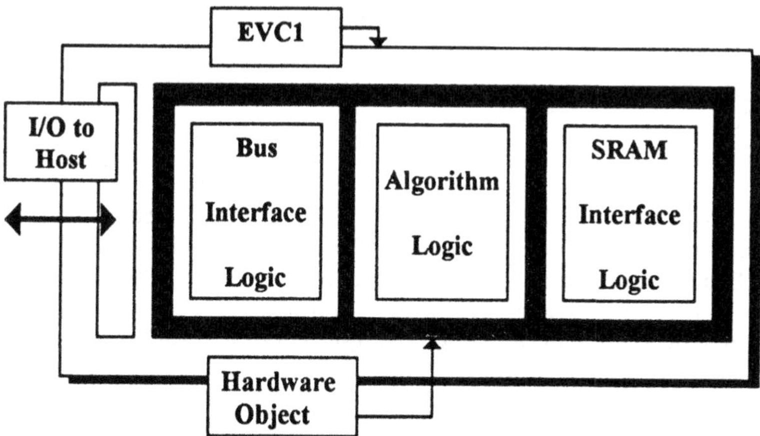

The **Verilog™** listing of the slave interface is provided in three modules. The top level **evc.v** calls out both the user module, "**user_module.v**" and the SBus interface module, "**sbus_if.v**". To create a new design, copy the top level (evc.v) to the current

project name. The top level module will call out the other modules as named. It is recommended that the three modules should be saved off to a back-up directory. As new projects are taken on, the modules can be copied into a working directory and renamed and modified as required without having to modify a prior project's modules.

3.3 Using a Hardware Object in a 'C' Application Program

The following functions need to be included in the 'C' language application program [10,11]. These routines include initializing the Hardware Object, downloading the Hardware Object into the FPGA and clearing or resetting of the FPGA.

1- The design's converted rawbits file Hardware Object:

#include "<header file name>"
The file name is the array name of the design - created by the **r2h** command used earlier in the design development cycle.

The **.h** header file is the compiled Hardware Object. It can be used and reused on-the-fly. Virtual Processing Libraries consists of sets of Hardware Objects.

2- The Hardware Object downloading function:

int (*EVCdownload());
EVCdownload(<filename>, Xilinx Type (0=4010, 1=4013), Slot number

For example: **EVCdownload(my_design,1,2)** downloads the Hardware Object called **my_design** into an EVC1s with a Xilinx 4013 plugged into the SBus slot #2.

The download function passes the Hardware Object's bit stream to the EVC and configures FPGA. An on-board PAL handles the FPGA configuration, SBus acknowledge and the function returns a pointer. This pointer is the base user address assigned by the workstation to the SBus. The pointer refers to ADR[0] in the SBus Interface Macro. A Read or a Write to this address activates ADR[0].

3.- The reset function:

int EVCreset();
Resets the EVC1 EVCreset(n) n=slot number of EVC1
For example: **EVCreset(2)**

The **EVCreset** function resets the FPGA, clearing the downloaded hardware object.

The downloading and reset functions are supplied with the EVC1s Utility Software. Besides the interfaces to Viewlogic, there are Verilog™ and VHDL interface macros.

4.1 The Virtual Random Number Generator - A Hardware Object

Random numbers generators have traditionally relied on determinastic algorithms. Presented in this paper is a method of producing random numbers that incorporates the non-determinstic behavior of the SBus. The following implementation of the non-determinstic random number generator serves as an example of design, execution and performance enhancements possible with Hardware Objects.

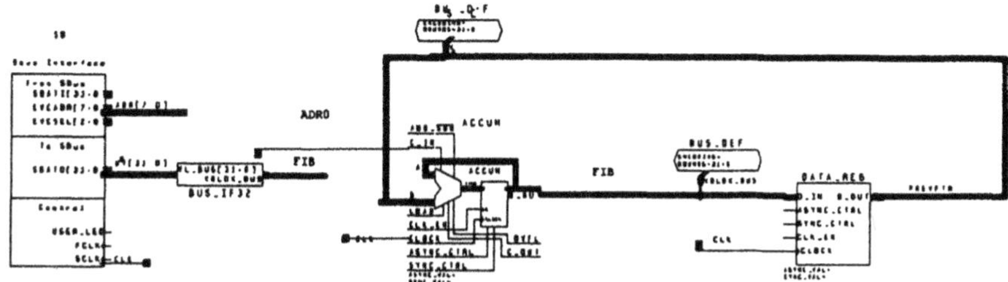

Fig. 4 Random Number Generator Schematic with SBus Macro

Once the design has been converted by VCC tools and inserted into an application, the Hardware Object is ready for use. By including print statements in the program, the developer is able to debug the design running in the EVC1s. The converted design goes into a Virtual Processing Library (VPL) as a Hardware Object. When compiling the application, use libvcc.a (which includes:**download, reset,& status routines**). To use the VRNG Hardware Object in an application, the random numbers need to be formatted in IEEE floating point format. The VRNG's performance (randomness) is compared to the Linear Congruential Method (LCM). We have used the VRNG in the NASA, NAS Embarrassingly Parallel Benchmark by replacing the LCM random number generator with the VRNG.

The following 'C' code contains two functions that load the Hardware Object and uses it to create random numbers between 0 and 1.

```
/* VRNG.C Virtual Random Number Generator Functions   */
#include <stdio.h>
#include <sys/types.h>
#include "vrng10.h"      /* RNG for XC4010 */
#include "vrng13.h"     /* RNG for XC4013 */
int (*EVCdownload());
int* vrng_config()
{
```

```
int     *return_val,i,j ;
int     numreps, slotnum, good_slot, good_probe;
int      probe, found_one;
..........................................................................
/*  This segment of code not shown due to limited space.
The routines locate EVC1 in host system */
..........................................................................
if (good_probe == 0)
return_val = EVCdownload(vrng10,0,good_slot); /* for XC4010 */
if (good_probe == 1)
return_val = EVCdownload(vrng13,1,good_slot); /* for XC4013 */
*(return_val) = 0;
for (i=0;i<100;i++)
        j = j+1;
return(return_val);
}
double vrng (evcbase)
int *evcbase;
{
double *number_ptr, return_val;
int result[2];
int *read_port;
read_port = evcbase;       /* reads from Hardware Object */
number_ptr = (double *) &result[0];
result[0] = *read_port;
result[1] = *read_port;
result[0] = result[0] & 0x3FFFFFFF;
result[0] = result[0] | 0x3FF00000;
return_val = *number_ptr - 1.;
return(return_val);
}
int vrng_array(evcbase,A,array_size)
int *evcbase;
double *A;
int array_size;
{
double *number_ptr;
int result[2],i;
int *read_port;
read_port = evcbase;       /* reads from Hardware Object */
number_ptr = (double *) &result[0];
for (i=0;i<array_size;i++) {
        result[0] = *read_port;
        result[1] = *read_port;
        result[0] = result[0] & 0x3FFFFFFF;
        result[0] = result[0] | 0x3FF00000;
        A[i] = *number_ptr - 1.;
        }
return(0);
}
```

Fig. 4 VRNG Hardware Object Function Library

4.2 An Example of a Benchmarking Application Program using the VRNG

The following code tests the Hardware Object VRNG against a software version of random number generator.

```
/* TESTVRNG.C Random Number Generator Test Program  */
#include <stdio.h>
#include <sys/types.h>
#include <math.h>
```

```
#include <sys/time.h>
#include <sys/resource.h>
#define Aconst 1220703125.0
#define S 271828183.0
#define REPS 1000000
double t1,tt;
double X[REPS];
int* vrng_config();
double vrng();
float timer();
main() {
int i, j[11], k, *base_ptr;
double random_number, max, min;
float time1,time2;
tt=S;
max=0.0;
min=1.0;
time1 = timer();
base_ptr = vrng_config(); /* Uses the VRNG.C Library Function */
vrng_array(base_ptr,X,REPS);
time2 = timer();
printf("time to generate virtual random numbers = %f\n",time2-
time1);
for(i=0;i<11;i++) j[i]=0;
for(i=0;i<REPS;i++) {
        random_number=X[i];
        if (random_number > 0.0 && random_number <= 0.01)j[10]++;
...............................................................
 /*  This segment of code abbreviated due to limited space.  The
routine calculates random numbers from Hardware Object in
formatted as shown */
...............................................................
        if (random_number > max) max = random_number;
        if (random_number < min) min = random_number;
}
k=0;
printf("non-deterministic virtual random numbers\n");
for (i=0;i<10;i++) {
k = k + j[i];
printf("numbers > %.1f and < %.1f %d\n",.1*i,.1*(i+1),j[i]);
}
printf("Total number of random numbers %d\n",k);
printf("max = %.60f\n",max);
printf("min = %.60f\n",min);
printf("numbers > 0 and < .01 %d\n\n\n",j[10]);
EVCreset(1); /* Clears and resets the EVC1)
max=0.0;
min=1.0;
k=0;
/* Uses Software Version of Random Number Generator */
time1 = timer();
        vranlc(0, &t1, Aconst, X);
        vranlc(REPS, &tt, Aconst, X);
time2 = timer();
printf("time to generate random numbers = %f\n",time2-time1);
...............................................................
 /*  This segment of code abbreviated due to limited space. */
...............................................................

/* This routine generates N uniform pseudorandom real*8 numbers
in  the range (0, 1) by using the linear congruential generator.
David H. Bailey October 26, 1990 */
```

4.3 Benchmark Results

The benchmarks were run on a SPARC2 Workstation with an the EVC1-13 (XC4013). The figures below came from two separate runs each generating 1,000,000 random numbers. The code profile for the software version is listed. The two top functions (**_aint** & **_vranlc**) were replaced with a Hardware Object. The Hardware Object saved 50.2% of the run time.

Run 1 Random Numbers

```
numbers > 0.000000 and < 0.100000
99979
numbers > 0.100000 and < 0.200000
100238
numbers > 0.200000 and < 0.300000
100190
numbers > 0.300000 and < 0.400000
99991
numbers > 0.400000 and < 0.500000
100290
numbers > 0.500000 and < 0.600000
99362
numbers > 0.600000 and < 0.700000
100268
numbers > 0.700000 and < 0.800000
99627
numbers > 0.800000 and < 0.900000
100384
numbers > 0.900000 and < 1.000000
99671

Total random numbers = 1000000
max = 0.999998027735670
min = 0.000000566747618
```

Embarssingly Parallel Benchmark

Using Sparc Station 2 for all

```
CPU TIME = 632.6700
           684.9300 (complied w/-p)

N = 2^24
NO. GAUSSIAN PAIRS = 13176389.
COUNTS:
0       6140517.
1       5865300.
2       1100361.
3         68546.
4          1648.
5            17.
6             0.
7             0.
8             0.
9             0.
```

Code Profile

%time	cumsecs	#call	ms/call	name
26.3	179.741	.....		_aint
23.9	343.380	257	636.73	_vranlc
15.1	446.781	0.01	0.00	_sqrt
14.2	543.680	1	96900	_MAIN_
13.7	637.730		0.00	mcount

Run 2 Random Numbers

```
numbers > 0.000000 and < 0.100000
100082
numbers > 0.100000 and < 0.200000
100478
numbers > 0.200000 and < 0.300000
100391
numbers > 0.300000 and < 0.400000
100174
numbers > 0.400000 and < 0.500000
99763
numbers > 0.500000 and < 0.600000
99686
numbers > 0.600000 and < 0.700000
99779
numbers > 0.700000 and < 0.800000
100166
numbers > 0.800000 and < 0.900000
99272
numbers > 0.900000 and < 1.000000
100209

Total random numbers = 1000000
max = 0.999998301519109
min = 0.000000095615990
```

Embarssingly Parallel Benchmark

Using EVC1 just for random numbers

```
CPU TIME = 353.8800

N = 2^24
NO. GAUSSIAN PAIRS = 13177271.
COUNTS:
0       6138931.
1       5865486.
2       1101640.
3         69558.
4          1634.
5            22.
6             0.
7             0.
8             0.
9             0.
```

Code Profile Notes

The top two functions (**_aint** & **_vranlc**) were replaced by a Hardware Object, which saves 50.2% (26.3% + 23.9%) of the run time.

5 Conclusion

Hardware Object Technology (H.O.T.) can virtually eliminate a subroutine from a programs' performance profile. Since Hardware Objects can be swapped out at anytime, a large portion of a program's running time can be accelerated.

[1] Kella Knack,
Prototyping with FPGAs, ASIC & EDA, December 1993, p.10.

[2] R. Larus,
Loop-Level Parallelism in Numeric and Symbolic Programs, IEEE Trans. on Parallel and Distributed Systems, Vol 4, No. 7 July 1993, pp. 812-826.

[3] S. Casselman
Virtual Computing and the Virtual Computer™, Proceedings of IEEE Workshop on FPGAs for Custom Computing Machines, April 1993

[4] Steve Casselman, Michael Thornburg, John Schewel
H.O.T (Hardware Object Technology) Programming Tutorial, Virtual Computer Corp. January 1995.

[5] D. Bailey, E. Barszcz, J. Barton, D. Browning, R. Carter, L. Dagum, R. Fatoohi, S. Fineberg, P. Frederickson, T. Lasinski, R. Schreiber, H. Simon, V. Venkatakrishnan, and S. Weeratunga
The NAS Parallel Benchmarks, RNR Technical Report RNR-94-007, NASA Ames Research Center. March 1994.

[6] Mike Thornburg, John Schewel, Steve Casselman
Users Guide EVC1s, Virtual Computer Corporation, 1994.

[7] *The XC4000 Data Book* Xilinx, August 1992

[8] *Workview Manual*, Viewlogic Systems Inc., 1991.

[9] James L. Lyle
Sbus: Information, applications and experience, Springer-Verlag, 1992.*Reconfigurable Silicon and Optical Fiber Memory*, Proceedings of IEEE Workshop on FPGAs for Custom Computing Machines, April 1993.

[10] S. Harbison & G. Steele Jr.
Programming in C, Hayden Book Company, 1991.

[11] Stephen Kochan.
C A Reference Manual Third Edition, Prentice Hall, 1991.

Rapid Hardware Prototyping of Digital Signal Processing Systems Using Field Programmable Gate Arrays *

L.E. Turner and P.J.W. Graumann

Department of Electrical and Computer Engineering
University of Calgary
Calgary, Alberta, Canada T2N 1N4
email: turner@enel.ucalgary.ca

Abstract. The design and hardware implementation of two Digital Signal Processing subsystems is described. Rapid design evaluation, algorithm verification and hardware prototyping are facilitated using the serial pipelined hardware description language DFIRST, the event driven simulator *dsim* and the gate compiler *trans* . The two subsystems are individually hardware prototyped and tested using Xilinx and Actel FPGA devices.

1 Introduction

Due to their low cost and user programmability, Field Programmable Gate Arrays (FPGAs) are particularly suited to the implementation of digital logic designs for hardware prototyping purposes or low volume applications. The use of schematic capture based Computer Aided Design (CAD) tools is a common approach to the design of custom logic devices using FPGAs. Using a hierarchical approach, the designer implements the required digital logic from a set of logic elements supplied by the integrated circuit manufacturer. The design is simulated at the logic gate level to verify that it has been constructed correctly.

In DSP applications, arithmetic circuitry for operations such as addition, subtraction and multiplication are commonly required. These arithmetic circuits can be designed and implemented using sub-circuits which can be re-used. However, as these designs can only be simulated at the logic gate level, it is difficult to verify the functional performance of the algorithm being implemented. It is particularly awkward to determine the potential undesirable side-effects of finite precision arithmetic, as this may require that large data sets be simulated and translated from numerical values to logic levels and vice-versa.

The process of designing and implementing a DSP system or subsystem can be time consuming when using a combination of schematic capture and logic level

* Partial support for this work has been provided by the Alberta Microelectronics Centre, Bell Northern Research, the Canadian Microelectronics Corporation, Micronet, NSERC and the University of Calgary. The contributions of students in ENEL 619.70 and ENEL 611 who implemented designs using the tools are gratfully acknowledged.

simulation. In addition, the specification of the DSP algorithm may have been verified using custom programs or discrete event simulators [1] suited to DSP. This introduces the need for a time consuming and error prone translation of the specification into digital logic. Once this translation is complete and the digital logic functions correctly, the designer is not inclined to make significant changes to the implementation.

One method of increasing the range of architectural solutions that a designer may explore in a reasonable time is to specify the DSP system with a Hardware Description Language (HDL); verify the finite precision performance of the system using a simulator for the HDL; and then automatically translate the HDL into a gate logic level netlist for implementation using vendor supplied FPGA tools. This approach allows the designer to detect errors in the implementation of an algorithm early in the design process using the HDL simulator and to rapidly generate prototype hardware implementations for evaluation.

The use of a HDL and an event driven simulator to facilitate the design and implementation of two DSP subsystems using Xilinx and Actel FPGAs is described. The subsystems implemented were taken from initial specification to final implementation (using Actel 1280 FPGA devices) over a period of four months. Each of these designs were the responsibility of three students enrolled in the course ENEL 619.70, Application Specific Integrated Circuit Design, Implementation and Testing.

Starting from an initial specification which was provided by a "Customer"[2], the students provided feedback on their progress through:

- Design Specification review (Convince the "Customer" of what might possibly be implemented).
- Design Implementation review (Convince the "Customer" to accept the design and pay for fabrication).
- Oral and written Design Presentation.
- Demonstrate the working Actel 1280 FPGA.

At the beginning of this course, formal lectures are held to introduce a variety of ASIC design concepts including, design for testability, meta-stability, asynchronous inputs, clock skew and FPGA technology. Students admitted to the course are expected to have a basic knowledge in digital logic design as a pre-requisite.

2 TOOLS: *trans* and *dsim*

The HDL used is a Register Transfer Language (RTL) which is designed for bit-serial and digit-serial pipelined arithmetic systems. This RTL, known as DFIRST [2, 3] is an extension of the bit-serial language FIRST [4]. The use of serial pipelined arithmetic is particularly effective in FPGA implementations because it does not overload the limited routing resources available. Bit-serial systems are also known to be efficient implementations when compared to bit-parallel designs using an area-speed product measure [4].

[2] The authors of this paper.

An event driven simulator *dsim* [2, 3] is used to verify that the DSP algorithm (as defined by the RTL DFIRST description) functions correctly using finite precision arithmetic. In addition to providing essential timing verification for the bit-serial and digit-serial design, *dsim* also converts the two's complement serial data into signed integer values to facilitate the input and display of data signals. Any internal signal may be identified and displayed at either the data word, data bit or gate timing level.

When the algorithm implementation is determined to be correct through simulation using *dsim* , the translation of the DFIRST RTL description into a gate level description can be performed using the gate compiler program *trans* [5]. *trans* converts a DFIRST netlist into the gate level logic netlist format appropriate for use with Xilinx or Actel FPGAs. This gate compiler implements each of the DFIRST arithmetic elements using generic logic elements and applies a series of reductions and mappings in order to obtain an effective implementation for the target FPGA.

The ViewLogic [6] CAD tools, ViewDraw and ViewGen, and the Xilinx [7] and Actel [8] FPGA place and route tools are also used to generate the logic block placement and routing for each FPGA. Reprogrammable Xilinx 3020, 3090, 4005 and 4010, FPGA devices are used to implement hardware prototypes of the DSP subsystems for evaluation, prior to programming the one time programmable Actel 1280 FPGA.

3 DSP SUBSYSTEMS

The two subsystems implemented made use of the DFIRST RTL language to specify the required DSP algorithms and *dsim* was used to verify the performance of these algorithms. The *trans* gate compiler provides an estimate of the number of FPGA cells required to implement each device in the target FPGA technology, allowing designers to quickly evaluate the size of different designs. The two designs implemented were a vector graphics controller for a 640 by 200 pixel Liquid Crystal Display (LCD) and a multiple data stream, single instruction stream, pipelined bit-serial DSP processor.

3.1 Vector LCD Controller

The TSC1500[3] LCD Controller implements a vector graphics interface for an Epson 640 by 200 dot matrix LCD. In addition to providing the required timing signals and interface to the raster display memory, the controller is capable of drawing (with user defined line styles) the graphics primitives of point, line, box and circle. Specialized arithmetic processors are implemented for drawing lines, circles and boxes. A block diagram of the LCD controller functional units is given in figure 1.

[3] Trevor Akitt, D. Sisay Yirga and Clinton Lam designed and implemented this LCD vector controller.

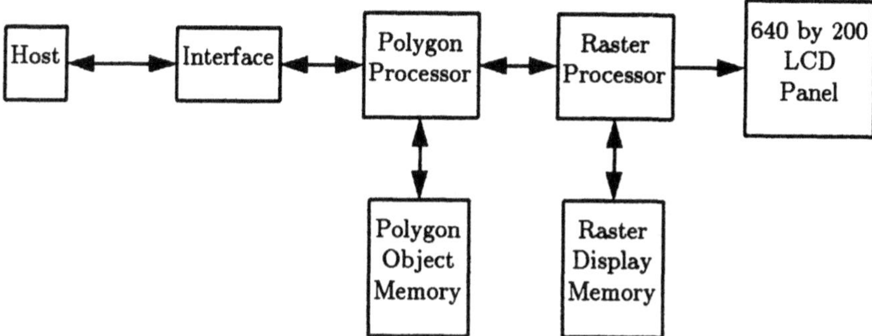

Fig. 1. LCD Vector Controller Block Diagram

LCD Controller Design The line processor uses Bresenhams' incremental technique [9] and the two end points of the line to determine the pixel locations for the line. The box processor calculates the pixels for a rectangular box given the locations of two corners by incrementing in the vertical and horizontal directions. For a given center location and radius, the circle processor performs the pixel calculations using a Midpoint algorithm [9] which is derived from Bresenhams' incremental technique. In addition, circle symmetry is taken advantage of by calculating only pixels for one eighth of the circle and mirroring these values about the horizontal and vertical axes to complete the circle.

Pixel locations are specified using a 15 bit two's complement data format. A clipping processor, which implements Sutherland's Clipping Algorithm [9], allows elements to be drawn which are partly or completely outside the 640 by 200 LCD display space. The origin at the lower left hand corner of the display.

As the pixel locations for lines, boxes and circles are calculated, these pixel locations are added to the object memory according to a specified eight bit pixel pattern. The pixel pattern can be different for each object drawn. The object memory contents are, on command, copied into the display memory using individual pixel OR, AND, XOR and XNOR operations. The raster processor continues to display the raster memory contents while the object memory contents are being copied and the display is not disrupted by this copying. When the object memory has been copied into the display memory, the object memory is then cleared. All address, data and control signals for the 16K by 8 bit static RAM object and display memories are provided by the controller.

The user interface for the LCD controller uses a 15 bit bi-directional data bus, a read-write signal and three register select lines. The three register select lines are used to access four data registers, an instruction register, a status register and an output register. The data registers are used to hold the vectors which define the objects to be drawn. The control register is used to initiate the drawing of each object and set the line style (pixel pattern). The status register is used to signal completion of each drawing operation and the output register allows the contents of any pixel location to be read.

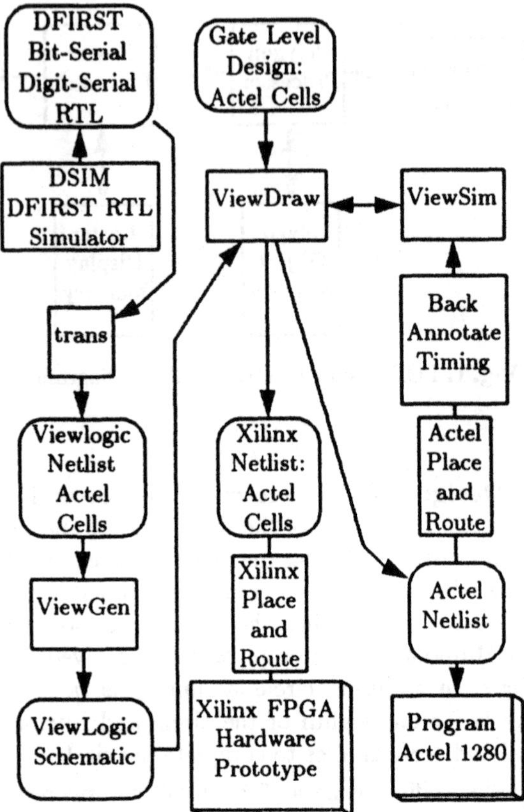

Fig. 2. LCD Vector Controller Design Flow

LCD Controller Implementation The design of this LCD controller required the use of both full custom logic for the raster controller and DSP algorithms defined using the DFIRST RTL language. Because of this requirement, the custom logic for the raster controller was implemented using the ViewDraw schematic capture tool and the Actel logic library. The object processors were specified using the DFIRST language and simulated using the event driven simulator *dsim* . The gate compiler *trans* and the ViewLogic tool ViewGen were used to generate Actel logic cell based instances of the object processor designs. ViewSim was used to simulate the complete Actel gate level description of the object and raster controllers. The complete design flow is shown in figure 2.

After verifying the functionality of the complete design from the ViewSim simulation of the Actel logic cells, a Xilinx netlist implementation of the design was generated using ViewDraw and wir2xnf. This was possible because only four low level logic elements were used in the ViewDraw representation of the Actel design. These four Actel elements were translated into equivalent Xilinx components with the gate compiler *trans* . Applying the Xilinx XACT [7] tools to the Xilinx netlist and the Xilinx models of the four actel elements, the complete design was placed and routed into three Xilinx FPGAs (two 4005s and one 3090). These three FPGA devices were connected on a prototype board and the system functioned at a clock rate of approximately 3 MHz. At this stage in the design process, all state machines and counters could be connected serially for testing purposes. In addition, a selection of signals between the major components were brought out to pins in order to provide observability and controllability of these components. These testability features were removed once the design was found to function correctly.

An error in the supply of the four bit raster data to the LCD display was detected at this stage. The two, four bit pairs of raster data which are used internally in the controller as eight bit values were being supplied in the reverse order required for correct display. This error was not detected during logic level simulation as the LCD panel was not included in the simulation. The output was interpreted incorrectly by the designers. This error was easily corrected at the top level circuit design and a revised implementation was tested using the Xilinx FPGAs. Full functionality was observed at the 3 MHz. clock rate.

An Actel netlist description was now generated for the corrected top level design and the Actel place and route tools were used to generate the design for the Actel 1280 FPGA. Routing difficulties for long line connections were corrected by adding additional signal buffering to complete the design. Once the design was placed and routed, additional timing information due to the actual interconnection wirelengths was back annotated to the Actel netlist and re-simulated using ViewSim. Race conditions detected at this stage were corrected by reducing the fanout of critical nets and the new design was placed and routed, back annotated and resimulated to verify that the errors had been corrected.

The Actel 1280 implementation operated correctly at a maximum clock rate of 12.8 MHz. This implementation used 1228 logic cells out of an available 1232 cells resulting in a 99.7% device utilization. The probability of routing success reported by the Actel software was 94%. Power consumption at the 10 MHz. operating rate was measured to be 700 milliwatts.

3.2 Bit-serial DSP Processor

The bit-serial multiple data stream, single instruction stream DSP processor[4] was designed to implement a channel equalization algorithm based on a QR-Decomposition Least-Squares Adaptive Decision Equalizer[10]. The equalizer was designed for use in a digital radio channel where each packet of data is preceded by a short training sequence. At the receiver, the training sequence, modified by the channel characteristics, is applied to the QR-Decomposition and back substitution algorithm to generate the coefficients for a fifth order Finite Impulse Response (FIR) filter. The data following the training sequence is then processed by the FIR filter and thus errors due to channel distortion are reduced. Arithmetic operations of multiplication, division, square root as well as

[4] Kenneth Chow, Brian Ebel and Garry Funk designed and implemented the bit-serial DSP processor.

addition and subtraction are required to implement the equalization algorithm. Because of the complexity of the algorithm, a programmable DSP processor, named APLUS (Arithmetic Programmable Logic Using Serial Architecture) was designed specifically for the equalization algorithm. The APLUS DSP processor was implemented using one Actel 1280 FPGA plus additional program and data memory devices.

Bit-serial DSP Processor Design A programmable processor implementation was chosen after careful consideration of the computational requirements of the QR-Decomposition equalizer. An attempt to generate a non-programmable special purpose bit-serial processor was made. However, due to the need for complex control timing sequences and the hardware size limitation of the Actel 1280 FPGA, a micro-programmable architecture was selected. This decision effectively shifted the design of the complex control signals from hardware to an external Programmable Read Only Memory (PROM). This partitioning allowed the finite precision algorithm development to proceed in parallel with the programmable hardware development.

The comparison of a finite precision simulation of the QR-Decomposition algorithm with an infinite precision simulation removed the need for a square root operation. Instead of using two multiplications, an addition and a square root operation a magnitude approximation [4] was implemented. The magnitude approximation made use of a magnitude ordering operator (AORDER), shifting, addition and subtraction operators. The AORDER operator has two inputs and two outputs. This operator performs a sorting operation based on the magnitude of the input signals. One output will always be the larger magnitude signal and the other output will be the smaller. This operator is substantially smaller in size than a square root operator.

The APLUS DSP processor Arithmetic Logic Unit (ALU) includes two multipliers, a divider, an AORDER operator, an adder, a subtracter, a conditional equality operator and a conditional less-than operator (for conditional branching). These pipelined operators function in parallel on a 16 bit serial data word. Each operator outputs a new value each 16 clock cycles. The divide operator requires two word times (32 clock cycles) to generate the output associated with a specific input. An external Random Access Memory (RAM) interface is used to load or store values to, or from, nine recirculating on chip ALU registers. A block diagram of the processor is given in figure 3.

The APLUS DSP Processor uses a 64 bit microcode word allowing each element of the ALU to be controlled independently. To reduce the number of input-output pins required, each microcode instruction is supplied in four 16 bit slices. Because of the pipelined operators, care must be taken to account for the word delay when implementing conditional branching operations. An assembler was written to facilitate programming the APLUS.

Bit-serial DSP Processor Implementation Finite precision simulations of the QR-Decomposition equalization algorithm were compared to infinite pre-

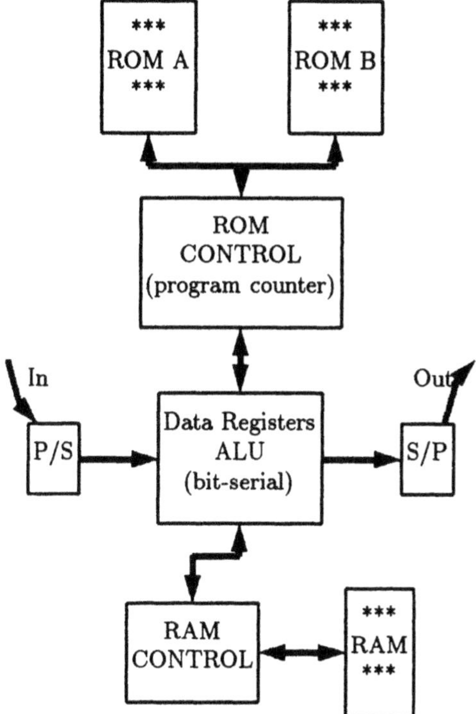

Fig. 3. Block diagram of the APLUS multi data stream bit-serial DSP processor.

cision simulations to determine the effects of finite precision arithmetic on the performance of the algorithm. These simulations, written in the "C" programming language verified that a 16 bit wordlength would be acceptable.

The APLUS ALU was implemented in the DFIRST RTL language and a functional simulation was performed using the *dsim* simulator. However, as *dsim* did not model the external RAM and ROM devices, a complete design was generated using the gate compiler *trans* . This complete gate level design was simulated using the LOGSIM [11] event driven logic simulator. The microcode ROM values were generated using the assembler. The entire QR-Decomposition and back substitution algorithm was simulated for the 14 iterations required to generate the FIR filter coefficients. This simulation required several hours to complete. The FIR filter coefficients were compared to the finite precision "C" language simulations to verify the accuracy of the implementation.

At this stage the *trans* gate compiler was used to generate a Xilinx FPGA netlist of the APLUS DSP Processor which was compatible with the ViewLogic ViewDraw schematic capture system. The ViewDraw netlist files were then converted to Xilinx XNF format using the wir2xnf program and the Xilinx XACT tools were used to place and route and finally program a Xilinx 4010 FPGA. This hardware verification stage made use of carefully written test programs to verify the functionality of the hardware.

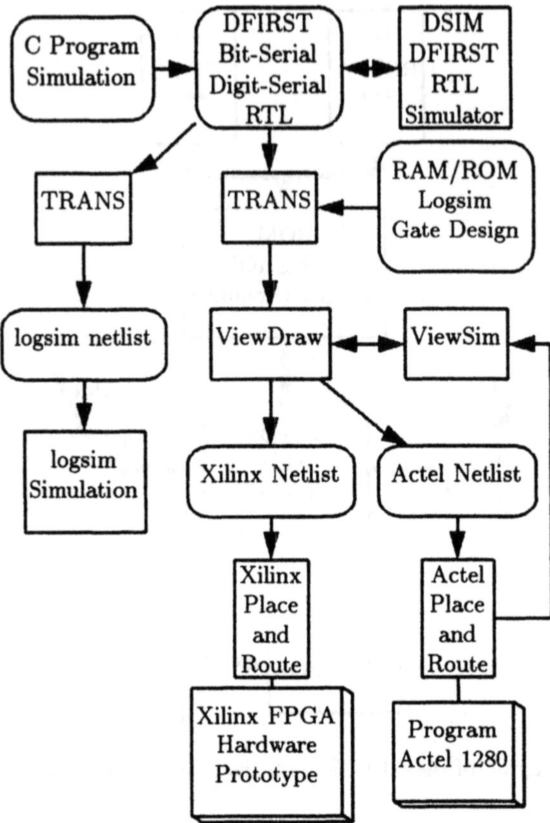

Fig. 4. Design flow for the APLUS DSP processor.

The ViewDraw system was then used to generate the an Actel netlist which was processed by the Actel tools to generate the programming file for an Actel 1280 FPGA. Because almost 100% of the available logic modules were utilized in the design, there was difficulty routing all nets on the device. Careful addition of buffers on critical paths identified by the Actel placement and routing software allowed all nets to be successfully routed. Using the timing information for the actual logic element placement and routing on the 1280 FPGA in a ViewSim simulation of the Actel design the actual device performance was determined. It was necessary to add buffers to internal nets which had a fanout in excess of 15.

All 1232 logic modules available on the Actel 1280 FPGA were used. Unfortunately, it was not possible to allocate specific I/O pins to nets and still route all internal nets. The I/O pins assignments were selected by the place and route software. The final ViewSim simulation using the actual timing back annotated from the placement and routing information indicated that the design would operate properly with at 10 MHz. clock. The design flow for the APLUS device is shown in figure 4.

The APLUS DSP processor implemented on the Actel 1280 FPGA used all

1232 logic modules available as well as 92 out of 142 I/O modules. The device functioned correctly at a 10 MHz. clock rate and consumed approximately 950 mW of power.

4 CONCLUSION

In this paper, the rapid prototyping of DSP algorithms for a vector graphics LCD controller and a multi data stream bit-serial DSP processor were described. Both reprogrammable (Xilinx) and one time programmable (Actel) FPGAs were used to test and implement these designs which use pipelined bit-serial arithmetic. A bit-serial RTL (DFIRST) was used to specify the DSP algorithms. The event driven simulator (*dsim*) was used to simulate the DFIRST specification before hardware implementation was attempted. The gate compiler *trans* was used to convert the DFIRST specification into Xilinx and Actel netlists prior to FPGA programming. The ViewLogic CAD tools ViewDraw, ViewSim and ViewGen were used to display, simulate and modify the form of the netlists used.

References

1. L.E. Turner D.A Graham and P.B. Denyer. The analysis and implementation of digital filters using a special purpose cad tool. *IEEE Transactions on Education: Special Issue on Teaching and Research in Circuits and Systems*, CAS-23(No. 3):pp. 287–297, 1989.
2. P.J. Graumann and L.E. Turner. Implementing dsp algorithms using pipelined bit-serial arithmetic and FPGAs. *First International ACM/SIGDA Workshop on FPGAs*, pages 123–128, 1992.
3. P.J. Graumann and L.E. Turner. Specifying and hardware prototyping of dsp systems using a register transfer level language, pipelined bit-serial arithmetic and fpgas. *2nd Candian Workshop on Field Programmable Devices*, 1994.
4. P. Denyer and D. Renshaw. *VLSI Signal Processing: A Bit Serial Approach.* Addison-Wesley Publishing Company, 1985.
5. L.E. Turner P. Graumann and S. Barker. TRANS User's Guide. *Department of Electrical and Computer Engineering, University of Calgary, Internal Report*, 1992, 1993.
6. ViewLogic. *VIEWlogic Reference Manual.* VIEWlogic Systems, Inc, 1991.
7. XILINX. *The Programmable Logic Data Book.* Xilinx Inc., 1994.
8. ACTEL. *FPGA Data Book and Desgin Guide.* Actel Corporation, 1994.
9. J. Foley A. Van Dam S. Feiner and J. Hughes. *Computer Graphics - Principles and Practice.* Addison-Wesley Publishing Company, 2 edition, 1990.
10. A. Sesay and M. Patton. Qr decomposition decision feedback equalization and finite precision results. *IEE Proceedings Part F, Radar and Signal Processing*, 1993.
11. B. Kish L.E. Turner J.M Bauer and R. Wheatley. Logsim users guide. *Department of Electrical and Computer Engineering, University of Calgary, Internal Report*, 1989.

Delay Minimal Mapping of RTL Structures onto LUT Based FPGAs

A R Naseer M Balakrishnan* Anshul Kumar

Department of Computer Science and Engineering
Indian Institute of Technology, Delhi
New Delhi - 110 001

Abstract. This paper presents an approach for mapping data paths onto FPGAs minimizing delay. The approach exploits the regularity of data path components. It involves slicing the components and generating "realizable cones" from slices of one or more connected components. The objective in delay minimization is to cover the RTL structure with realizable cones in such a way that the maximum path length is minimized. A parameter called delay benefit is defined for each cone based on its potential to reduce the path delay. A greedy heuristic is employed to cover the current critical path with cones having maximal delay benefit. Experimental results are shown to demonstrate delay reduction obtained by this approach.

1 Introduction

The Field Programmable Gate Array(FPGA) is a relatively new technology that allows circuit designers to produce ASIC chips without going through the conventional fabrication process. Short turnaround time and low manufacturing cost have made FPGA technology popular for rapid system prototyping and low- or medium-volume applications. The increase in complexity of FPGAs in recent years have made it possible to implement data path oriented designs on FPGAs. It is easy to visualize that the high level synthesis techniques would be increasingly used in the FPGA context, more so as the fast turnaround time and low design costs are the key benefits of this technology.

We have been looking at the problem of mapping data paths generated by high level synthesizers directly onto FPGAs[1, 2, 3, 4]. Recently, a number of technology mapping techniques have been proposed for delay optimization in LUT based FPGA designs, with the objective of minimizing the number of Configurable Logic Blocks (CLBs) in the longest path in a boolean network. Important among previous work for delay optimization include Chortle-d[5], Improved Mis-pga [6], Flow_Map [7], etc. Most of these approaches are based on adopting the logic synthesis techniques with the flexibility that k-input LUTs can synthesize any k-input function. These approaches, though applicable to map data paths onto FPGAs, ignore a key feature of all the data path components; namely,

* Currently at University of Dortmund, Germany on leave from IIT Delhi

their regular and iterative structure. We exploit this regularity to slice the components into bits and map slice of one or more "connected" components onto CLBs. A technique for cost minimal mapping of RTL structures onto FPGAs was reported in [3].

This paper presents an approach (part of a system called FAST[1]) for mapping data paths in such a manner that delays are minimized. FAST forms a backend to a Data path Synthesizer[8] and is integrated to IDEAS[9]. Starting from a VHDL like behavioral description language and a global time constraint RTL data path is generated automatically. As the RTL modules are generic in nature, they cannot be directly mapped onto a single CLB. A dynamic slicing technique based on the iterative structure of RTL modules is used to partition them into component parts. Each RTL module is viewed as consisting of slices of one or more bits. Closely connected slices of different modules are considered together and mapped onto one or more logic blocks. At each stage an attempt is made to reduce the number of CLB levels in the paths corresponding to the combinational logic and interconnections between the primary inputs/register outputs and register inputs /primary outputs.

The rest of the paper is organized in five sections. Section 2 presents the preliminary definitions and terms used in this paper. In section 3, we present the expressions for computation of delays of individual nodes, which are used in section 4 to build expression for path delays. Algorithm for delay minimal mapping of RTL structures onto LUT based FPGAs is given in section 5. In section 6, results of delay minimal mapping for some high level synthesis benchmarks on XILINX devices [10], alongwith conclusions are presented.

2 Definitions and Terminology

The input network is a data path RTL structure obtained from a high level synthesizer. This network is represented as a directed graph $G(V, E)$ where each node represents a *module* which could be either a register, a functional unit (such as ALU, Adder) or an interconnection element(such as MUX) and directed edges represent connections between the modules.

A *cell* is an indivisible part of a module that is iterated to form a *module*. A *slice* is an array of contiguous cells of a *module*. A *cone* is a set of slices of nodes which lie on paths converging on a particular node called *apex* of the cone. A *realizable* cone is one that fits in a CLB. Similarly, slices of a node which can be directly realized by a CLB are known as *realizable slices*.

Let *width(n)* and *width(s)* represent width (i.e., the number of cells) of a node n and slice s respectively. In our approach, the slice width is not decided a-priori, rather it is dynamically determined during the mapping process. We refer to this as *Soft-slicing*. The maximum slice width of a node which is a realizable cone by itself, is referred to as *max_slice_width*.

Let n_k represent slice of node n with width k, then

$$max\_slice\_width(n) = \max k \mid n_k \textit{ is realizable in a CLB} \qquad (1)$$

A cone is said to be *trivial* or *simple* if it contains only slices of the same node. A cone which contains slices of different nodes is termed to be *non-trivial* or *compound*. Note that all slices are *trivial* form of cones and all slices of node n upto *max_slice_width(n)* are *realizable* cones.

For a slice s, $node(s)$ denotes the corresponding node. We define *width* of a cone C to be the width of the maximum width slice in the cone and is given by

$$width(C) = \max_{s \in C} width(s) \tag{2}$$

During the technology mapping process, we need to examine the data paths originating at primary inputs or register outputs and ending at primary outputs or register inputs. We call these paths *d_paths*. To facilitate examination of *d_paths*, we split each register node r of G into two nodes - ro, a register output node which carries with it all the outgoing arcs of r and ri, a register input node which carries with it all the incoming arcs of r. Now every directed path from a source node to a sink node is a *d_path*. Let DP be the set of all *d_paths*.

3 Delays of nodes

To analyse delay of a node in garph G, we consider the network of cells corresponding to the RTL component represented by the node n. We define *cell_levels(n)* to be the number of cells in the critical path in this network. Considering a node n alone, its delay is minimum when it is realized using largest possible trivial slices. This delay is given by the following expression :

$$\left\lceil \frac{|cell\_levels(n)|}{max\_slice\_width(n)} \right\rceil \cdot D_{cell}(n) \tag{3}$$

where $D_{cell}(n)$ is the cell delay of node n which is an integer multiple of D_{CLB}, the delay of a CLB which is nothing but the delay of the function generators or LUTs.

Example 1 : For a 16-bit adder as shown in figure 1, $max\_slice\_width = 1$, $cell\_levels = 16$, $D_{cell} = 1{*}D_{CLB}$ and total delay $= 16{*}D_{CLB}$, whereas for a 4-input 16-bit wide mux, $max\_slice\_width = 1$, $cell\_levels = 1$, $D_{cell} = 2{*}D_{CLB}$ and total delay $= 2{*}D_{CLB}$.

When the RTL components are connected to form a data path, we need to take into account the direction of signal propagation within the components. In the adder, the carry signal propagates from LSB to MSB. There are instances of modules(e.g., comparator) in which signals propagate from MSB to LSB. Based on the direction of signal propagation through the cells, the type $type(n)$ of a node n is either "+" (e.g., for adder) or "-" (e.g., for comparator) "0" (e.g., for Mux).

To facilitate the computation of delays of the *d_path* containing different types of nodes, we split the delay of a node n into two parts : *vertical delay* $vd(n)$ representing the delay of a single cell and *horizontal delay* $hd(n)$ representing

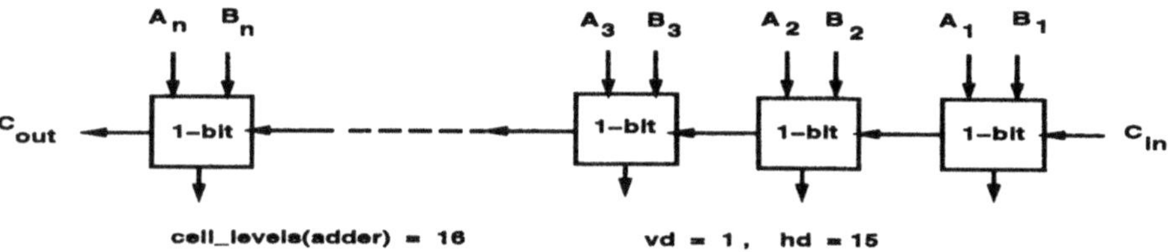

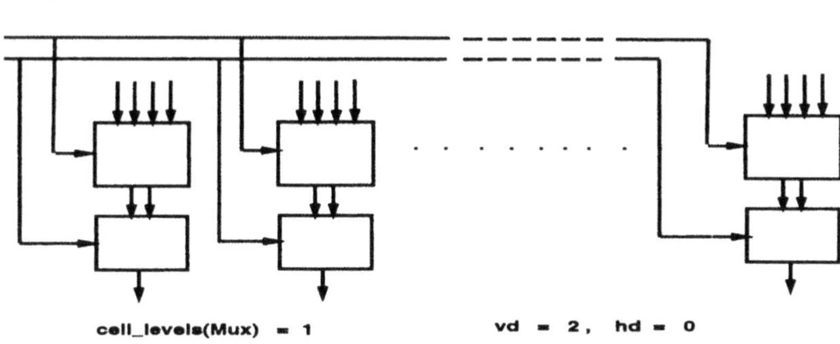

Fig. 1. (a) A 16-bit adder (b) A 16-bit 4-input Mux

the additional delay encountered due to signal propagation within the node as follows.

$$vd(n) = D_{cell}(n) \tag{4}$$

$$hd(n) = \left(\left\lceil \frac{|cell\_levels(n)|}{max\_slice\_width(n)} \right\rceil - 1 \right) . D_{cell}(n) \tag{5}$$

These two parts of the node delay combine differently when path delays are computed.

4 Delays of d_paths

Let *node_set*(*dp*) represent the nodes in a *d_path dp*, then the minimum delay of *dp* is obtained by adding the horizontal and vertical parts of the minimum delays of nodes in *node_set*(*dp*) appropriately. All the vertical delays are added unconditionally whereas the horizontal delays are added conditionally, depending upon the directions of signal propagation in the adjoining nodes.

$$min\_delay(dp) = \sum_{n \in node\_set(dp)} vd(n) + \sum_{n \in node\_set(dp)} hd(n) * p\_flag(n, dp) + source\_delay(dp) + sink\_delay(dp) \tag{6}$$

where *source_delay*(*dp*) represent either the input pad delay or register propagation delay, *sink_delay*(*dp*) represents either the output pad delay or register setup and p_flag is a 0-1 flag which determines whether the horizontal delay of a node n to be included in the *d_path* delay computation. This decision is taken based on the types of node n as well as type of its predecessor and successor in the *d_path* using the following theorem.

Theorem : Let the predecessor and successor of node n in dp be denoted by p and s respectively. Then the $p\_flag(n, dp)$ has a value true under any of the following four conditions :

i) type(n) $\neq$ type(s) and hd(n) $>$ hd(s)
ii) type(n) $\neq$ type(p) and hd(n) $\geq$ hd(p)
iii) type(n) $\neq$ type(p) and type(n) $\neq$ type(s)
iv) hd(n) $>$ hd(s) and hd(n) $\geq$ hd(p)

We omit the proof of this theorem here and illustrate the basic idea through an example.

Example 2 : To illustrate the delay computation of a *d_path* containing different types of nodes, let us consider a portion of datapath containing five 6-bit nodes connected as shown in figure 2 a). Cell level structure corresponding to these nodes is given in figure 2 b). The values of vertical, horizontal delays and propagation flag for each of the nodes are listed in table 1, along with the computation of the *min_delay* of the *d_path dp*.

A *d_path* is critical if it has the largest *min_delay* among all the *d_paths*. The

node no	max_slice_width	vertical delay vd	horizontal delay hd	propagation flag p_flag
1	2	1	2	1
2	3	1	1	0
3	2	1	2	1
4	2	1	2	1
5	3	1	0	0
min_delay(dp)		$(5) + (2 + 0 + 2 + 2 + 0) = 11^{*}\ D_{CLB}$		

Table 1. Delay computation of a *d_path*

critical path delay of a graph G is given by

$$critical\_delay(G) = \sum_{dp \in DP} min\_delay(n) \tag{7}$$

5 Minimal Delay Mapping Algorithm

Our algorithm is based on packing slices from multiple nodes into a single CLB. Initially, all nodes are considered to be covered by maximum width slices, forming trivial cones. Then the cones are identified. We consider only those cones

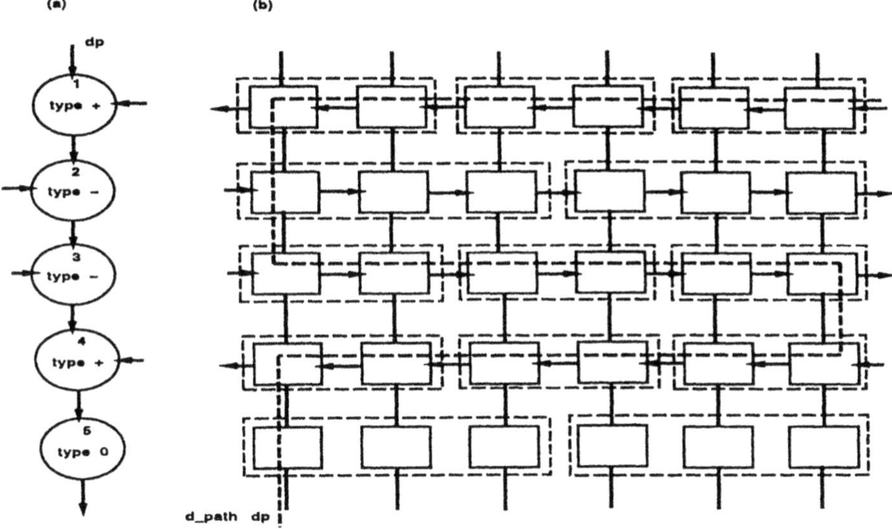

Fig. 2. (a) A portion of a 6-bit data path (b) Its cell level structure

which are beneficial (i.e. those which reduce the number of CLB levels). Due to differences in widths of slices in C, some of the nodes in $node\_set(C)$ may not be completely covered by that cone. The uncovered parts of partially covered nodes are considered to be covered by maximum width slices, forming trivial cones. The number of instances of C required to cover the critical path of cells within the node n is given by

$$cvr\_inst\_cnt(n, C) = \left\lceil \frac{cell\_levels(n)}{width(C)} \right\rceil \tag{8}$$

The number of instances of max_slice_width slices required per instance of a cone(for covering the uncovered part of n) is given by

$$uncvr\_inst\_cnt(n, C) = \left\lceil \frac{width(C) - width(s)}{max\_slice\_width(n)} \right\rceil \tag{9}$$

where $s \in C \mid node(s) = n$. Let $node\_set(C)$ denote the set of nodes whose slices are included in C. Let dp/C denote the portion of the path dp intersected by cone C (i.e. node_set(dp) $\bigcap$ node_set(c)). Let us denote by $delay(dp, CS)$ the delay of the path dp after selection of non-trivial cone set CS. Then delay of a d_path dp before selecting any non-trivial cone is given by

$$delay(dp, \phi) = min\_delay(dp) \tag{10}$$

Now delay of a *d_path* dp after selecting a non-trivial cone C is given by

$$delay(dp, CS + C) = delay(dp, CS) - delay\_benefit(C, dp, CS) \quad (11)$$

where $delay\_benefit(C, dp, CS)$ represents the reduction in delay or benefit due to covering of path dp/C by cone C which is given by

$$delay\_benefit(C, dp, CS) = \sum_{n \in node\_set(dp/C)} (\Delta vd(n, C, CS) - \Delta hd(n, C, CS).p\_flag(n, dp) - D_{CLB}) \quad (12)$$

where Δ vd is the change in vertical delay which is given by

$$\begin{aligned} \Delta vd(n, C, CS) &= vd(n) \;\; if \; CS + C \; covers \; n \; fully \\ &= 0 \;\; otherwise \end{aligned} \quad (13)$$

and Δ hd is the change in horizontal delay in the path dp due to covering by cone C which is given by

$$\begin{aligned} &\Delta hd(n, C, CS) \\ &= ((cvt\_inst\_cnt(n, C)\,(1 + uncvr\_inst\_cnt(n, C)) - 1)\,.D_{cell}(n)) - hd(n), \quad (14) \end{aligned}$$

if n is neither partially nor fully covered by CS
$= 0$ otherwise.

The cones are selected on the basis of their potential benefit, given by

$$cone\_benefit(C) = \max_{dp \mid dp \text{ is critical } d\_path} delay\_benefit(C, dp, \phi) \quad (15)$$

The algorithm described in figure 3 shows the major steps involved in minimal delay mapping of RTL structures onto FPGAs. In step 1, for each node in the graph, we compute the *min_delay* i.e., the number of CLB levels required to realize slices in the critical path of the nodes. Next we determine the *d_path* delays and select set of nodes in the critical paths. Step 2 generates and identifies realizable and beneficial cones. We traverse the network backwards starting from sink nodes and generate 'realizable' cones with non-negative 'delay-benefit' by considering various soft slicing options and merging them till no more merger is feasible or source nodes are reached. The realizability of these cones are checked as they are generated and only 'realizable' delay beneficial ones are retained. Step 3 finds a cover which minimizes the CLB levels or delay in the critical path following a greedy approach. Minimal delay cone cover procedure given in figure 4 follows a greedy approach. In this step, we start with choosing a cone from the set of beneficial cones generated in step 2, which reduces the longest path delay by maximum value while covering uncovered or partially covered nodes. We update delays on all the paths which contain the nodes forming this selected cone. Next, we determine the new critical.path in the updated graph. This process of choosing a delay beneficial cone , updating the path delays and determining the new critical path is repeated until all the nodes in the graph are completely covered.

```
Algorithm FAST_DMAP

 Input  :  RTL Data path structure
 Output :  Delay minimal interconnected CLB map

1. Computation of Dealy upper bound
     1.1  for each node n ∈ node set V of graph G
          1.1.1  compute  (i)    max_slice_width(n)
                          (ii)   no_of_levels(n)
                    and   (iii)  min_delay(n)
     1.2  compute d_path delays
     1.3  identify the critical paths
2. Realizable Cone generation
     2.1  for all 'feasible slices' s of nodes in V do
          2.1.1  generate all candidate cones
                 with s as apex using softslicing [3]
          2.1.2  identify realizable cones by
                 FAST decomposition [1]
          2.1.3  compute cone benefit (eqn. 15)
          2.1.4  Select realizable cones with cone
                 benefit ≥ 0
     2.2  coneset = list of selected cones in decreasing
                    order of their cone benefit
3. Minimal delay cone cover
     3.1  Generate complete cone sets which covers the
          entire network G with minimum delay
          in a 'greedy' manner
```

Fig. 3. Algorithm for Delay Minimal Mapping of RTL Structures onto FPGA's

6 RESULTS AND CONCLUSION

The package FAST has been implemented on a SUN Classic workstation. We have synthesized four structures corresponding to High level synthesis benchmarks : GCD, Diff_eqn, AR_filter, Elliptic filter. The mapping onto XC2000, XC3000 and XC4000 device CLBs has been performed and is reported in table 2.The allocation used in each example is listed in second column of the table. The CLB count and critical path delays obtained using simple and compound mapping are given in the last four columns of the table. For benchmarks containing multipliers, we have assumed that multipliers are external to the design and are realized separately.

To conclude, we have presented an approach for delay minimal mapping of RTL structures onto FPGAs. The technique is primarily meant for data path of the design and effectively utilizes iterative structure of the data path components. The slices of well connected components are generated and are called cones. These functions form the inputs to the decomposition process. The approach is

```
Procedure Minimal_delay_cone_cover
  {
      CS = φ
      Repeat
      {
              (i)     Choose a cone C ∈ coneset of
                      maximum delay_benefit for some critical path
                      while covering uncovered or partially
                      covered nodes. Include C in CS
              (ii)    Update delays on all d_paths dp ∈ DP (eqn. 11)
              (iii)   Update critical path set
      }
      Until (all nodes ∈ V are completely covered)
  }
```

Fig. 4. Minimal delay cone cover

flexible and can handle various families of XILINX FPGAs. The synthesized CLB boundaries correspond to RTL component boundaries which would imply ease in testability and simulation.

Acknowledgements

This work is partially supported by Department Of Electronics(DOE), Govt. of India under IDEAS project and Ministry of Human Resource Development (MHRD), Govt. of India under QIP programme. Visit of M. Balakrishnan to University of Dortmund under the Konrad Zuse programme (DAAD) was sponsored by Prof. Peter Marwedel.

References

1. A. R. Naseer, M. Balakrishnan and Anshul Kumar, *FAST : FPGA targeted RTL structure Synthesis Technique*, Proc. IEEE/ACM 7th Int. Conf. on VLSI Design'94 January 1994, pp. 21-24
2. A. R. Naseer, M. Balakrishnan and Anshul Kumar, *A technique for synthesizing Data Part using FPGAs*, Proc. IEEE/ACM/SIGDA 2nd Int. Workshop on Field Programmable Gate Arrays, Berkeley, February 1994.
3. A. R. Naseer, M. Balakrishnan and Anshul Kumar , *An efficient technique for Mapping RTL structures onto FPGAs*, Proc. 4th Int. Workshop on Field Programmable Logic and Applications, Prague, September 1994, Lecture Series in Computer Science, Springer Verlag, vol. 849, pp 99-110.
4. M. Balakrishnan, A. R. Naseer and Anshul Kumar , *Optimal Clock Period for synthesized Data Paths*, Fachbereich Informatik Technical Report No. 574, University of Dortmund, Germany, April, 1995.

Benchmark example	Allocation	FPGA device family	CLB Count simple Mapping	CLB Count compound Mapping	Critical Path Delay simple Mapping	Critical Path Delay compound Mapping
GCD	$\{1alu, 1cmp\}$	XC2000	160	80	18	17
		XC3000	80	56	18	16
		XC4000	72	48	10	9
Differential Equation	$\{1+, 1<, 2*, 1-\}$	XC2000	468	324	19	18
		XC3000	282	210	19	17
		XC4000	234	178	11	9
	$\{1+, 1<, 1*, 1-\}$	XC2000	432	320	22	20
		XC3000	280	216	22	19
		XC4000	200	152	14	11
Elliptical Filter	$\{3+, 2*\}$	XC2000	784	576	22	20
		XC3000	504	360	22	19
		XC4000	352	280	14	11
	$\{3+, 1*\}$	XC2000	688	544	22	21
		XC3000	440	328	22	20
		XC4000	320	272	14	12
	$\{2+, 1*\}$	XC2000	688	496	23	20
		XC3000	424	336	23	19
		XC4000	296	232	15	11
AR Filter	$\{2+, 4*\}$	XC2000	752	512	21	20
		XC3000	480	336	21	18
		XC4000	320	256	13	10
	$\{1+, 2*\}$	XC2000	672	448	22	20
		XC3000	400	280	22	18
		XC4000	296	228	14	11
	$\{1+, 1*\}$	XC2000	544	368	21	20
		XC3000	312	224	21	18
		XC4000	256	176	13	9

Table 2. Results of mapping for HLS Benchmarks

5. R. J. Francis, J. Rose, Z. Vranesic, *Technology Mapping of Look- up Table- Based FPGAs for performance*, Proc. Int. Conf. on CAD, 1991, pp.568-571.
6. R. Murgai et al., *Performance-Directed Synthesis for Table Look-up Programmable Gate Arrays*, Proc. Int. Conf. on CAD, 1991, pp. 572-575.
7. J. Cong and Y. Deng, *FlowMap : An optimal Technology Mapping algorithm for Delay Optimization in LookUp-Table based FPGA designs*, IEEE Trans. in CAD, vol. 13, No. 1, January 1994, pp. 1-12.
8. M. V. Rao, M. Balakrishnan and Anshul Kumar, "DESSERT : Design Space Exploration of RT Level Components", Proc. IEEE/ACM 6th Int. Conf. on VLSI Design'93, January 1993, pp. 299-303.
9. Anshul Kumar et al. "IDEAS : A Tool for VLSI CAD ", IEEE Design and Test, 1989, pp.50-57.
10. Xilinx Programmable Gate Array Users' Guide, 1994 Xilinx, Inc.

Some Notes on Power Management on FPGA-based Systems

Eduardo I. Boemo, Guillermo González de Rivera*,
Sergio López-Buedo, and Juan M. Meneses

E.T.S.I. Telecomunicación, Universidad Politécnica de Madrid,
28040 Ciudad Universitaria. Madrid - España, e-mail: ivan@die.upm.es

* Dept. de Ingeniería Electrónica, Universidad Alfonso X El Sabio,
28691 Villanueva de la Cañada, España, e-mail: gdrivera@uax.es

Abstract. Although the energy required to perform a logic operation has continuously dropped at least by ten orders of magnitude since early vacuum-tube electronics [1], the increasing clock frequency and gate density of the current integrated circuits has appended power consumption to traditional design trade-offs. This paper explore the usefullness of some low-power design methods based on architectural and implementation modifications, for FPGA-based electronic systems. The contribution of spurious transitions to the overal consumption is evidenced and main strategies for its reduction are analized. The efectiveness of pipelining and partitioning inprovements as low-power design methodologies are quantified by case-studies based on array multipliers. Moreover, a methodology suitable for FPGAs power analysis is presented.

1 Introduction

The general advantages of power consumption reduction are well-known: it allows expensive packaging to be avoided, the chip life operation to be increased, cooling to be simplifyed and the autonomy of battery powered systems to be extended (or their weight to be reduced). Even in fast prototyping, excessive consumption can be inconvenient: CMOS delays increase 0,3 % per ºC [2], as well as synchronous circuits can exhibit current peaks so they affect apparently independent variables like PCB features. Analogous to throughput or area occupation, the reduction of consumption can be achieved at any level of hierarchy; however, in this paper attention will be focused on architectural-implementation transformations, availables to FPGA end-users. Additionally, these approaches are not aggressive and can be applied in conjunction with any other strategy.

The main consumption in CMOS technology correspond to dynamic power: the energy per clock cycle involved in the charge/discharge of all circuit node capacitances. This power component can be modelled by:

$$P = \sum_{all\ nodes} C_n f_n V_{DD}^2 \qquad (1)$$

where f_n is the effective frequency of each circuit node (usually different from the system clock), C_n is the output capacitance of each node, and V_{DD} is the power supply voltage (Eq.(1) assumes that all capacitances reach V_{DD} after the loading period). Thus, setting aside V_{DD} manipulations, the power consumption can be modified by varying: the topology (that influences all the variables); the data (that vary f_n); and finally, the interconnection network, which affect C_n, but also f_n. However, the estimation or control of the effective frequency f_n of each node is difficult due to the appearance of glitches. Although the glitches do not produce errors in a well-designed synchronous systems, they can significantly increase the circuit activity.

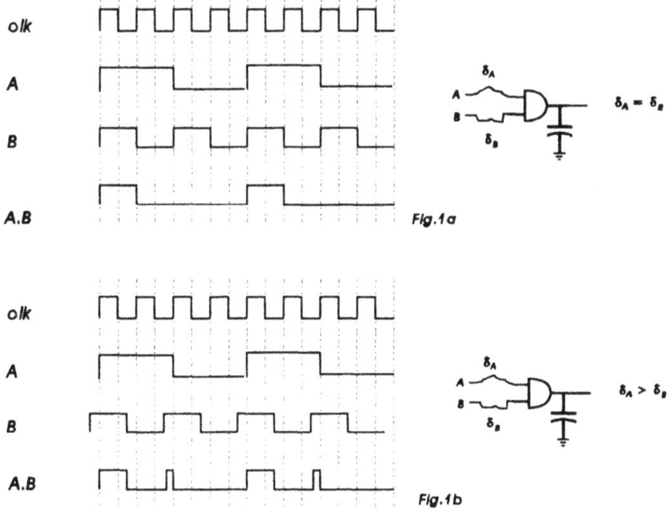

Fig.1: The effect of net unbalance on node activity.

The effect of glitches is illustrated on Fig.1: in the above graph the delay nets are equalized and the spurious activity level is zero; however if an unbalance between the paths exists (Fig.1b), glitches appear and power consumption increases. Depending on the circuit topology, these spurious transitions can progress across the following stages, producing an avalanche effect on power consumption. For example, combinational array multipliers with automatic placement-routing utilized in this work gave rise to arround 25 to 40 intermediate values before reaching the correct result, meanwhile a manually path-equalized version of the same circuit just exhibed 5 to 8 intermediate values.

FPGA user has three ways to diminish glitches: pipelining, partitioning improvements and path delay equalization. Pipelining, a popular way to speed up circuits also allows power consumption to be reduced [3]-[4]. Its usefulness is based on a marginal effect of the intermediate pipeline registers: the obstruction of the propagation of spurious (asynchronous) transitions. Pipelining also affects power consumption by the modification of datapath wiring loads: global lines (which usually broadcast the input data into the array) are split into a subset of lightly loaded lines, reducing the overall capacity. The second way to disminish spurious activity is to pack critical parts into look-up tables (LUTs) by using a manual partitioning process: thus glitches, wiring and nodes can be reduced. Finally, path delay equalization also reduces spurious, as well as can conduce to wave pipelines or maximum-rate circuits [5]; the application of this technique on FPGAs is analyzed in [6].

In the next section, a methodology suitable for the analysis of power consumption on FPGAs is presented. In section 3, the effect of spurious transitions on datapath power and the efectiveness of pipelining and partitioning improvements is quantified by a set of case-studies; additionally, the magnitude of off-chip and clocking consumption is evaluated.

2 Power Budget on FPGAs

Dynamic consumption on FPGAs can be separated into three parts: datapath, synchronization, and off-chip power. The first component corresponds to the combinational blocks and associated interconnection power; the second part is the consumption by registers, clock lines, and buffers; and finally, off-chip power, is the fraction dissipated in the circuit output pads (where the capacitances are several times larger than those for conventional microelectronics). Knowledge of the relationship between these components for a given FPGA technology is fundamental: it allows the effectiveness of any particular power reduction method to be determined *a priori*. For example: partitioning improvements and "cold scheduling" [7] would be superfluous if datapath power is relatively small; Gray Code counters for addressing external circuits would be useful only if heavily loaded buses exist; self-timed synchronization, wave pipelining, DET registering [8], or stoppable clocks strategies would be effective provided synchronization power is dominant.

Datapath power measurements require the definition of a test vector set. Random sequences allow average consumption to be determined (and thus battery life operation to be predicted). Meanwhile special vector sequences that try to maximize the toggle of the circuit nodes allows the peak of dissipation to be deduced (and thereby stablishing the power supply requierements or testing off-chip power characteristics). In this work, circuits have been tested using a 2^{16}pseudo-random data as well as a reduced set of 16 data, which toggle near the 93 % of the outputs in each cycle, rather than the 56% toggled by the random sequence. Because current pipelined circuits can run faster than affordable pattern generators, both test sequences were produced by another FPGA, a XC3120-3, allowing a low-cost 140 MHz pattern generator to be obtained.

In several circuits synchronization power consumption can be determined by clocking the circuit while maintaining the input data constant. Thus, there is no activity neither in the datapath nor in the I/O pads; then, the measured power can be assigned to registers, clock lines and buffers. Finally, although off-chip power is strongly application-dependent, for a given FPGA-based system it can be easily determined simply by measuring chip power twice: first in normal operation, and then activating the 3-state output pad option. The difference between the values, allows the designer to diagnose if off-chip power reduction is necessary.

The average value of the power components can be indirectly determinated by measuring average FPGA input current; or subtracting average system power values measured with and without clocking the FPGA chip under test (all the circuits should include registered I/O). Both methods require the voltage on the FPGA chip to be held constant.

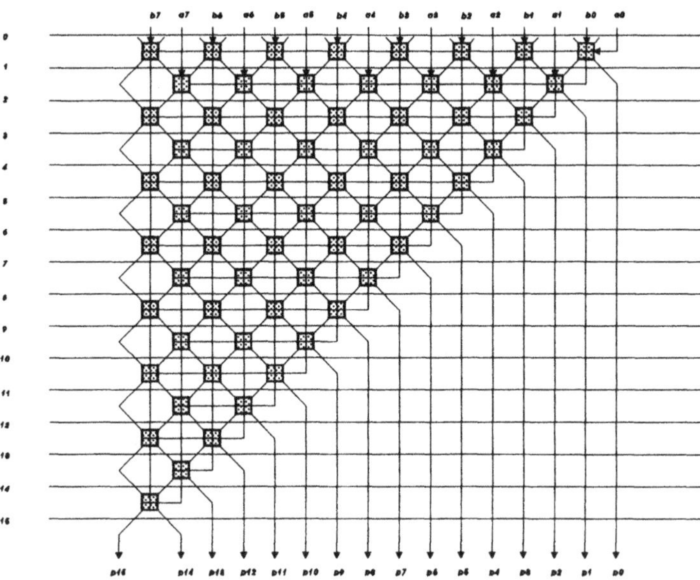

Fig.2: 8-bit Guild Multiplier. Equitemporal lines.

3 Experimental Results

The relationship between pipelining, partitioning and power consumption has been quantified by a set of 8-bit Guild pipelined array multipliers [9] implemented using a XC3090PC84-100 and a XC4005PC84-6 Xilinx FPGAs. Each array family includes versions pipelined with five different granularities β, defined as the maximum number of elementary processors (EP) between successive register banks [10]: β=1 (all EP I/O registered), β=2 (data registered in even lines of Fig.2), β=4 (data registered in lines 0, 4, 8, ...), β=8 (data just are registered in lines 0, 8 and 15), and finally β=15, a combinational array with registered I/O. In order to assess the effect of efficient LUT utilization, two versions for each circuit has been constructed for the XC3090 chip: a non-optimized (default APR) implementation; and another corresponding to a manual partitioning optimization. Additionally, some full manual high-optimized prototypes has been also developed.

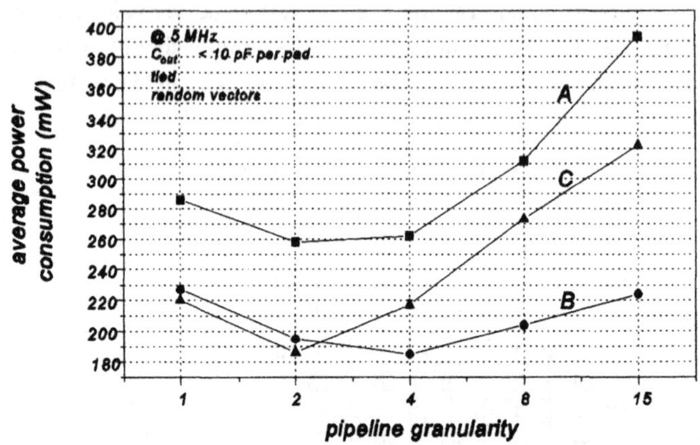

Fig.3: 8-bit Guild Array. Average power consumption vs granularity. XC3090 default PPR (curve A) and optimized partitioning (curve B), and XC4005 default PPR (curve C)

3.1 Pipelining as a Low-Power Strategy: Fig.3 shows the average power consumption of three pipelined array sets versus pipeline granularity (measured at 5 MHz, a frequency at wich all prototypes can be compared). The off-chip power quota has been maintained as low as possible in order to avoid masking datapath power effects; thus, each pad supports just the 10 pF (max.) logic analyzer probe load. Despite the hardware overhead, fine grain pipelines not only ran faster than combinational versions (β=15), but also exhibed lower consumption if operated at the same frequency. In all cases the minimum power dissipated corresponded to logic depth from two to four LUTs between registers. Thus, pipelining allows the designer to trade power consumption for additional logic blocks and latency. For example, for default implementation conditions, the consumption of a 5 MHz β=15 multiplier can be reduced by 33 % (XC3090) or by 58 % (XC4005) if it is β=4 pipelined. In both cases the number of registers required would increase from 32 to 104, and the latency from one to four clock cycles.

3.2 Power Reduction via Partitioning Improvements: This strategy not only diminishes CLB occupation and speed up circuits but also reduces power consumption. It can be evaluated comparing curve A and B on Fig.3. Note that, for each β, both versions have the same synchronization and off-chip power; thus, the difference corresponds exclusively to datapath power consumption. Note that the benefits of partitioning improvements increase with the logic depth.

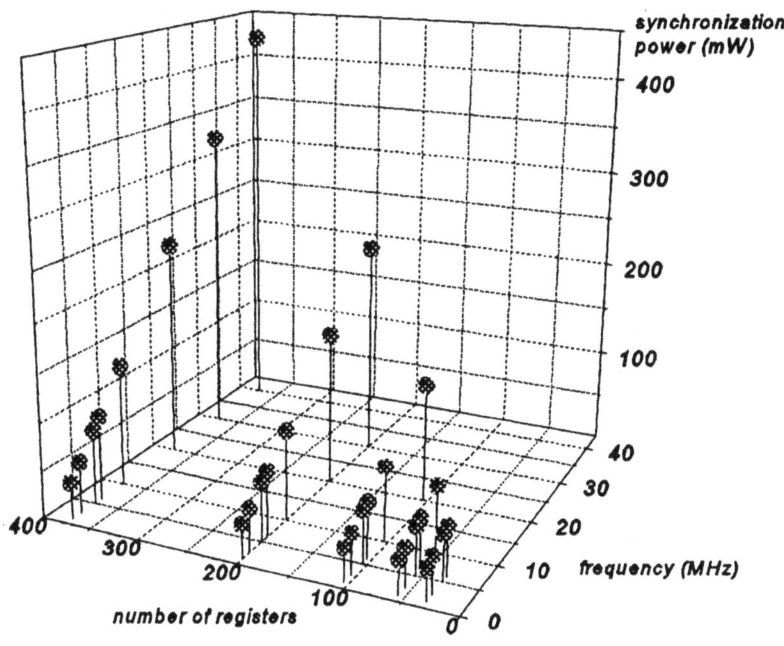

Fig.4: Synchronization power vs frequency and number of registers.

3.3 Synchronization Power: This component can be modeled by measuring multipliers with different numbers of registers. In Fig.4 the value of the power consumption has been plotted, versus frequency and number of register. Thus, the following model has been derived for 3090PC84-100 synchronization power (tied option) as a function of the number of registers (NR) and frequency:

$$Synch.\ Power\ (\ mW\ ) \approx (\ 2{,}7 + 0{,}019\ NR\ )\ frequency\ (MHz)\ +\ 31\ mW$$

Note that the off-set of 31 mW includes overall chip consumption at low frequency. This model has been successfully utilized to predict the results of other tied arrays with different topologies and numbers of registers[1]. Error for different partitioning and placement have been estimated near 7% (circuits that make use of both CLB registers exhibed less sinchronization power).

Although for high frequency operation, the synchronization power fraction is not excessive, it can not be reduced by using the CLB clock enable facility. Thus, the application of techniques like stopable clocks to FPGA-based systems, must block the clock signal at chip input pin in order to be effective. Finally, in Fig.5 an example is shown of the three components of total power for a full manual PPR, 8-bits-240CLBs-70MHz, β=1, pipelined Guild array multiplier on a 3090PC84-100.

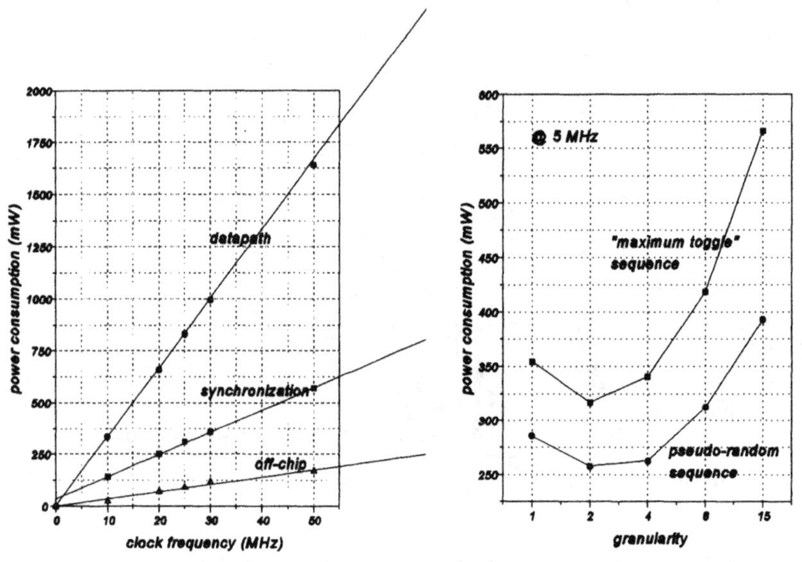

Fig.5: Power components. High-optimized 8-bit β=1 pipelined array multiplier

Fig.6: Data-dependence of power consumption. Non-optimized part.

3.4 Data-dependence of Power Consumption: In Fig.6 the power consumption is shown for the same family of circuits when they process different input data. Note that the sequence of sixteen vectors for maximum output toggle produces a significant increase in power consumption, even for the small off-chip capacitance values. However, this effect can be utilized in a reverse mode: for a particular application, a subset of data that minimizes toggle output would be effective to disminish datapth power consumption. In

[1] All the experiments presented in this paper have been repeated using a Hatamian array [11], providing similar results.

another application, a similar idea is utilized in [7], where the internal activity is reduced by minimizing it at the circuit inputs.

3.5 Correlation between Occupation, Bandwidth and Power Consumption: In spite of architectural trade-offs, from an implementational point of view, occupation, bandwidth and power consumption can be improved simultaneously. In Fig.7 is plotted power consumption @ 5 MHz versus minimun clock period of a set β=15 multipliers implemented on a XC3090PC84-100 using the "aprloop" facility. Although there are some exceptions, the faster circuit runs, the smaller is the power consumption for a given frequency.

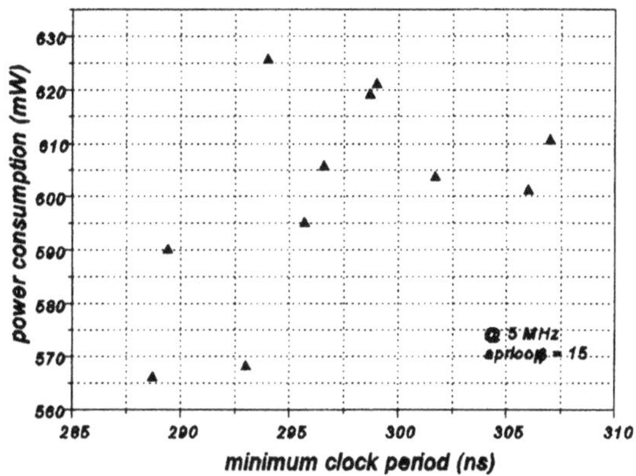

Fig.7: Aprloop results from a power consumption perspective.

4 Conclusions

The effect of pipelining as a low-power design technique on FPGAs has been quantified; a model for synchronization power has been proposed; and a general methodology to characterize others applications or FPGAs has been presented. The results show that pipelining can produce a reduction of power consumption by about 25% - 40 %, and nearly 15% - 45 % can be achieved simply by improving partitioning. It can be stated as a rule of thumb, that circuits than run faster, use less CLBs and dissipate less power. The common origin of these improvements is the reduction of the interconnection network influence.

From a research or educational point of view, it has been demonstrated that RAM-based FPGAs exhibit important advantages over other technologies in terms of power analysis; their layout editors combined with the changeable structure of logic blocks allow circuit modifications like: inserting/deleting registers without altering the routing; modifying the routing without affecting the logic or placement; isolating any block from the system clock;

confining critical parts to LUTs; using positive or negative edge registers; disconnecting the outputs pads, etc. Additionally, the fast design cycle and reprogramability of this technology allows prototypes to be builded and measured without significant cost. However, the FPGA net information based on delays rather than node capacitances make modelling of the consumption difficult.

Acknowledges

This work has been supported by the CICYT of Spain under contract TIC92-0083. The authors wish to thank Seamus McQuaid for his constructive comments.

References

1. R. Keyes, "Miniaturization of electronics and its limits", *IBM J. of Res. Develop.* Vol.32, nº1. January 1988.
2. Xilinx, Inc., *Technical Conference and Seminar Series,* 1995.
3. Z. Lemnios y K. Gabriel, "Low-Power Electronic", *IEEE Design & Test of Computers*, pp. 8-13, winter 1994.
4. A. Chandrakasan, S. Sheng y R. Brodersen, "Low-Power CMOS Digital Design", *IEEE Journal of Solid-State Circuits*, Vol.27, Nº4, pp.473-484. April 1992
5 D. Wong, *"Techniques for Designing High-Performance Digital Circuits Using Wave Pipelining"*, Technical report No. CLS-TR-92-508. Stanford University, february 1992.
6. E. Boemo, S. López, G. González and J. Meneses, "On the usefulness of pipelining and wave pipelining as low-power design technique", *Proc. 1995 PATMOS Conf.* (in press).
7. C. Su, C. Tsui y A. Despain, "Low Power Architecture Design and Compilation Techniques for High-Performance Processors", *Proc. Sprint COMPCON 94*, pp.489-498. IEEE Press, 1994.
8. R. Hossain, L. Wronski y A. Albicki, "Low Power Desing Using Double Edge Triggered Flip- Flops", IEEE Trans. on VLSI Systems, Vol.2, Nº2, pp.261-265. June 1994.
9. H.H. Guild, "Fully Iterative Fast Array for Binary Multiplication and Addition", *Electronic Letters*, pp.263, Vol.5, Nº12, June 1969.
10. C. Hauck, C. Bamji and J. Allen, "The Systematic Exploration of Pipelined Array Multiplier Performance", *Proceeding ICASSP 85*, pp.1461-1464. New York: IEEE Press, 1985.
11. M. Hatamian and G.Cash. "A 70-MHz 8-bit x 8 bit Parallel Pipelined Multiplier in 2.5-um CMOS". *IEEE Journal of Solid-State Circuits*, August 1986.

An Automatic Technique for Realising User Interaction Processing in PLD Based Systems

Keith Dimond
Electronic Engineering Laboratories
University of Kent at Canterbury
Canterbury Kent UK CT2 7NT
k.r.dimond@ukc.ac.uk

1 Introduction

With the availability of large fpgas it becomes possible to realise complete digital systems on one or two devices. Quite often the digital system will have a number of user inputs to control the operation of the system. These may consist of simple push-buttons or they may be more complex numerical key pads. As the system complexity increases the range of different modes of operation also tends to increase and hence the complexity of the user interaction. This paper describes a method of automatically designing that part of the system which processes the user interaction.

2 Overall Structure of the system

We can consider the overall system to be divided as follows

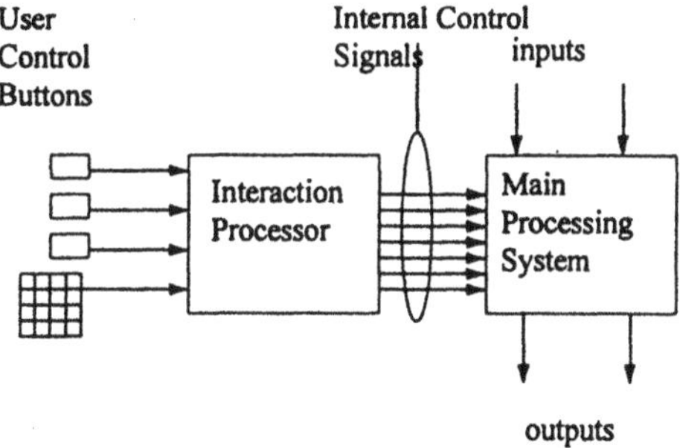

Fig. 1.

The interaction processor takes the signals from the control buttons and the numerical input devices. After processing these input signals a set of internal control signal are produced. These then determine the function to be carried out in the rest of the system, i.e. how the system inputs are mapped to the outputs etc. With complex systems the interaction is complex and techniques are needed to design the interaction processor automatically.

3 Representation of the user interaction

In this approach we represent the user interaction by a grammar in much the same way as a programming languages described by a grammar. Much work has been carried on automating this process and parser generators or compiler-compiler systems are available to help. The compiler-compiler that we have used is called YACC (1) which uses a modified BNF grammar to produce what is technically called a shift-reduce parser. The parser is a table driven automaton, similar to a finite state machine hence it is relatively straightforward to implement in hardware.

Hardware Implementation of the parser

The shift reduce parser is relatively simple in operation and requires a push-down automaton. The push-down automaton (pda) is a finite state machine where the state register forms part of a push-down stack. The push-down stack allows the sequence of states to be stored, which is necessary to allow recursive grammar definitions to be processed.

To describe the structure and operation of the hardware implementation of the parser it is sensible to consider a simple example. We are designing the interaction processor for an all-digital sinusoidal signal generator. This has a number of push buttons on the front panel which controls the magnitude and frequency of the output signal

The control buttons are as follows

ampl which designates the start of the amplitude specification sequence
volts which selects a voltage output
amps which selects a current output
freq which designates the start of the frequency specification sequence
cps which terminates the frequency specification sequence
digit a single digit decimal value

Typical sequences of key presses would be

ampl 2 5 volts

freq 2 0 0 cps

The first designates the output magnitude to be 25 volts and requires four key presses. The second designates the frequency specification and requires five key presses. The YACC grammar which represents this interaction is shown in table 1.

```
%start script
%token ampl volts amps freq cps digit
```

```
%%
script:   config   | script config ;
config:  amplitude  | frequency ;
amplitude:       ampl avalue units;
frequency:       freq fvalue cps;
avalue:          value;
units:           volts | amps ;
fvalue:          value;
value:           digit | value digit;
%%
```

Table 1

The identifiers in the line %token correspond to the buttons which the user presses. Each production has a label followed by a colon. The body of the rule or production is then defined on the right hand side. The vertical bar designates an alternative rule. Due to the limitations of space it is not possible to discuss this further. Interested readers can obtain more information from (2).

Having written the grammar YACC is used to process it to yield the moves that the push-down automata has to make. As has been said before the parser has two types of moves, one is a shift and the other is a reduce. The parser initialises to the zero state. Then as inputs signals - button presses are applied the parser shifts or moves from state to state. As an example consider the valid input sequence

ampl 2 5 volts

Each button press requires the parser to perform a shift with the state being stored on the stack. When the complete set of button presses have been applied the parser is in a state which signifies that the valid sequence has been input and performs a reduce operation. The grammar above is a little more complex to enable arbitrary number of digit presses.

4 The Parser Operation

We now consider the detailed operation of the parser so that the hardware structure can be developed. When the grammar rules have been processed YACC produces a diagnostic file - y.output which details the moves of the parser. These take the form of a description of the actions which are to occur in each state, i.e. the shifts and reduces. Thus the parser output starts with a description of what happens in the state 0.

```
state 0   $accept : _script $end
          ampl  shift 5
          freq  shift 6
          . error
          script  goto 1
```

config goto 2
amplitude goto 3
frequency goto 4

Table 2. Parser moves for state 0

When in state 0 the pressing of the ampl button causes the parser to shift to state 3, whilst if the freq button is pressed then the parser moves to state 6. If any other button is pressed signified by '.' then this is an error and the parser will attempt to restart at a suitable point by flushing the stack.

The next set of moves are referred to as the goto actions. These come into play when a reduce action is performed. An example of a reduce move is given in state 17, the final state of the parser.

state 17 frequency : freq fvalue cps_ (6)
reduce 6

In this state the parser will have seen the complete sequence of inputs which define the frequency command, i.e. the button press freq, a numerical sequence defining the frequency and the button press cps. The reduce action removes from the stack all the states associated with the grammar rule number 6. In doing this the parser represents all the inputs associated by this rule by the so called non-terminal frequency. The number of items which are moved from the stack is defined by the number of elements contained in the grammar rule, in this case 3. When these are removed, the stack will still contain a value, this will be either state 0 or state 1 depending upon the previous inputs. If we assume it is state 0 then we can look at the goto actions in state 0 to determine what to do. If we refer to these it is seen that when reduce has occurred producing the non-terminal frequency then the action is goto 4 - goto state 4. Although it is referred to as a goto it is essentially a shift to state 4. The parser will then continue parsing according to the moves associated with state 4.

To implement the parser we need to have a stack. To manipulate the stack we need to have logic to implement the shift, goto and reduce actions.

5 Implementation using PLDs

Previous work (3) has investigated architectures for the hardware parser suitable for custom silicon implementation. The approach which was adopted was to use a conventional implementation of the fsm by means of a PLA and state-register. The PLA was personalised by analysing the y.output file to build a matrix of 0's and 1's defining the PLA. This was fed into a minimisation program and hence to a PLA generator to produce the masks. The other components were connected together using standard CAD tools.

This approach is not directly applicable to the realisation in the PLDs. We are aware that with most fpgas the fan-in of the logic functions tends to be limited. It is for this reason that it is normal to implement finite-state machines using the one-hot technique. The remainder of the paper is devoted to a description of the novel one-per-state implementation of the push-down automaton.

Central to this realisation is the implementation of the stack. In a conventional stack we expect to employ a block of read/write memory and an index register to implement the structure. The index register can be incremented or decremented and values stored in the RAM. The one-hot stack on the other hand has a radically different structure. This is shown in Figure 2.

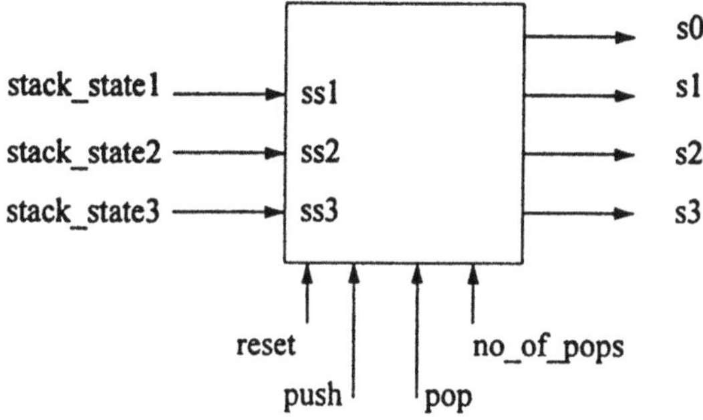

Fig. 2. Block Diagram of a four state one_hot stack

This shows the outline of a four sate stack. When reset is activated the stack goes to state 0, the output s0 will be 1 and all the other outputs 0. When moving to state 2 then s2 output will be 1 and all the other outputs will be 0. The fact that the state was in state 0 is stored. After a series of shift moves the parser will perform a reduce, this is then achieved by applying to the no_of_pops input the number of states to be removed and activating the pop input. This has been implemented in a modular form.

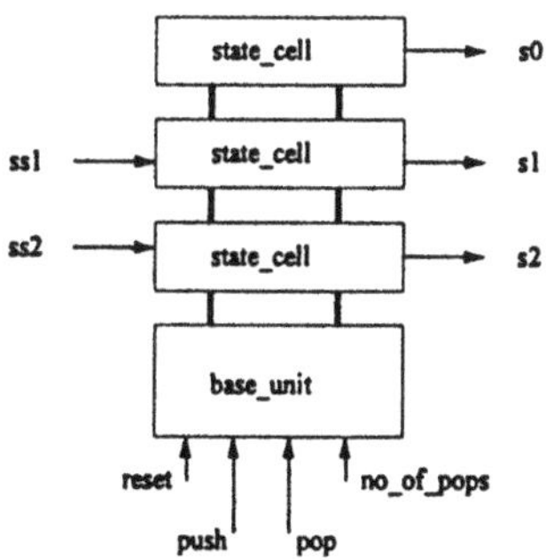

Fig. 3 Internal structure of the one_hot stack

In this structure there is a base unit which maintains details of the current stack level. For each of the states in the stack a state_cell is included. Signals couple the base_unit to all of the state_cells. The basic operation of the stack is to maintain in the base unit a counter, which determines the current stack level. This current stack level is applied to all the state cells. Each state_cell has an internal register, this register can be loaded with the value of the stack level. It may also be compared with the current stack level. When a push operation is performed the stack level is incremented and loaded into the state_cell register whose stack_state input is set to 1, and a flip-flop is set to indicate the state is set. Logic is provided in each state_cell to compare the current stack level with the state_cell register. If the two are equal then the output of the state_cell flip-flop is output, indicating the stack state. If another state is pushed then again the same procedure is adopted causing the new state output to got to 1. All previously stacked state outputs are zero since the state output is only 1 if the state_cell register contents equals the current stack level.

For the popping operation, the number of states to be removed is then applied to the no_of_pops input and when the pop control signal is activated the stack level is decremented by the appropriate amount This means that the state_cell output of the state uncovered will go to 1. Another feature of the state_cell is that if the value in the state_cell register is greater than the current stack level, then the flip-flop in the state_cell is set to zero, indicating that the state has been discarded.

6 Control of the Stack

Having discussed the structure for the stack we now have to consider how the stack_state inputs of the stack are driven. It will be remembered that although in the description of the parser operation we have identified two mechanism, the shift and the goto, in reality they are almost the same operation. If we analyse the output of the parser we can generate a table which identifies all the shifts or gotos. This is shown below in table 3.

current state	token	nonterminal	next state
0	ampl		5
0	*freq*		*6*
0		script	1
0		config	2
0		amplitude	3
0		frequency	4
1	ampl		5
1	*freq*		*6*
1		config	7
1		amplitude	8
1		frequency	4
5	digit		10
5		avalue	8
5		value	9

6	digit		10
6		fvalue	11
6		value	12
8	volts		14
8	amps		15
8		units	13
9	digit		16
11	cps		17
12	digit		16

Table 3. Shift and Goto Operations of the PDA

Looking at this table we see that there are two rows where the next state is state 6. This state is reached from state 0 and an input of freq or from state 1 with the same input signal. The way in which the move to state 6 is implemented employs the following iterative structure shown in figure 4 (entries for state 6 are italicised in table 3). Each block has an input for the state signal and the token or signal from the button. The output of one block is passed to the next so that it may be combined internally with an OR operation. In addition to these inputs there is a further signal which passes through all the shift blocks so that it is possible to determine if one of the shift actions has occurred. If no shift operation has occurred then sh_en_out will be 0.

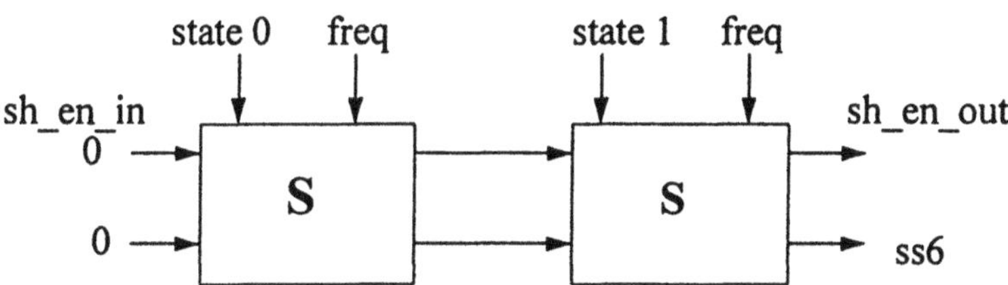

Fig. 4 Connection of shift blocks for stack state 6

7 The Reduce Operation

Previous sections have described the reduce operation. In summary this is where a number of states are popped from the stack and a non-terminal signal corresponding to the rule is generated. This non-terminal signal is then used in the goto operation. Table 2 show a summary of the reduce operations.

current state	no_of_pops	nonterminal
2	1	script
3	1	config
4	1	config
9	1	avalue
10	1	value
12	1	fvalue
13	3	amplitude
14	1	units
15	1	units
16	2	value
17	3	frequency

Table 4. Reduce Operations

This gives the state for each reduce rule, the number of states to be removed from the stack and the non-terminal signal produced. To consider the hardware implementation we see from Table 2 that the non-terminal config results from reduces in either state 3 or state 4. Hence the logical structure in Figure 5.

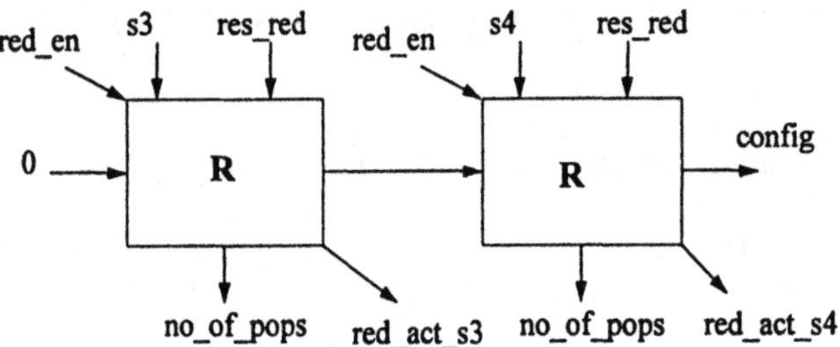

Fig. 5. Interconnection to produce the non-terminal config

This block is basically a latch with some extra logic. When a reduce occurs the 1 from the state output is stored in the reduce block. In the case of the non-terminal config this can be produced from two states, state 3 or state 4, so these two reduce blocks are cascaded together. The reduce action happens when red_en is activated. The other output of the reduce block is a signal which can be used to produce the required value of no_of_pops. The form of this block is shown in figure 6

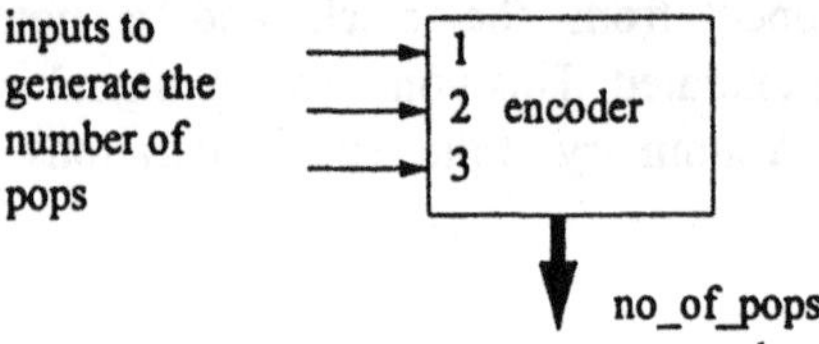

Fig. 6. The no_of_pops encoder

This has a number of inputs one input for each of the different values required for the no_of_pops input of the stack. Thus if the pop input is set to 1 then the output of the block will be the binary value 1 and similar for all the other inputs.

8 The Complete Parser

The important feature of this implementation of the parser is that all the blocks are made of simple logic functions which map on to the basic cells of the fpgas easily. A complete design can be realised by interconnecting a small number of basic blocks. The very important and novel feature of this approach is that the application is defined solely by the interconnections of these simple blocks. Thus to realise the system one needs to put down the required number of blocks and then interconnect them in the appropriate way. In the studies which have been made we have employed the Xilinx fpga. In this technology it is possible to specify the interconnection of basic blocks in terms of the .xnf file. This is a textual file specifying how the basic gates are connected together. Representations of the blocks using .xnf files are linked together using the standard Xilinx tools. This composite .xnf file is produced automatically from the y.output file. Figure 7 shows a part of the parser defined in stylised blocks. This shows how the inputs from buttons are connected to the shift blocks and the non-terminal signals connected to the goto blocks. It must be stressed that this diagram is only to illustrate the final design, in practice only a .xnf file would be produced.

9 How to Control the Processing Sub Block

The paper so far has concentrated on the way in which the parser processes sequences of user commands. The way in which these signal are communicated to the rest of the system is application specific. In general terms the red_act output of the reduce blocks can be used to control the rest of the system. A relatively simple example is concerned with the input of numeric values. With the user interface specified these are input via the digit token and associated with this signal will be the numeric value of the key pressed. Additional arithmetic processing units will take key values and accumulate an overall value. The rule which governs this is

value: digit | value digit;

This consists of two productions, and it is the red_act signalfor each of these productions which can be used to control the processing. The first production controls a single digit input, whilst the second ensures that all subsequent digit values are added with a suitably scaled running total to produce the final accumulated value.

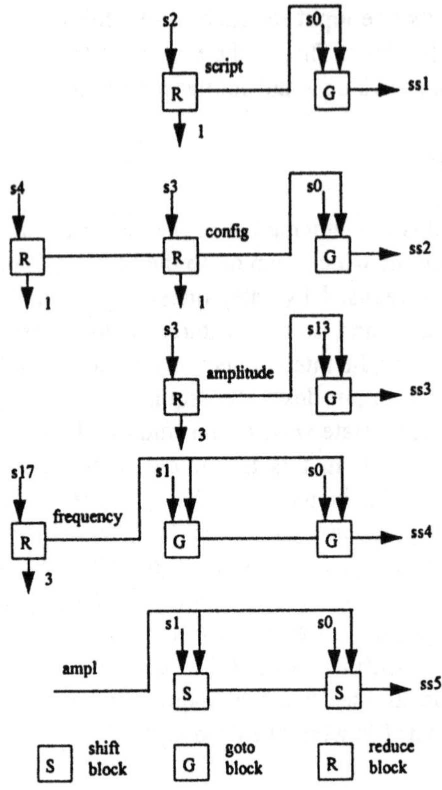

Fig. 7. Part of the logical Structure of the Parser

10 Conclusions

This paper has described how a shift-reduce parser can be implemented in hardware. Of more significance is the way in which the architecture has been tailored to the specific logical structures of fpgas. The paper then describes how the hardware can be constructed automatically by analysing the output of the compiler-compiler. Although the example cited in the paper very simple the method has been used to develop sophisticated user interfaces for domestic electronic equipment. Whilst the Xilinx fpga is used in this paper studies are continuing to examine other types. With the current implementation this design takes approximately 90 CLBs to realise, it is expected that finer grain fpgas could offer some interesting advantages.

11 References

1 YACC yet another Compiler-Compiler Chapter 10 Programming Utilities and Libraries, Sun Micro Systems.

2 Aho and Ullman - Principles of Compiler Design, Addison Wesley.

3 Dimond, K. R. Integrated Circuit Structures for Processing Man-machine interactions. pp 347-350 IEEE ISCAS 1987 Philadelphia.

Proper Use of Hierarchy in HDL-Based High Density FGPA Design

Carol A. Fields
Xilinx Inc.— 2100 Logic Drive, San Jose CA USA 95124

As the density and complexity of FPGA-based designs has increased to 10,000 gates and beyond, the use of high-level design languages (HDLs) is rapidly supplanting schematic entry as the preferred design entry format. However, to obtain the best results, the hierarchical design techniques already familiar to schematic users can be even more critical in an HDL-based design. Furthermore, the choice of partition size can be critical to meeting capacity and performance goals, as demonstrated by the implementation of a 15,000 gate design.

Introduction

In today's highly competitive market, system designers are faced with the conflicting challenges of greater system complexity and the need for short, efficient design cycles. These complexity and time-to-market pressures continue to reshape the art of designing electronic systems, and have led to the emergence and growing acceptance of top-down design methodologies and logic synthesis tools. With top-down design, engineers start work at a higher-level of abstraction than with traditional, gate-level design techniques. The designer manipulates logical or functional abstractions, and uses logic synthesis tools to produce the gate-level implementation. Over the past decade, top-down designs has become increasingly popular for the design of gate array and custom cell devices. As the density and complexity of FPGA-based designs has increased to 10,000 gates and beyond, users are now employing the same techniques and similar tools for FPGA design. The use of logic synthesis for FPGA design will continue to accelerate as ever-larger FPGA devices are introduced.

The use of high-level hardware design languages, such as VHDL and Verilog-HDL, allows designers to create and manage larger designs. As a result, there may be a temptation to disregard the importance of the hierarchical structure. Current design methodologies tend to target ASIC devices. 10,000 to 20,000 gates ASIC devices can often be compiled disregarding the hierarchical structure. This temptation to disregard the importance of designing the hierarchy should be avoided. Proper use of hierarchy in high density FPGA designing is critical in achieving desired device utilization.

The benefits of hierarchical design basically remain unchanged regardless of the design entry method; the use of hierarchical design techniques adds structure to the design process, ease debugging, allows for 'mixed-mode' design (wherein different design entry methods can be used for different portions of the design), provides a mechanism for dividing the design task among members of the design team, and facilitates the creation of libraries of reusable functions that, in turn, ease evolutionary product development.

Due to the basic functionality and interaction of FPGA synthesis and implementation tools, the actual size of each hierarchical module - in terms of the number of resulting logic blocks used in the target FPGA architecture - can have significant effect on the efficiency and performance of the resulting circuit. With current tools, synthesizing a design as one large, flattened module can result in designs that are very difficult to route. The placement algorithms in FPGA 'place and route' programs typically are based on a cost function that attempts to minimized the total net length of the resulting block interconnections. As a results, interconnected logic blocks are placed as close together as possible. With very large designs, this strategy can reach the point of diminishing returns, in that the logic condenses into one region of the FPGA device causing local routing congestion. On the other hand, the use of a multiplicity of very small modules can results in wasted logic capacity; since each module is synthesized separately, logic functions are not optimized across module boundaries.

Empirical evidence suggest that, give the current state-of-the-art of FPGA synthesis and implementation tools, partitioning a large design into modules in the 3,000 to 5,000 gate range, and employing floorplanning techniques to govern the relative placement of those modules, results in the most-effective combination of FPGA device capacity and performance. In this manner, the logic is spread more evenly throughout the device, but related logic within a given module is still placed in an optimal topology for meeting performance requirements. The benefits of such an approach include the following:

- Since each module is reasonably large, gate utilization is not overly diminished by the inability to optimize logic across module boundaries. Of course, careful partitioning of the modules, keeping related logic functions in a common module, further minimizes any deleterious effects caused by not optimizing across module boundaries. For example, if a design contains several small 4-bit incrementors, resource sharing would occur if these incrementors are in the same VHDL 'process'. If they are not in the same process and they were implemented in gates they could be combined by the optimizer further reducing the gate utilization.
- The design's routability is improved by grouping the modules and specify their floorplan according to the design's hierarchy and data flow. In this process, the designer has effectively added structure to the design and has passed this information onto the placement and routing tools. Floorplanning these modules into regions of the device evenly spreads-out the logic.
- Routing times are reduced; since modules are constrained into regions of the device, the automatic placement and routing tools has less area to evaluate. The designer has done some of the work by specifying the structure of the design to the placement and routing tools.
- Logic can be added or changed easily, since changes to one module can be made without effecting the placement and routing of other modules by using the re-entrant place and route programs.
- Making the design easier to debug. The design's modules are isolated into a region of the device. The contents of the modules and the location are defined by the designer.

A Design Example

The design methodology described in this paper was applied to several difficult to route 5,000 - 20,000-gate designs. The design example used to illustrate this technique is a transceiver receiver circuit for a telecommunication product that was targeted for an XC4025 FPGA device (-5 speed grade, 299-pin PGA package). This design was synthesized with the Synopsys FPGA Compiler using three different design methodologies. First, the design was compiled as one flat module. Second, it was compiled using the design's original hierarchical structure. Third (and recommended), it was compiled in 6 mid-size modules. The Xilinx Floorplanner was used to define the location of the modules. The Xilinx placement and routing tool was use to implement the design into the XC4025pg299-5 device and to evaluate the utilization and timing results.

This called TOP contained a lower level module called CORE as shown in Figure 1. The "CORE" level of this design consist of two large modules, R0 and X0, and two smaller ones, UP0 and DD0. The two larger modules consist of over 30 sub-hierarchical modules. The sizes of these modules range from 4 to 591 configurable logic blocks (CLBs). Note: The names of the modules have been re-named in order to protect the confidentiality of the design.

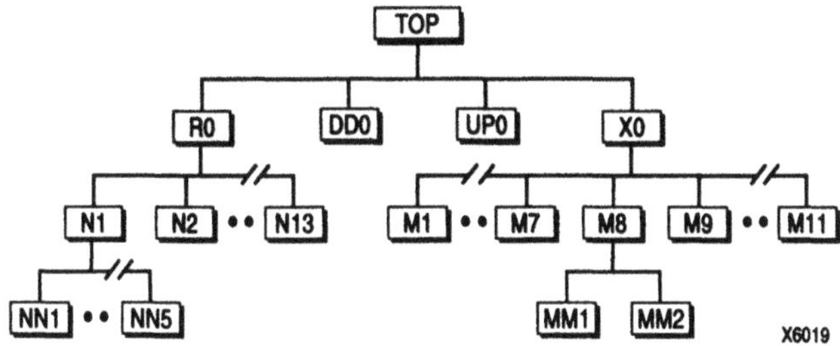

Figure 1 Original Hierarchy of Top Design

Compiling the Flatten Design

The design was initially compiled as one module using the Synopsys FPGA Compiler's "compile -ungroup_all" command on the CORE level. This design was unroutable, even though it only utilized 737 out of 1024 (71%) in the 1,024 CLBs available in the XC4025 FPGA. The logic of a large design compiled as one flat module often concentrates into one region of the device, creating a design that is difficult or impossible to route, as illustrated in Figure 2.

Next, the design was compiled using the design's existing hierarchy containing 5 levels of hierarchy and over 30 lower level modules. The logic block utilization increased from 737 CLBs to 1024 CLBS, a 29% increase in gates.

PPR was run on the design with and without floorplanning the X-BLOX RPMs. The design that was not floorplanned had ~8,000 unrouted nets. Once again the logic concentrated into one region of the device, causing local routing congestion similar to the design shown in Figure 2.

In order to route the design using the X-BLOX DesignWare library, the RPMs were floorplanned in a manner that away as to forced the logic to spread out in the device. The placed and routed design is shown in Figure 3. Overconstraining the design or poor floorplanning can make the design unroutable.This design methodology allows PPR to place the unconstrained cells anywhere in the device, making critical paths difficult to debug. Any design changes then would require an entirely new placement and routing, and, possibly, further floorplanning. If the design is difficult to route, a design change may cause the design to be unroutable.

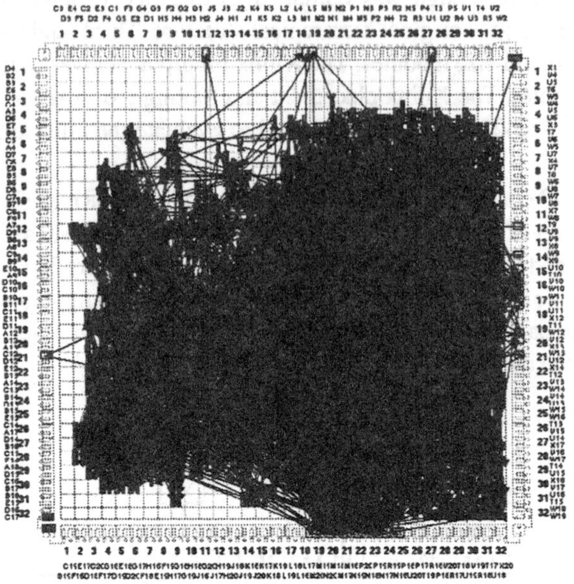

Figure 2 Ratsnets of Design Example Compiled Flat Compiling Using the Existing Hierarchy

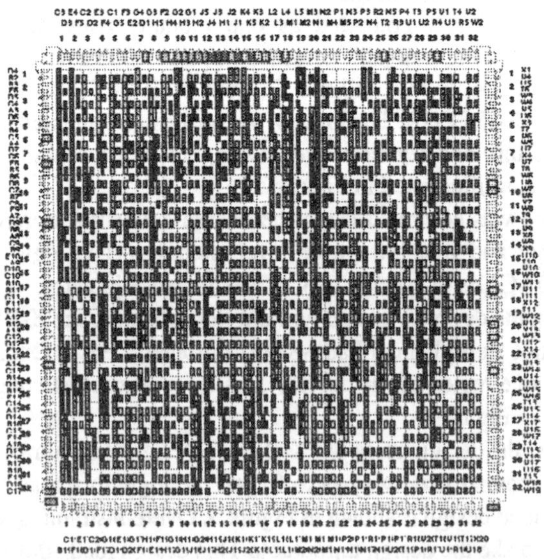

Figure 3 Placement Cells (without Ratsnets) for Top Design Compiled Using the Original Hierarchy with the RPMs Floorplanned

An 8 Mhz internal clock speed was required for this design. Using the existing design's hierarchy with X-BLOX, the longest constrained clock-to-setup delay, 143.5 ns, exceeded the requirement of 125 ns. All constrained pad-to-clock met the requirements as shown in Table 1.

Re-Grouping the Hierarchy

The recommended methodology for a large HDL design is to re-group the design's hierarchy into mid-size modules. The original design hierarchy consists of 4 major blocks at the "CORE" level. The estimated CLB utilization was determined using the Synopsys FPGA Compiler's "report_fpga" command. The block "R0" has an estimate of 591 CLBs, "X0" has 342 CLBs, UP0 has 25 CLBs and DD0 has 4 CLBs. Since the ideal module size is around 100 - 200 CLBs, this design was 're-grouped' to create a better hierarchical structure for the placement and routing tools.

The design's original hierarchy is shown in Figure 1. "R0" was separated into 4 modules and "X0" into 2 modules as shown in Figure 4.

The new grouping of the original modules was based on the sizes of the modules and the modules' interconnections with surrounding modules. An ideal grouping of modules will reduce the gate count and reduce the numbers of nets routed between the top level modules. The ideal module size is between 150 - 250 CLBs. The modules are grouped into new modules as shown in Figure 4.

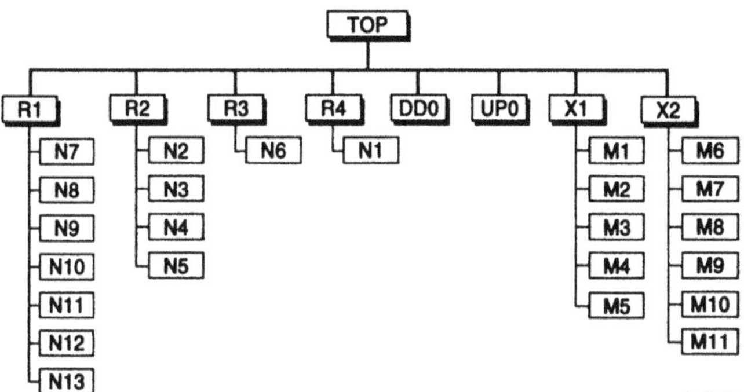

Figure 4 New Design Hierarchy

The two smaller modules, UP0 and DD0, were not combined with any other modules since these modules have an equal amount of interconnects with all of the new modules.

The Synopsys "group" command was used to define the new hierarchy. For example the module RO was regrouped into four smaller module R1 - R4 using the following commands:

```
current_instance = R0
group {N7,N8, N9, N10, N11, N12, N13} -design_name R1 -cell_name R1
group {N2, N3, N4, N5} -design_name R2 -cell_name R2
group {N6} -design_name R3 -cell_name R3
group {N1} -design_name R4 -cell_name R4
```

Each group was then compiled individually using the "compile -ungroup_all" command. A new script file was created that defined the new hierarchical groups, compiled the new groups, and created the XNF file for the CORE level. The lowest level modules were compiled before running this script and saved into a db file (e.g., N1.db). The script for the top level module reads in the top level, reads in the CORE level, assigns the I/Os, and writes out the design to an XNF file, top.sxnf.

A capacity reduction of 9 CLBs (30 packed CLBs) was achieved from compiling and flattening larger groups of logic together. An additional reduction of 67 CLBs (115 packed CLBs) was achieved when the Synopsys DesignWare modules were used in place of the small bit-width RPMs. (The 46 RPMs where 4-6 bits wide).

Floorplanning the Modules into Regions

Next, the modules were constrained into regions of the device as shown in Figure 5 using the Xilinx Floorplanner as shown in Figure 6. Each region must be large enough to fit the module and provides the placement and routing tools room to route the module. The height of the regions must be tall enough to accommodate the tallest structure in the module. For example, an 8-bit adder would need the region to be at least 5 CLBs tall. The locations of the regions were selected based on data flow.

The two smaller modules UP0 and DD0 were not constrained to allow the placement tools to determine the best location.

A constraint file for the placement and routing tools (PPR) was created specifying the regions which each module is constrained by using the Xilinx Floorplanner.

This design was then placed and routed using the constraints file and the timing specification passed from the Synopsys FPGA Compiler.

```
ppr top placer_effort=4 router_effort=3 cstfile=top_des
```

The placement of the resulting design is shown in Figure 7. Constraining the CLBs into regions of the device assists PPR in spreading the logic evenly throughout the device.

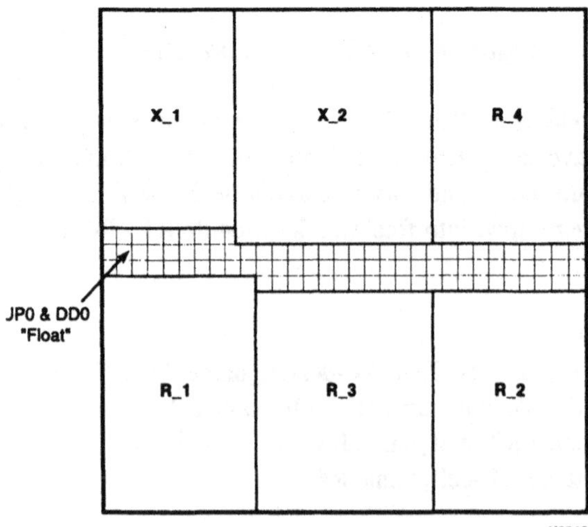

Figure 5 Floorplanning Modules into Areas

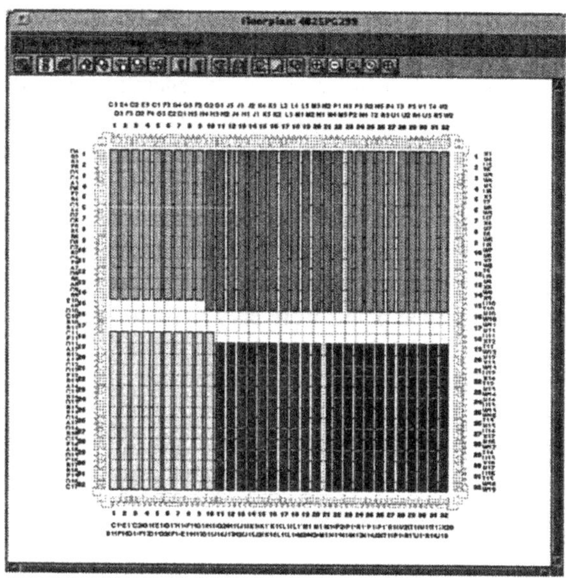

Figure 6 Floorplanning Modules into Areas

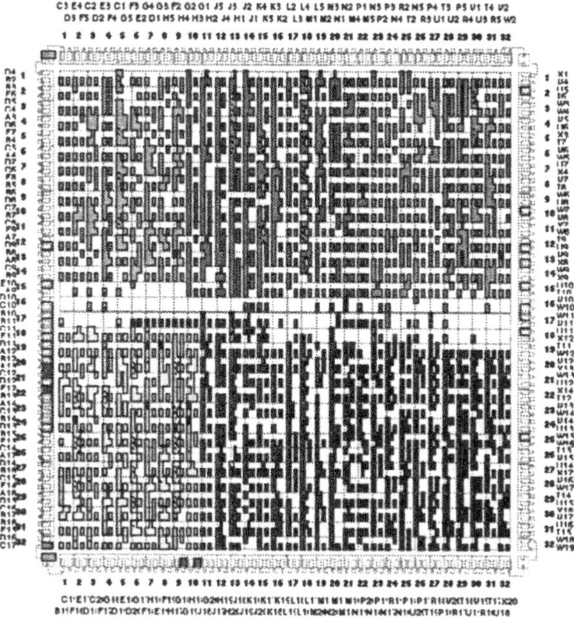

Figure 7 Placement of New Design Hierarchy with Region Constraints

Debugging a Design

It is common to connect internal signals to unused I/Os in order to debug a design. If the I/O pins are constrained, design changes are generally more difficult. The Xilinx Design Editor's (XDE) "Defineprobe" command can be used to specify an unused IOB as a probe point. The "Assignprobe" command is used to route an internal net to a probe point.

Two probe points were defined for both the design with the original hierarchical structure and the design with the re-defined hierarchical structure. In the design with the original hierarchy, the probe points were unroutable. However, in the design with the re-defined hierarchy, the probe points were easy to route, since this design contained unused logic in the center of the device.

Design Methodology Comparison

The flattened design used the least amount of resources. However, this design was packed so densely that it was unroutable. In addition, a flat design tends to be difficult to floorplan since the hierarchical structure is lost.

Using the existing design hierarchy required floorplanned to assist PPR in routing the design. This design utilized 100% of a XC4025. The longest constrained clock-to-setup delay was 143.2 ns and the longest constrained pad-to-clock delay was 100.1 ns. A timespec of 100 ns was specified when the design was placed and routed. However, the actual time requirement was 125 ns. 'Place and route' execution time approached 16 hours. Any small changes would require that the design be re-synthesized and re-compiled, resulting in another 16 hours to place and route.

Re-grouping the design's hierarchy did not require that individual cells be Floorplanned, but used the re-defined hierarchy to distribute the logic in the device. The design utilized 93% of a XC4025. The longest constrained clock-to-setup delay was 113.5 ns and the longest constrained pad-to-clock was 112.7 ns. This design did meet the system time requirements. 'Place and route execution time was reduced to 8 hours. Isolating the modules into regions reduces the run time, since less of the device has to be considered. A small design changed made using this design methodology only requires that the changed module be re-synthesized and re-placed and routed, significantly reducing the placement and routing time. Iterative design changes are easier to make. Logic can be added to unused portions of the device. For example, in Figure 7, additional logic can be added to the center of the device.

Efficient HDL Coding

Efficient FPGA design starts at the HDL code, before the design is placed and routed. No amount of re-design of the hierarchy can compensate for a poorly written HDL code. Like designs entered using schematic capture tools, highly structured, synchronous designs will route easier and perform better than unstructured, asynchronous designs. Devices with limited routing resources require that the designer consider the data flow of the designs during coding. Utilization and system speed can also be improved by using the system features of the FPGA architecture. HDL tends to abstract the design process, causing designers to lose sight of the implementation of the device. Using the correct construct will create a better implementation. A good understanding of the

Table 1. Comparison of Design Methodologies

Design Methodology XC4025pg299-5 PPR V5.1.0	Packed CLBs	Clock ToSetup Rising Edge	Pad ToSetup	PPR Run Time (CPU Time)	
Flat Design (no X-BLOX)	619 60%	n/a*	n/a*	n/a*	n/a*
Original Design Hierarchy; no Floorplanning	745 72%	n/a*	n/a*	n/a*	n/a*
Original Design Hierarchy; with Floorplanning	745 72%	143.2 ns	100.1 ns	Partition Placement Routing Total	01:13:41 02:05:16 12:53:22 16:14:36
Re-Group Design Hierarchy with X-BLOX	715 69%	106.7 ns	108.8 ns	Partition Placement Routing Total	01:05:42 01:34:49 08:07:08 10:49:37
Re-Group Design Hierarchy without X-BLOX	630 61%	113.4 ns	107.6 ns	Partition Placement Routing Total	01:02:10 05:39:46 04:29:02 11:12:55

synthesis tool's capability will further improve the design's utilization and performance.

Summary

Synthesis tools provide HDL users with the capability to reduce logic when modules are optimized together. Flattening large designs often creates unroutable designs. Retaining an existing design hierarchy with a multiplicity of very small modules can result in wasted logic capacity and can often leads to a design that does not fit in the target device. Re-structuring the design's hierarchy into mid-size modules (i.e., 100 - 200 CLBs) not only reduces the area utilization, but also allows the user to make small changes to the design and to easily locate logic for debugging of critical paths. This also reduces place and route execution times.

This methodology was tested on six additional designs. Half of the designs were implemented using the XC4013 devices and the other half using XC4025. All six designs exhibited similar results. Each of these designs did not route prior to regrouping the design's hierarchy. In all cases the system speed doubled, design changes were easier to make and execution time was significantly reduced. In one design their was significant utilization improvement to allow the designer to add additional functionality to the device. Additional studies where performed on XC4013 and XC4025 designs which did not require re-structuring of the design's hierarchy. These designs tend to have a significant amount of structure and module boundaries which were registered.

This methodology of proper use of hierarchy will lead designers into creating easier-to-route structured design by assisting designers in considering the data flow in the device.

Reference

Xilinx - Xilinx Synopsys Interface Guide (1994)
Synopsys - VHDL Compiler Reference Manual
Synopsys - Solv-it
Xilinx - Floorplanner's User Guide
Xilinx - High Density Application Note

Compiling Regular Arrays onto FPGAs

W. P. Marnane, C. N. Jordan and F. J. O' Reilly

Department of Electrical Engineering and Microelectronics
University College Cork, Ireland

Abstract. Many DSP functions can be implemented as arrays of simple Processing Elements (PEs) connected to their nearest neighbours in a regular manner. Field Programmable Gate Arrays consist of an array of user-configurable logic blocks and a matrix of user configurable interconnection between the logic blocks. Thus FPGAs are prime candidates for implementing regular arrays. In this paper we present FPGA Regular Array Description Language (FRADL) which will map the regular array into a FPGA.

1 Introduction

A regular array can be defined as a circuit consisting of several identical PEs with identical inter-connections between PEs throughout the array. These interconnections can result in a linear array or a rectangular mesh. The flow of data and control information is also regular. The PEs are generally only connected to their nearest neighbours on a two dimensional surface. Each PE can be purely combinational or contain both combinational and sequential logic blocks. The class of circuits we call regular arrays contains many sub-classes, such as linear iterative logic arrays (a simple example being a ripple carry adder), two dimensional iterative logic arrays (such as a carry save multiplier), and systolic arrays. In a systolic array, long combinational paths are eliminated by pipelining the circuit with registers on the inter-connections. Thus the propagation of data through the array is synchronized to a single clock. In signal processing, many of the Matrix and Vector operations can be implemented using regular arrays. In particular, applications such as convolution, correlation, matrix multiplication, filtering and Fourier transforms have all been implemented as systolic arrays [8, 6, 10].

A Field Programmable Gate Array (FPGA) consist of an array of identical but user-configurable Logic Blocks and a Matrix of user-configurable interconnections between the CLBs. The Xilinx 3000 series FPGA [13] contains Configurable Logic Blocks (CLB), Input/Output Blocks (IOB), and inter CLB connections. Each CLB in the array consists of a combinational logic section and two flip-flops. The external interface pins (IOBs) of the Xilinx FPGA can be programmed to perform as input or output pins. Communication between CLBs and IOBs on the FPGA is achieved through general purpose interconnect, direct connections or longlines.

Bit level regular arrays consist of a regular array of Processing Elements with typical complexity of a gated full adder. The interconnections between PEs are nearest neighbour. These two features make FPGAs a suitable device for implementing an array design.

2 Synthesis

Prototyping and implementing designs on FPGAs is becoming more popular. This has resulted in improved CAD support for the design process. A typical design route is to capture the design using a hardware description language such as VHDL and then to use a synthesis tool to translate this design into an FPGA implementation. These synthesis tools are primarily aimed at glue-logic, the initial market identified for FPGAs. Thus these tools would not preserve the inherent regularity present in a bit-level regular array and hence do not give area and time efficient designs.

However we are not only interested in synthesizing bit-level arrays, but also word level arrays where the PE has a complexity of a parallel multiplier or divider. Thus rather than VHDL we require a system where we can capture the algorithm description and from this using proven transformations, determine a parallel processor implementation. Such a system is provided by ALPHA [12, 5]. The ALPHA language has as its underlying model the system of affine recurrence equations [11]. It is well suited for the expression of regular algorithms and therefore provides a framework for the transformation of the algorithm specification into an architecture.

2.1 Synthesizing Bit Level Arrays

The first step in the synthesis process is the algorithm. For example a N tap FIR filter is the convolution sum given by:

$$y_n = \sum_{k=0}^{N-1} x_{n-k} \times h_k$$

where y_n is the filter output and x_n is the filter input at time n, and h_k is the filter coefficients. However $y_n = (y_{n,m-1}, y_{n,m-2}, \ldots, y_{n,0})$ is an m-bit integer, and if we consider the problem at the bit level from the start, an expression for the jth bit of the result $y_{n,j}$ takes the form:

$$y_{n,j} = \sum_{k=1}^{N-1} \sum_{i=0}^{m-1} x_{n-k,i}.h_{k,j-i} + carries$$

One way to form the bit $y_{n,j}$ is to sum over k index first and then accumulate over i. We can describe the algorithm in ALPHA and apply suitable timing functions and allocation functions to transform the algorithm description into the description of a bit level systolic array [5]. The resulting array PE illustrated in figure 1 performs this bitwise accumulation on the serial bit stream $x_{0,0}, x_{0,1} \ldots$ [7, 9]. Each PE of this filter stores the filter coefficients $h_{k,i}$ and is a gated full adder. The array is fully pipelined with the systolic latches on each data line.

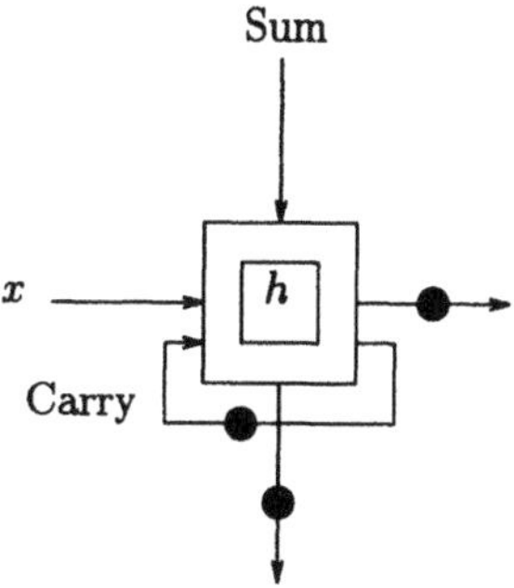

Fig. 1. Inner Product PE

2.2 Synthesizing Word Level Arrays

A typical problem in the area of spectral estimation is the requirement to solve a set of linear equations[1]. One algorithm to do this is LU decomposition. The description of this algorithm in ALPHA is given in figure 2. Any array transformed from this description will require a divider PE as well as a multiply accumulate PE, to operate on a data word.

```
system LU (a : {i,j|1<=i<=8;1<= j<=8} of real)
  returns (l : {i,j |j<i;i<=8;1<=j<= 8} of real;
           u : {i,j |1<=i<= 8;i<=j;j<=8} of real);
  var
    F : {i,j,k|1<=i<=8;1<=j<=8;0<=k<i;k<=j} of real;
  let
    F[i,j,k] = case
          {|k=0}: a[i,j];
          {|i>j; j=k}: F[i,j,k-1] / F[k,j,k-1];  #Division
          {|i>k; j>k}: F[i,j,k-1] - F[i,k,k] * F[k,j,k-1]; #MAC
    esac;
    l[i,j] = F[i,j,j];
    u[i,j] = F[i,j,j-1];
  tel;
```

Fig. 2. ALPHA Description of LU Decomposition

An example of the divider circuit necessary in the LU array is illustrated in figure 3. This bit level array takes a dividend ($N = n_0.n_1n_2n_3n_4n_5n_6$) and a divisor ($D = d_0.d_1d_2d_3$) to produce a quotient ($Q = q_0.q_1q_2q_3q_0$) and a remainder ($R = 0.0000r_3r_4r_5r_6$) [3]. Each PE of this array is a controlled add-subtract. The PE interconnections are regular and to nearest neighbours.

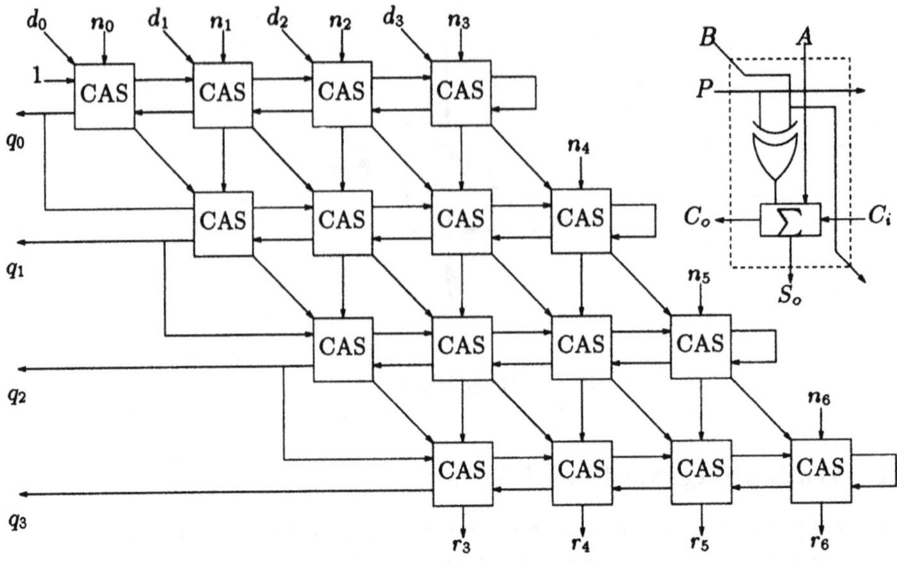

Fig. 3. Nonrestoring Array Divider

2.3 Synthesizing to FPGAs

As we have illustrated in the previous two sections we can synthesis algorithms to both word and bit level arrays. We now wish to consider implementing these arrays using FPGAs. In the case of bit level arrays we can implement them directly on FPGAs [5]. In the case of word level arrays an array of FPGAs is required, but in general an iterative logic array such as the divider or parallel array multiplier will be implemented as a PE on the FPGA [4].

In order to map regular arrays to FPGAs we can identify three requirements:

- It should be possible to specify processor sections in the FPGA and to tile them across it.
- Complete and detailed control over the placement of cells and routing resources must be available.
- Major design modifications should be attainable in a few minutes.

The Logic Description Generator (LDG) [2] is a hierarchical systolic description language, supporting the systolic unit from a high level. Thus hierarchical designs can be mapped into FPGAs. LDG supports the above three requirements. However LDG is transformed to the XNF netlist description as input to Xilinx FPGAs. The XNF netlist does not allow absolute control at the switching transistor level of the routing resources available in the FPGA. Moreover LDG does not support proven transformations as ALPHA and is dedicated to a specific hardware platform.

In order to exploit the capabilities of ALPHA and to have more exact control of the routing resources we present a FPGA Regular Array Description Language (FRADL).

3 FRADL

FRADL is designed to map the layout of a regular structure into a FPGA. FRADL operates specifically as an automation tool for FPGA layout. As such it is designed to be viewed and driven from the FPGA chip level resources. This gives the advantage that designs and routes can be aggressively chosen and optimized for performance. These optimized designs can then be replicated across the chip giving a tightly controlled level of performance and area efficiency throughout the FPGA. To achieve this control the LCA format is used as the input to the Xilinx design environment XACT and converted to a bit stream for programming the FPGAs.

FRADL uses high level structural concepts such as 'rectangle' and 'linear', to map designs in a rectangular or linear fashion respectively, but also contains FPGA specific design details. This high level input language is then compiled into the full LCA format by the compiler which expands structures and maps the design across the FPGA chip, allowing one instance of a design to be repeated numerous times. This technique allows the description of high level structures in a concise manner yet importantly maintain the FPGA specific speed advantages.

3.1 FRADL Commands

As has already been stated the routing resources available in the Xilinx 3000 series FPGAs are direct interconnect, general purpose interconnect and longlines. The general purpose interconnect consists of a grid of metal segments located between the rows and columns of CLBs. A transistor switching matrix is used to route signals through these metal lines. Direct routing is available for communication between adjacent CLBs. Longlines bypass the switching matrix and are available for broadcasting data to many CLBs.

For regular arrays we are primarily interested in the direct interconnect and longlines. The FRADL construct for routing is of the form:

BB.X - > BC.DI

Thus the X output from CLB BB is routed to the DI input of CLB BC. This routing is maximized for speed, by using a minimum number of switching transistors.

Longlines are particularly useful for routing global resets and broadcast data for regular arrays. The command:

longline: BB.C repeat 6 step 2

will draw a single longline, connecting the C input, starting at CLB BB and repeating 6 times, skipping a block each time.

As we are particularly interested in bit level systolic arrays we have a specific construct for the global clock. This allows us to connect the clock input (K) of CLBs together.

In order to define the function of a particular CLB, FRADL uses a **clbdefinition** command. The CLB description follows the Xilinx XACT "editblk" screen. The FRADL compiler needs to know which member of the Xilinx 3000 series devices is targeted, by using the **CHIPtype** command. As FRADL is designed specifically for repetitive or regular designs, we can define a pattern of repetition along with the pattern to be repeated. The function **rectangle** and **linear** will map the 2 dimensional or 1 dimensional array of the routing or CLB definitions given. For example, if the carry out of a full adder is output X of the CLB, and the carry-in is input B of the CLB, the FRADL command:

```
rectangle rectlayout1 Horiz repeat 7 step 1
                      Vert repeat 8 step 1
{
   route: BB.X -> BC.B ;
}
```

will generate the LCA code to route from CLB output X to CLB input B for a 7×8 array, starting a CLB BB.

3.2 FRADL and Regular Arrays

Using the above FRADL commands we now present the FRADL code necessary to generate the regular arrays. The FRADL code to generate the divider array is given in figure 4. The PE of the divider array is a controlled add subtract circuit which can be mapped directly onto one CLB. The function **clbdefinition** defines the function a particular CLB (a controlled add subtracted cell with the C CLB input the control line P of figure 3. The two CLB definitions in figure 4 are required to incorporate the feedback of the control line P to the carry-in of the first PE in each row.

The FRADL code necessary to generate the FIR Filter array is given in figure 5. In this array the PE requires 4 latches and a gated full adder. This requires two CLBs to implement a PE. Thus we require a step facility in our FRADL code as we have two distinct CLB designs.

4 LCA File

The FRADL is compiled directly into the Xilinx LCA description of the FPGA. An example of the routed divider circuit implemented on a XC3042 device is illustrated in figure 6. This array will take a 15 bit value and divide it by a 7 bit

```
chiptype = 3042;

clbdefinition clbdesign1
{
    F=((A@C)*E)+B*((A@C)+E)
    G=(A@C)@E@B
    X: F
    Y: G
}

rectangle rectlayout1 Horiz repeat 7 step 1
                      Vert repeat 8 step 1
{
    clb: clbdesign1 BC ;
    route: BB.X -> BC.B ;
    route: BB.Y -> CC.E ;
}

linear lin1 Horiz repeat 8 step 1
{
    longline: BB.C repeat 8 step 1 ;
}

clbdefinition clbdesign2
{
    F=((A@C)*E)+A*((A@C)+E)
    G=(A@C)@E@A
    X: F
    Y: G
}

linear lin2 Vert repeat 8 step 1
{
    longline: BB.A repeat 8 step 1;
    route: BI.Y -> CJ.E ;
    route: BI.X -> CI.A ;
    clb: clbdesign2 BB ;
}
```

Fig. 4. Divider FRADL Code

```
chiptype = 3020 ;
CLBDefinition clb1
{   F=(A*B)+(Ã*QY)
    G=QX*QY
    X: QX
    Y: G
    DX: DI
    DY: F
    CLOCK: K }
rectangle rect1 Horiz repeat 5 step 1
                Vert repeat 3 step 2
{   clb: clb1 BB ; }
rectangle rect3 Horiz repeat 4 step 1
                Vert repeat 2 step 2
{   route: BB.X -> BC.DI ; }
rectangle rect4 Horiz Repeat 5 step 1
                Vert repeat 2 step 2
{   route: BB.Y -> CB.A ;
    longroute: CB.Y -> EB.B ; }
CLBDefinition clb2
{   F=(A*B)+(A*QX)+(B*QX)
    Y: QY
    DX: F
    DY: G
    CLOCK: K }
rectangle rect2 Horiz repeat 5 step 1
                Vert repeat 5 step 2
{   clb: clb2 CB ; }
linear lin1 Horiz repeat 5 step 1
{   longline: BB.B -> FB.B step 2; }
linear lin2 Vert repeat 3 step 2
{   longline: BB.A -> BF.A ; }
globalclock
{   line: BB -> GB ;
    line: BC -> GC ;
    line: BD -> GD ;
    line: BE -> GE ;
    line: BF -> GF ; }
```

Fig. 5. FIR Filter FRADL code

value to give a 8 bit result. The regular array requires 56 PEs and the compiled design uses 56 CLBs.

The routed 3 by 3 PE FIR filter is illustrated in figure 7, using 3 by 6 CLBs. The FIR filter PE requires 4 latches and is partitioned into two CLBs. CLB1 is used to store the value of h in the QY flip flop, with logic block F used to initiate the PE with the appropriate value of h before the computation begins. The QX flip flop stores the x value as it is propagated from PE to PE across each row

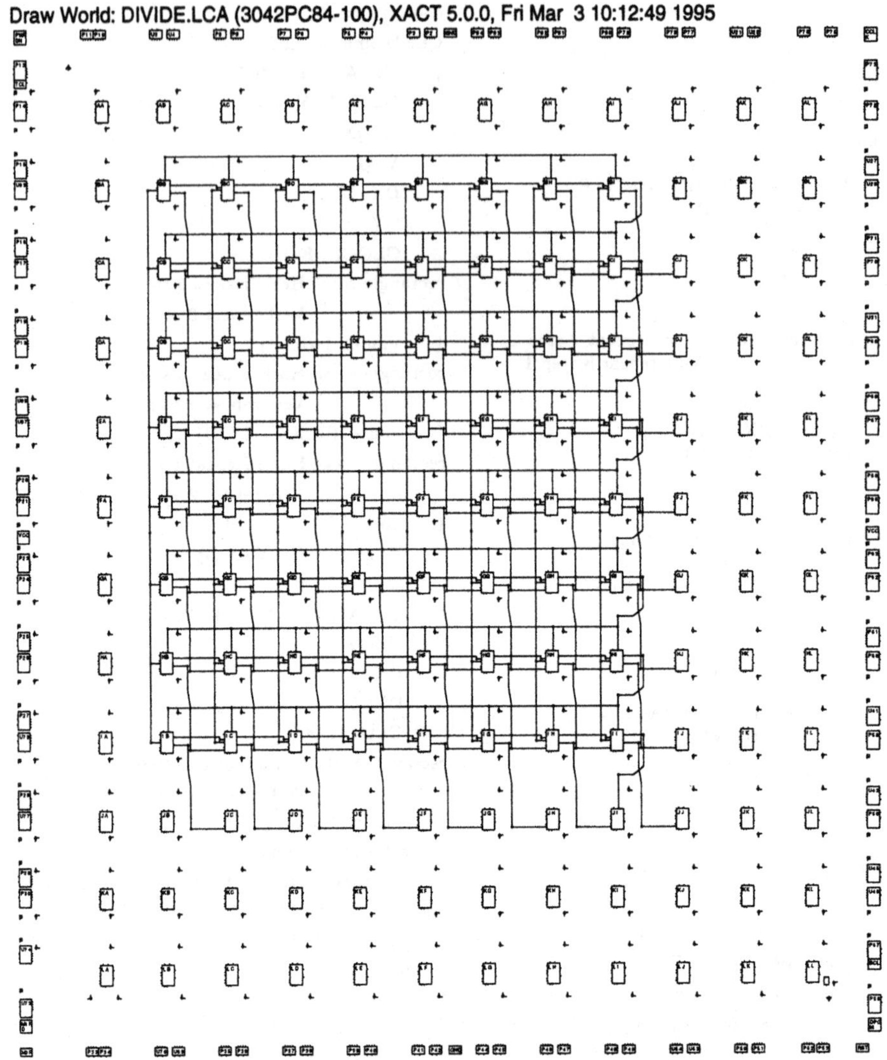

Fig. 6. Divider LCA

of the array. Finally we AND the contents of the two flip flops in logic block G and feed the result to the next CLB below on the Y output. CLB2 is used to calculate the sum and carry of each PE. The QX flip flop stores the carry which is generated in the F logic block, the QY flip flop stores the sum output before passing it on to the next PE down the column.

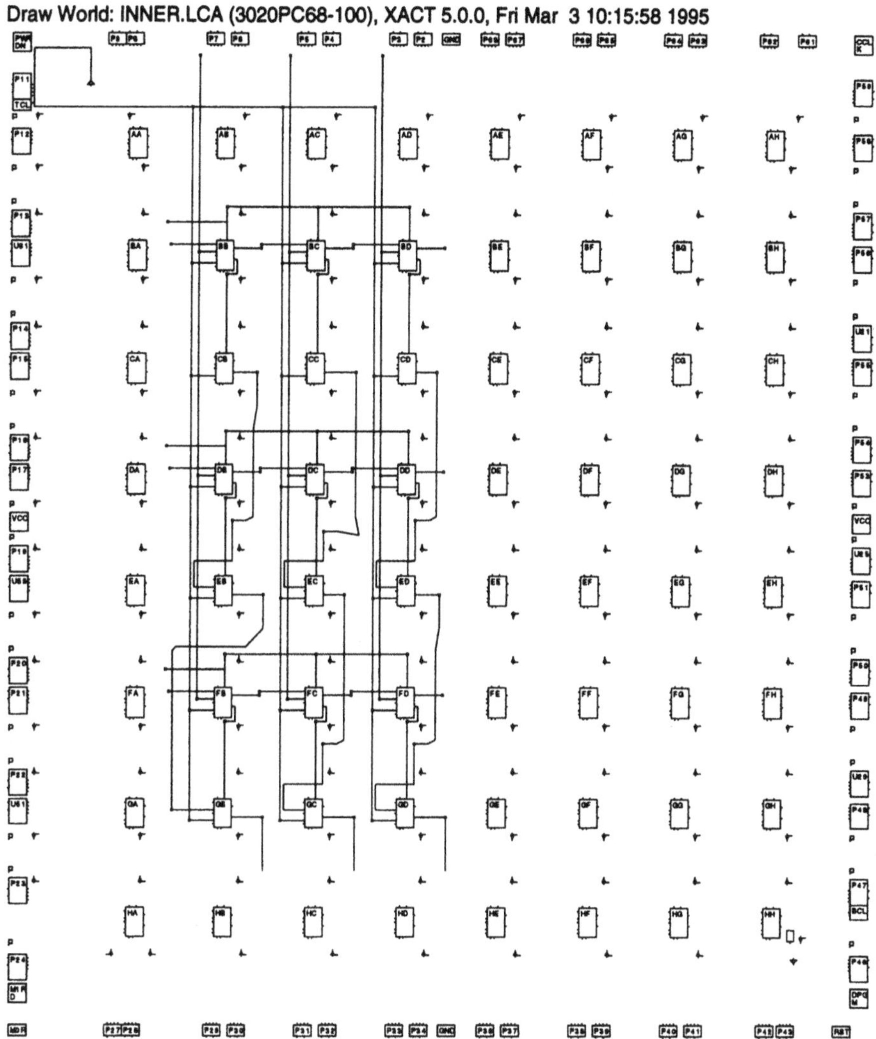

Fig. 7. FIR Filter LCA

5 Conclusion

In this paper we have presented a description language that allows us to compile a regular array directly into a FPGA. This approach preserves the inherent regularity present in the regular array to give us area and timing efficient designs in the FPGA. The motivation in the design of the FRADL language is to automate the layout of designs Xilinx FPGAs. Regular designs benefit most from this capability. However, as FRADL operates at a chip level, regularity in the design is not a pre-requisite. Non regular connections and mapping are supported in the syntax. The output of the FRADL compiler is the Xilinx LCA format, thus allowing the potential of utilizing FRADL within the existing XACT software environment.

It should be noted that our ultimate aim is to automate the process of mapping DSP algorithms to FPGA based implementations [5]. We have chosen ALPHA as our environment for capturing the algorithm and transforming it into a regular structure. We are currently automating the process of mapping an ALPHA description of an array into the FPGA through the FRADL description.

References

1. S. Bellis, W. Marnane, and P. Fish. Systolic Architectures for the Modified Covarience Spectral Estimator used with Ultrasonic Doppler Blood Flow Detectors. In *Proceedings of IRISH DSP and Control Colloquium*, pages 71–78, 1994.
2. M.B. Gokhale, S. Kopser, S.P. Lucas, and R.G. Minnich. The Logic Description Generator. In S. Y. Kung, E. Swartzlander, J. Fortes, and K. W. Przytula, editors, *Application Specific Array Processors*, pages 111–120. IEEE Computer Society Press, September 1990.
3. Kai Hwang. *Computer Arithmetic, Principles, Architecture and Design.* John Wiley & Sons, 1979.
4. C. Jordan and W. Marnane. Prototyping Systolic Arrays using Field Programmable Gate Arrays . In *Proceedings of IRISH DSP and Control Colloquium*, pages 79–86, 1994.
5. W. P. Marnane, C. J. Jordan, and F. J. O' Reilly. Synthesising a FIR Filter onto a FPGA. In *Proceedings of IRISH DSP and Control Colloquium*, June 1995.
6. J. V. McCanny, J. G. McWhirter, and E. Swartzlander, editors. *Systolic Array Processors.* Prentice Hall, 1989.
7. J. V. McCanny, K. W¿ Wood, J. G. McWhirter, and C. J. Oliver. The Relationship Between Word and Bit Level Systolic Arrays as Applied to Matrix × Matrix Multiplication. *Proceedings of SPIE Int. Soc. Opt. Eng.*, 495:114–120, 1984.
8. W. R. Moore, A. P. H. McCabe, and R. B. Urquhurt, editors. *Systolic Arrays.* Adam Hilger, 1987.
9. R. B. Urquhart and D. Wood. Systolic matrix and vector multiplication methods for signal processing. *IEE Proceedings Part F*, 131(6):623–631, October 1984.
10. M. Valero, S. Y. Kung, T. Lang, and J. Fortes, editors. *Application Specific Array Processors.* IEEE Computer Society Press, September 1991.
11. H. Le Verge, C. Mauras, and P Quinton. The ALPHA Language and its use for the Design of Systolic Arrays. *Journal of VLSI Signal Processing*, (3):173–182, 1991.
12. D. K. Wilde and O. Sié. Regular array synthesis using alpha. In *Proc. International Conference on Application-Specific Array Processors - ASAP'94*, pages 200–211, 1994.
13. Xlinx Inc., San Jose, California, USA. *The Programmable Gate Array Data Book*, 1991.

Compiling Ruby into FPGAs

Shaori Guo[1] and Wayne Luk[2]

[1] Computing Laboratory, Oxford University, Parks Road, Oxford OX1 3QD, UK
[2] Department of Computing, Imperial College, London SW7 2BZ, UK

Abstract. This paper presents an overview of a prototype hardware compiler which compiles a design expressed in the Ruby language into FPGAs. The features of two important modules, the refinement module and the floorplanning module, are discussed and illustrated. Target code can be produced in various formats, including device-specific formats such as XNF or CFG, and device-independent formats such as VHDL. The viability of our floorplanning scheme is demonstrated by a compiler backend for Algotronix's CAL1024 FPGAs. The implementation of a priority queue is used to illustrate our approach.

1 Introduction

Compiling selected parts of application programs into hardware, such as FPGAs, has recently attracted much interest. This method holds promise of producing better special-purpose systems more rapidly than existing techniques. A number of hardware compilers (see, for example, [8], [11]) have been developed for designs described in various languages into hardware netlists, which can then be mapped onto FPGAs by vendor software.

This paper presents an overview of two important modules, the refinement module and the floorplanning module, in a prototype compilation system. The system is based on Ruby [4], [9], a relational language for capturing block diagrams parametrically. There are mechanisms in Ruby for describing spatial and temporal iteration, allowing succinct and precise design specification. Moreover, the explicit representation of different forms of spatial iteration simplifies the production of layouts, and the declarative nature of the language allows designs to be refined by simple equational reasoning. Our aim is to exploit these features of Ruby to provide an efficient hardware compilation system.

The refinement module enables users to focus on the high-level structure of a design without being overwhelmed by details such as the size of individual datapaths. It is based on a constraint-propagation procedure. Given the size of inputs and a library of bit-level operators, it automatically constructs efficient low-level designs rapidly and in a provably-correct manner; this facilitates exploring architectures and evaluating the effects of different bit-level data representations.

Another important module, the floorplanning module, is devised to reduce the time to place and route a netlist produced by a hardware compiler. Since Ruby expressions carry information about the way a circuit can be assembled from primitive parts, our method is designed to exploit the structure of the source program in generating a layout. It is also possible for the user to guide the placement of components and to import layouts that are developed manually or by other tools. Much of our floorplanning procedure is syntax-directed and is therefore very efficient.

While our floorplanning scheme is largely device-independent, to demonstrate its viability a compiler backend has been developed for Algotronix CAL1024 FPGAs. The implementation of a priority queue will be used to illustrate this approach.

2 Ruby

Ruby is a language of functions and relations. It has been used in developing a wide range of designs including signal processing architectures [2] and butterfly networks [4], and it has also been used in producing implementations partly in hardware and partly in software [7]. Detailed descriptions of Ruby can be found, for instance, in [4] and [9].

In Ruby a design is captured by a binary relation R, which relates the interface signals x and y in the form of $x\ R\ y$. For instance the *max* operator, which produces the maximum of two numbers, can be described by

$$\langle x, y\rangle\ max\ (maximum(x,\ y)),$$

so $\langle 3,4\rangle\ max\ 4$ and $\langle 10,6\rangle\ max\ 10$. The *min* operator for finding the minimum of two numbers can be described in a similar way. The identity relation *id* is given by $x\ id\ x$. To select or regroup components of composite data, there are wiring primitives such as *fork*, π_1 and *rsh*, given by $x\ fork\ \langle x, x\rangle$, $\langle x, y\rangle\ \pi_1\ x$ and $\langle x, \langle y, z\rangle\rangle\ rsh\ \langle\langle x, y\rangle, z\rangle$. To reflect a component along its trailing diagonal, we can use the converse operator, given by

$$x\ R^{-1}\ y\ \Leftrightarrow\ y\ R\ x.$$

Complex designs in Ruby can be formed by composing simpler designs. For instance, two components Q and R with a compatible interface connected in series is denoted by $Q\ ;\ R$ (Figure 1a):

$$x\,(Q\ ;\ R)\,y\ \Leftrightarrow\ \exists s : (x\ Q\ s) \wedge (s\ R\ y).$$

The $\exists$ symbol means that, unlike x and y, s is not an interface variable of the composite and cannot be observed.

If there are no connections between Q and R, the composite design is represented by parallel composition $[Q, R]$ (Figure 1b), where

$$\langle x_0,\ x_1\rangle\,[Q, R]\,\langle y_0, y_1\rangle\ \Leftrightarrow\ (x_0\ Q\ y_0) \wedge (x_1\ R\ y_1).$$

Repeated compositions of n copies of Q can be described by Q^n or $\mathsf{map}_n\ Q$, so for instance $fork^4 = fork\ ;\ fork\ ;\ fork\ ;\ fork$ and $\mathsf{map}_3\ rsh = [rsh, rsh, rsh]$.

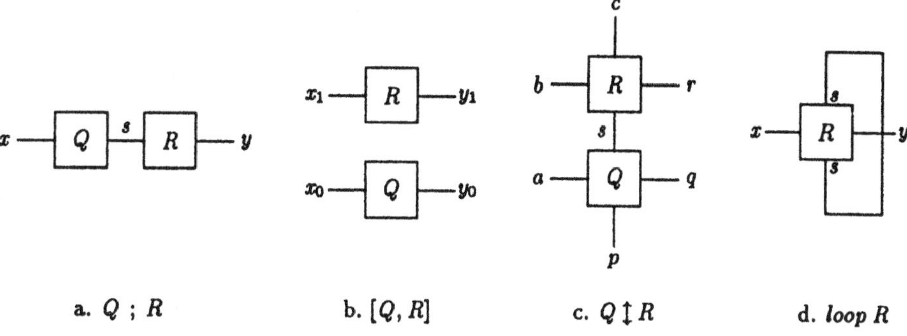

a. $Q \; ; R$ b. $[Q, R]$ c. $Q \updownarrow R$ d. *loop R*

Fig. 1. Some Ruby operators.

Components with connections on four sides can be joined together by the *beside* and *below* operators; *below* (Figure 1c) is given by

$$\langle\langle a, b\rangle, c\rangle \, (Q \updownarrow R) \, \langle p, \langle q, r\rangle\rangle \; \Leftrightarrow \; \exists s : (\langle a, s\rangle \, Q \, \langle p, q\rangle) \wedge (\langle b, c\rangle \, R \, \langle s, r\rangle).$$

To deal with designs operating on time-varying data, a relation in Ruby can be considered to relate an infinite sequence of data in its domain to another infinite sequence in its range; elements in these infinite sequences can be regarded as values appearing at an interface at successive clock cycles. Given that $\forall t$ denotes "for all values of t", a squarer can be described by

$$x \; sq \; y \; \Leftrightarrow \; \forall t : x_t^2 = y_t.$$

A latch can be modelled by a delay relation D, given by

$$x \; D \; y \; \Leftrightarrow \; \forall t : x_{t-1} = y_t.$$

A latch initialised to value i is denoted by $D\, i$.

Latches are used in designs with feedback to prevent unbuffered loops. A design Q containing an internal feedback path s can be modelled by the operator *loop* (Figure 1d):

$$x \, (loop \; R) \, y \; \Leftrightarrow \; \exists s : \langle x, s\rangle \, R \, \langle s, y\rangle.$$

3 Refinement

We can use Ruby to describe word-level designs, like the *max* or the *min* operator for integers. At bit-level, these operators can be built by logic gates which can also be captured in Ruby. The aim of our refinement system is to automatically produce the most efficient bit-level design from a high-level description.

Bit-level designs produced by the refinement system should satisfy constraints specified by the designer. Examples of constraints include the speed, size, latency and power consumption of a design, the maximum and minimum values of inputs and outputs, or a combination of the above. Of course, if the constraints are too strict, there may not be any bit-level design that satisfies them all. Our efforts so far have been concentrated on constraints specifying the maximum and minimum values of inputs for a circuit.

There may be many possible bit-level designs which can implement a given word-level design. Also each data representation (such as two's complement representation) will result in a specific family of bit-level implementations. The refinement system can refine a word-level design into several bit-level implementations, depending on the bit-level data representation.

The refinement module is based on a constraint-propagation algorithm. The maximum and minimum values of inputs are propagated across the circuit. For a given component, once all constraints on its inputs are known, the constraints on its outputs can be derived. Resolving the constraints fixes the size of the components and the width of the output data path. Given a library of parametrised bit-level operators and their sizes, our constraint-propagation procedure can be used to determine the widths of all the data paths. A bit-level Ruby design can then be constructed. As an example, consider a priority queue which can be specified in Ruby as follows.

$$N = 4. \tag{1}$$

$$pq = pqcell^{N}. \tag{2}$$

$$pqcell = loop\ ((sort2 \updownarrow mux2) \updownarrow ([id,\ D\,127]\ ;\ fork2)). \tag{3}$$

$$sort2 = fork\ ;\ [min,\ max]. \tag{4}$$

$$mux2 = fork\ ;\ [muxr\ 2,\ \pi_1]. \tag{5}$$

$$fork2 = {\pi_1}^{-1}\ ;\ [fork^{-1},\ fork]\ ;\ rsh. \tag{6}$$

Let us briefly introduce the correspondence between the Ruby program and the pictorial description of the priority queue; further details about possible designs and their development can be found in [9]. The Ruby descriptions for the word-level design (Figure 2) are shown above, which is implemented as a linear array of a repeating unit *pqcell* (expression 2), and the length of the array is 4 (expression 1). The repeating unit *pqcell* (expression 3) consists of three parts: an insertion sorter cell *sort*2 (expression 4), a selection unit *mux*2 (expression 5) and a data distribution unit *fork*2 (expression 6). There is an internal path in *pqcell* where the minimum output of the sorter is fed back while the maximum value is output to the next cell (expression 3). A latch (shown as a small triangle) is placed on the top of the feedback path, and it is initialised to the value 127.

Suppose the constraint specified by the designer is that the input data are natural numbers no larger than 127. Given that a bit is either T (True) or F (False), 127 is represented by $\langle T, T, T, T, T, T, T\rangle$. The refinement system produces a bit-level Ruby program with *min* replaced by *min_un_b* 7 7, where 7 represents the number of bits of the input. It also replaces *max* with *max_un_b* 7 7 in expression 3, *muxr* 2 with *muxr2_bit* 7 in expression 4, and *D* 127 with *map* 7 (*D T*) in expression 5. The bit-level descriptions include instantiations from a library of parametrised bit-level components, which contains, for instance, *max_un_b*, *min_un_b* and *muxr2_bit*, the bit-level implementations of

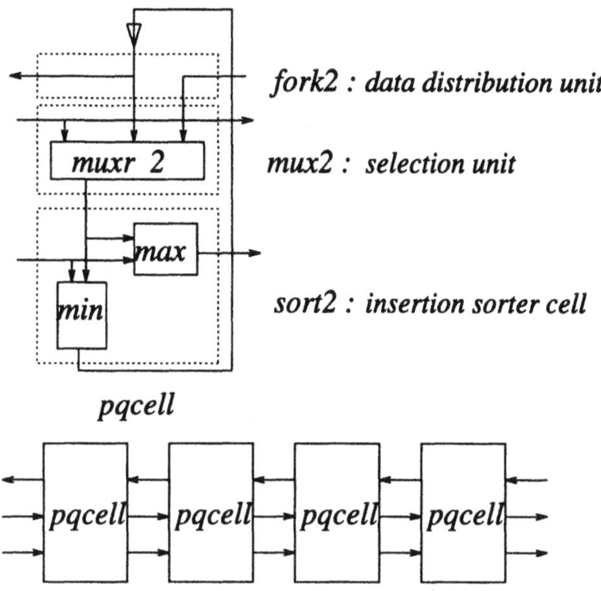

Fig. 2. A priority queue ($n = 4$).

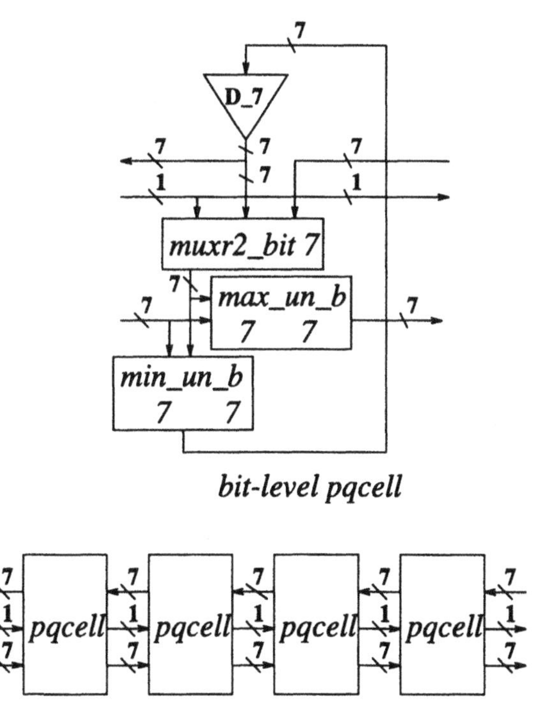

Fig. 3. Bit-level priority queue, with inputs not larger than 127 ($n = 4$).

max, *min* and the multiplexer *muxr* 2 operating on unsigned integers. The bit-level implementation of the priority queue is shown in Figure 3. Notice that the big triangle D_7 represents seven D latches in parallel.

There are compiler backends for converting a bit-level description into various formats, such as XNF (Xilinx Netlist Format) or VHDL. The physical mapping onto FPGAs can then be carried out using commercial tools. An alternative implementation path will be sketched in the next section.

4 Floorplanning

A major bottleneck in automatic hardware synthesis is the time to place and route the netlist produced by a hardware compiler. The aim of our floorplanning module is to expedite the placement and routing procedure by exploring the structure of the source descriptions. To achieve high quality layouts, our floorplanning scheme includes facilities which allow combination of layouts produced both automatically and manually.

The floorplanning procedure consists of two phases. The first phase is the global placement and routing, which is mainly device-independent. In this phase a design is modelled as a rectangular block with connecting points on its four sides. Our floorplanning scheme allows the variation of block sizes, so that connecting positions between two adjacent blocks match each other to minimise the routing between them. In the second phase, the detailed routing within the blocks and their interface will be determined.

Consider first the global placement and routing phase. A design in Ruby is represented by a binary relation, while in pictorial form it is modelled as a rectangular block. A convention is required for assigning the domain and range variables of a relation to each side of the block – this step is known as direction assignment. The following convention is chosen: the domain data will be mapped onto the western or northern side, while the range data will be mapped onto the southern or eastern side [4].

Following this convention, the layout of a relation with its domain in the form of a two-tuple $\langle x, y \rangle$ can be a block with x on the western side and y on the northern side, or both x and y on either the western or the northern side. Similarly, the layout of a relation with its range in the form of a two-tuple can be a block with some of its connecting points on the southern side and some on the eastern side, or all of them on either the southern or the eastern side. One can show that, for a relation with both its domain and range in the form of a two-tuple, there are nine possible layouts [3]. The choice of which layout to adopt is determined by context or by a default convention. For instance some combinators in Ruby carry contextual information about possible direction assignment; the *below* combinator requires two of its domain and two of its range connections to be horizontal (Figure 1c).

After direction assignment, we check the compatibility of the interfaces between connected components. Since polymorphism is allowed in the domain and range of some Ruby primitives such as *fork*, a simple structure comparison is insufficient. Instead a general unification algorithm was used to determine the most general substitution for the domain and range components, so that the interface constraints can be satisfied.

Sometimes information on direction of signal flow is necessary for certain devices, such as the cells used in Algotronix's CAL1024. In these cases we apply a constraint-propagation algorithm to determine the direction of signal flow for each Ruby wiring constructs.

The placement stages of our floorplanning system are not time-consuming because we exploit the structure of Ruby programs for placement. If we want to include a circuit which has been placed and routed manually or by other tools, we need to specify its size and the connection positions. Interface between the original and the imported layouts can then be produced by the compiler. A pair of curly braces are employed in the source Ruby program to indicate which part of the circuit should be laid out separately. The right curly brace is followed by a pair of parentheses which enclose the name of the manual layout file, so that the compiler can import this part of the layout and link it with others.

Further descriptions of our syntax-guided placement technique can be found in [3].

5 Device-Specific Mapping

While our approach to global placement and routing is largely device-independent, the detailed placement and routing flattens each block produced after global placement and routing, and it requires information specific to a particular device. To demonstrate the viability of our floorplanning scheme, a compiler backend has been customised for CAL1024 FPGAs developed by Algotronix (now Xilinx Development Corporation).

CAL1024 arrays are orthogonally connected structures obtained by replicating a basic cell which has one input port and one output on each of its four sides. An input port can be programmed to connect to one or more output ports, or to a function unit which can be programmed to behave as a two-input combinational logic gate or as a latch. The output of this function unit may also connect to one or more output ports. Hence a CAL cell may be used to perform processing and routing simultaneously. Figure 4 shows a CAL cell with its northerly output connected to its easterly input, and its easterly output is the Boolean conjunction of its westerly and northerly inputs.

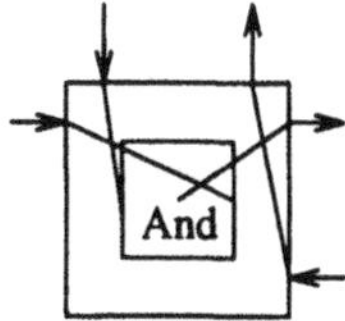

Fig. 4. CAL Cell.

During global placement and routing, two kinds of blocks are produced: blocks for combinational primitives such as *AND* and wiring primitives like *fork*. For combinational primitive blocks, we have developed a simple river routing algorithm to connect the connecting points on the four sides of the block to the cell performing the logic function of the primitive. A simple switch-box routing algorithm has also been devised to implement the detailed routing for the wiring blocks. The output of the floorplanner is a program in OAL [6], a variant of Ruby specialised for CAL devices. The OAL compiler can then be used to generate CFG files used for FPGA programming.

Although the floorplanner can perform the placement and routing fully automatically, the quality of the final implementation may be inferior to one produced by hand or by other tools. It is our intention to give the designer the flexibility to use our compiler for global placement and routing, while part of or all of the detailed placement and routing can be produced by other means. For instance, a designer may wish to develop by hand the repeating unit of an array-based circuit, since any inefficiency in the basic cell will be multiplied many times. The compiler can incorporate existing CAL designs into the implementation according to the annotations specified by the designer in the source program, as described in section 4.

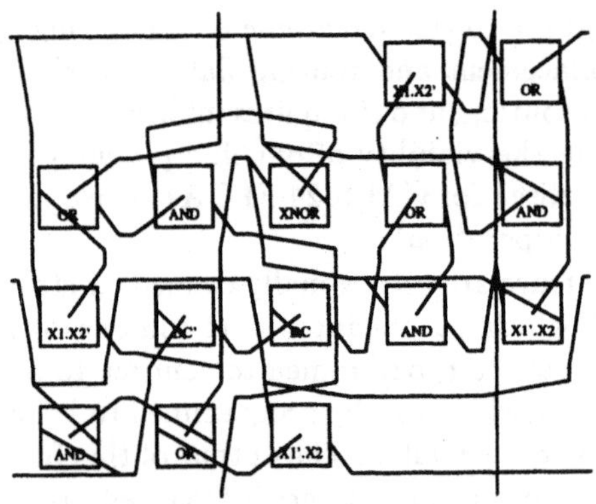

Fig. 5. CAL implementation of a bit-level priority queue cell.

Consider a priority queue implementation obtained by optimising the bit-level design in section 3 (see [9]). The bit-level repeating unit (Figure 5) was developed by hand and is highly optimised; this unit is then replicated vertically to form a column which corresponds to the core of a *pqcell* in Figure 3. The CAL implementation of the priority queue is shown in Figure 6. Note that the number and order of the interface connections correspond to those in Figure 3, except that the two bottom outputs of the rightmost *pqcell* are discarded.

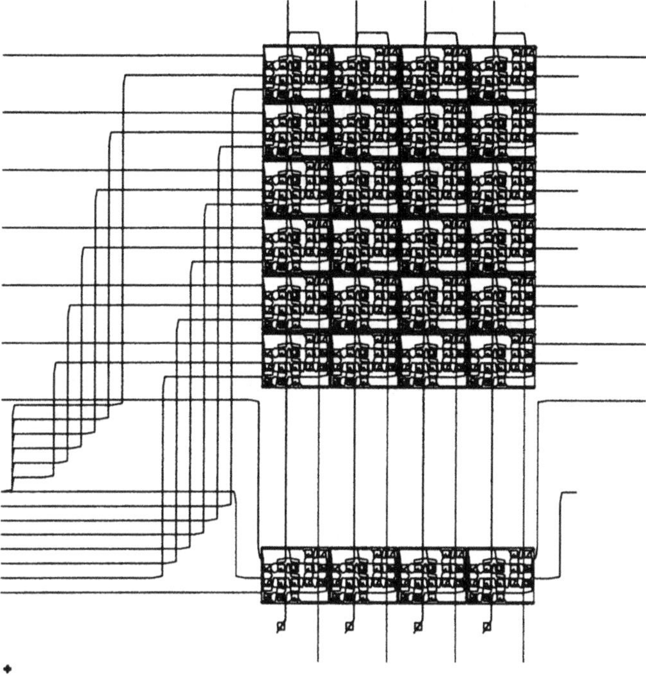

Fig. 6. CAL implementation of a bit-level priority queue (n = 4, m = 7).

6 Future Work

In the refinement module of our compilation system, we have focused on constraints specifying the maximum and minimum values of inputs for a word-level circuit. Our method can be extended to take into consideration other kinds of constraints: examples include critical path, latency or the number of a particular component. If no solutions exist that satisfy all user-specified constraints, we can choose the solution that satisfies most of the high-priority constraints.

The CAL backend of our compiler demonstrates the viability of our floorplanning module. We have not, however, optimised the switch-box routing or the river-routing algorithms, and the layouts produced automatically can become rather large. For better results, we can use methods like min-cut or simulated annealing hierarchically in placement and routing [10]. Device-specific compaction techniques should also be studied.

Much of our method for generating layouts is syntax-directed. The quality of the compiled implementation depends largely on the Ruby source program which describes the design; therefore source transformation can be adopted for optimisation. One way to automate this step is to have an accurate performance estimation procedure to drive the transformation engine.

It will also be interesting to extend our work to support partial and run-time reconfiguration of FPGAs, to support developing multi-chip systems, and to support implementing asynchronous and self-timed designs [1].

Acknowledgements

The support of Xilinx Development Corporation, Scottish Enterprise, Department of Computing, Imperial College and Oxford University Hardware Compilation Research Group is gratefully acknowledged. S. Guo thanks the Sino-British Friendship Scholarships Foundation for their support.

References

1. E. Brunvand, "Using FPGAs to implement self-timed systems", *Journal of VLSI Signal Processing*, vol. 6, 1990, pp. 173-190.
2. S. Guo, W. Luk and P. Probert, "Developing parallel architectures for range and image sensors", in *Proc. IEEE Int. Conf. on Robotics and Automation*, IEEE Computer Society Press, 1994, pp. 2205–2210.
3. S. Guo and W. Luk, "Producing design diagrams from declarative descriptions", to appear in *Proc. Fourth Int. Conf. on CAD and CG*, SPIE, 1995.
4. G. Jones and M. Sheeran, "Circuit design in Ruby", in *Formal Methods for VLSI Design*, J. Staunstrup (ed.), North-Holland, 1990, pp. 13–70.
5. W. Luk, "Analysing parametrised designs by non-standard interpretation", in *Proc. Int. Conf. on Application-Specific Array Processors*, S.Y. Kung, E. Swartzlander, J.A.B. Fortes and K.W. Przytula (eds.), IEEE Computer Society Press, 1990, pp. 133–144.
6. W. Luk and I. Page, "Parameterising designs for FPGAs", in *FPGAs*, W. Moore and W. Luk (eds.), Abingdon EE&CS Books, 1991, pp. 284–295.
7. W. Luk and T. Wu, "Towards a declarative framework for hardware-software codesign", in *Proc. Third International Workshop on Hardware/Software Codesign*, IEEE Computer Society Press, 1994, pp. 181–188.
8. W. Luk, D. Ferguson and I. Page, "Structured hardware compilation of parallel programs", in *More FPGAs*, W. Moore and W. Luk (eds.), Abingdon EE&CS Books, 1994, pp. 213–224.
9. W. Luk, "A declarative approach to incremental custom computing", in *Proc. IEEE Workshop on FPGAs for Custom Computing Machines*, D.A. Buell and K.L. Pocek (eds.), IEEE Computer Society Press, 1995.
10. M. Newman, W. Luk and I. Page, "Constraint-based hierarchical hardware compilation of parallel programs", in *Field-Programmable Logic: Architecture Synthesis and Applications*, LNCS 849, Springer-Verlag, 1994, pp. 220–229.
11. M. Wazlowski et. al., "PRISM II: compiler and architecture", in *Proc. IEEE Workshop on FPGAs for Custom Computing Machines*, D.A. Buell and K.L. Pocek (eds.), IEEE Computer Society Press, 1993, pp. 9–16.

The *CSYN* Verilog Compiler and Other Tools

David Greaves

University of Cambridge, Computer Laboratory, Cambridge, UK. djg@cl.cam.ac.uk

Abstract. The CSYN Verilog compiler was written by Dr Greaves in early 1994 as a vehicle for research in logic synthesis algorithms and to support experimental extensions to the Verilog language to test high-level specification techniques. A basic version of CSYN is in use at a number of local companies for industrial FPGA design. This paper describes CSYN and its use with Xilinx devices for teaching. To extend this work, we are defining formal semantics for Verilog, both for simulation and compilation into hardware. This paper reports the performance of CSIM, an X-windows Verilog simulator based on the formal simulation semantics and expresses the desire for a general purpose semantics for Verilog, which can help prove the equivalance of different implementations of a module.

1 Background

The Systems Research Group of the University of Cambridge Computer Laboratory has for ten years or more owned and used a set of CAD tools based around the 'Cambridge HDL' heirarchic net list format [1]. The tools were first written to support internal research projects, but they have also been commercialised from time to time and used by a dozen or so local companies and they have been used extensively in teaching.

In recent years, limited logic synthesis capabilties were added to the tools using enhanced ABEL as the input specification language. ABEL's infix operators for addition and complex combinatorial expressions so became available. In addition, a local language FDL (functional definition language) was added to allow behavioural specification of standard, non-synthesised parts, such as RAMs, FIFOs or clock modules and test wrappers. Hence the human input to the CAD system was split over three source languages which necessarily resided in different source files, owing to different parsers and front-end processing.

A few years ago, the author became interested in the Verilog language and so the process of moving the tools to Verilog began. Like VHDL, Verilog is able to integrate these three forms of specification using a common syntactic structure, and indeed the forms can be intermixed at the fine grain of individual declarative statements within a Verilog 'module'. The module is the heirachic building block, and can be anything form a gate to a microprocessor or complete system.

Today the tools run on Unix or other Posix machines (such as Linux) using an X-windows interface for the simulators and command line or makefile driven interfaces for the remainder. About half of the designs generated with the tools are targeted at Xilinx devices and the remainder include standard cell,

PCB board models, MACH devices or abstract designs with their own research content.

The following programs have been written:

- CSYN - Verilog compiler.
- CSIM - Verilog simulator.
- XSIM - Cambridge HDL/FDL simulator.
- CVAUX - Verilog flattener.
- CVXNF - Verilog to Xilinx net list format convertor.
- LCATOV - Xilinx logic cell array to Verilog convertor.
- CPAL - Verilog to JDEC pal compiler.
- TRANS - Verilog multi-level logic minimiser.

2 CSYN Verilog Compiler

CSYN is described in the CSYN user manual [2]. The input to CSYN is a set of Verilog files, including the design proper and standard libraries. The libraries contain reusable building blocks, such as a Huffman or Manchester coder, which can be compiled for any target technology, and target technology-specific definitions such as input/output pads or special buffers which drive things such as the Xilinx global clock nets.

The output from CSYN is a heirarchic Verilog net list known as a `vnl` file where each module in the output corresponds to a module in the input source (CSYN can also generate Cambridge HDL). A `vnl` file is the subset of Verilog language obtained by deleting the RTL and behavioural declarations: `assign`, `initial` and `always`, and disallowing the entity declarations `integer`, `reg`, `time`, `event` etc.. This leaves only the `wire` and `tri` data types and the only significant construct is structural instantiation of a submodule.

The heirarchy in a `vnl` file can be flattened when necessary using an extra program called 'CVAUX'. A related program, CVXNF will flatten and generate Xilinx net list format (xnf) for input into the Xilinx tools. In this mode, it takes an additional command line argument to specify a technology library, such as 'xi4000', which contains appropriate macros which are expanded to generate the xnf.

Logic synthesised by CSYN is not mapped onto a specific target technology library. Instead, CSYN converts all RTL and behavioural constructs into a fixed set of gates, which are AND2, OR2, INV, DFF, TLATCH, MUX2. It is up to the software which processes the target `vnl` file to understand these gate types and fit the logic to the target device. However, the instantiated modules output from CSYN do not only include these gates, since the source Verilog may contain structural instantiations of library or other leaf modules such as the pads etc. mentioned earlier. Attempts to actually synthesise these leaf modules by CSYN is prevented by flagging their definition with a small extension to the Verilog language syntax: the keyword `primitive` may be introduced before the word `module` in each such definition, or alternatively the line `primitive everything` may be added to a file to mark as primitive all textually subsequent modules.

For primitive modules, CSYN ignores the module's body, if present, and does not flag 'output not driven' warning messages when the body is absent. However CSYN does cross-check the `input`, `output` and `inout` port direction statements for compatibility at each instantiation of the primitive module.

Using CSYN, the designer has control of the logic that will be synthesised through the way he uses continuous assignments and parenthesis. If part of the circuit is critical, it must be essentially hand coded in the source file using the more simple constructs of Verilog which have predictable synthesis paths, rather than using esoteric `for`, `case` or `while` constructs. Such control might be termed `in-band` when compared with other logic synthesisers which accept complex `out-of-band` annotations for each input and output, expressing a desire for low load or late arrival etc.. Advanced minimisation of logic functions using techniques based on *Expreso* are not considered appropriate within CSYN since the best optimisations are target technology specific and best implemented in subsequent vendor fitter tools.

Infact, CSYN does have a small set of command line parameters which modify slightly the way it synthesises structures such as adders, but these are generally left unused. Their use is to trade speed of operation against gate count.

2.1 CSYN operation

CSYN first parses the input Verilog files to build a lisp parse tree. It then selects the subset of sourced modules that will be present in the output (i.e. are being used) using a recursive tree walk from a command line specified root name, which must be one of the sourced modules.

It then uses simple algorithms to convert the supported behavioural constructs into continuous and behavioural assignments, generating for them a parse tree structure similar to that which they would have if they had been specified in this form to start with. D-type flip-flops or broadside registers are generated and inserted into the structure as though they had been structurally instantiated in the source file.

An algorithm which expands all busses into individual signal nets is then applied, and at the same time this algorithm rewrites the tree to convert all of the 'higher-order' logic functions, such as addition, bus comparison or unary reduction, into a small set of operators: namely AND, OR, XOR, INVERT and MUX2.

An algebraic simplifier then runs over the tree, reducing simple tautologies and applying other Boolean identities, but without applying cubic division or other more complicated minimisation.

The final stage is a recursive gate generating algorithm which walks over the tree replacing each node with a gate instantiation and allocating a signal identifier for its output. Before generating each gate, the gate generator checks whether it has already produced a gate of the same type and input connection pattern, and if so, uses the output from that instead. To make this associative search cheap, the gates are given a normalised input ordering and then hashed to provide a simple array index.

After this stage, the circuit is composed entirely of structural instantiations of modules and leaf gates. A port checker cross-checks the input/output/inout port direction across all submodule instantiations and checks that each signal has exactly one output driving it, unless it is of type `tri`.

2.2 CSYN Synthesisable Constructs

No contemporary Verilog compiler can convert every behaviourally expressable Verilog construct into gates. At the RTL level, CSYN supports the subset of the Verilog combinatorial operators shown in table 1 together with the unary reduction operators, bus concatenation and dynamic subscripting of busses.

<table>
<tr><th>Symbol</th><th>Function</th><th>Resultant width</th></tr>
<tr><td>¬</td><td>monadic negate</td><td>as input width</td></tr>
<tr><td>–</td><td>monadic complement (*)</td><td>as input width</td></tr>
<tr><td>!</td><td>mondaic logic not (*)</td><td>unit</td></tr>
<tr><td>*</td><td>unsigned binary multiply (*)</td><td>sum of arg widths</td></tr>
<tr><td>/</td><td>unsigned binary division (*)</td><td>difference of arg widths</td></tr>
<tr><td>%</td><td>unsigned binary modulus (*)</td><td>width of rhs arg</td></tr>
<tr><td>+</td><td>unsigned binary addition</td><td>input width plus one</td></tr>
<tr><td>–</td><td>unsigned binary subtraction</td><td>input width plus one</td></tr>
<tr><td>>></td><td>right shift operator</td><td>input - shift amount</td></tr>
<tr><td><<</td><td>left shift operator</td><td>input + shift amount</td></tr>
<tr><td>==</td><td>net/bus comparison</td><td>unit</td></tr>
<tr><td>! =</td><td>inverted net/bus compare operator</td><td>unit</td></tr>
<tr><td><</td><td>bus compare operator</td><td>unit</td></tr>
<tr><td>></td><td>bus compare operator</td><td>unit</td></tr>
<tr><td>>=</td><td>bus compare operator</td><td>unit</td></tr>
<tr><td><=</td><td>bus compare operator</td><td>unit</td></tr>
<tr><td>&</td><td>diadic bitwise and</td><td>minimum of both inputs</td></tr>
<tr><td>↑</td><td>diadic bitwise xor</td><td>maximum of both inputs</td></tr>
<tr><td>↑ ¬</td><td>diadic bitwise xnor (*)</td><td>maximum of both inputs</td></tr>
<tr><td>|</td><td>diadic bitwise or</td><td>maximum of both inputs</td></tr>
<tr><td>&&</td><td>diadic logical and</td><td>unit</td></tr>
<tr><td>||</td><td>diadic locical or</td><td>unit</td></tr>
<tr><td>? :</td><td>conditional expression</td><td>maximum of data inputs</td></tr>
</table>

Table 1. Verilog Operators in order of Binding Power. Asterisked operators are not supported in current release of CSYN.

In Verilog, behavioural statements must be included within an `initial` declaration or an `always` declaration. CSYN (release cv2) ignores the contents of `initial` statements, but they may be present and are vital when simulating the source file with a simulator (to generate resets and so on).

CSYN compiles only the following form of the Verilog `always` construct. It has the syntax

```
always @( <sensitivity-list> ) <behavioural-statement>
```

This statement causes execution of the behavioural statement each time an event in the sensitivity list occurs. The behavioural-statement is typically a `begin-end` block containing other statements. The only sensitivity lists supported are the `posedge` or `negedge` of a single clock net or the full *support* list of the following statement (block).[1]

CSYN compiles behavioural constructs using a function $CC_{I\sigma}$ which takes a section of source code whose extent is free from event control statements (such as `posedge`), a substitution set σ and a condition expression I and returns a new set of substitutions σ'. A substitution set is a mapping of each variable (`integer`, `wire` or `reg` etc.) that appears in the block to an expression. At entry to a source block (i.e. after any previous event control) the initial σ passed to $CC_{I\sigma}$ is set to map each variable directly to the register, input pin or other source that drives it. I is set to true. CSYN transforms the σ returned by $CC_{I\sigma}$ into an appropriate RTL assignment of new values for the variables. $CC_{I\sigma}$ will handle `begin-end` sequencing, `case`, `if-then-else` and both blocking and non-blocking assignments as follows:

- $CC_{I\sigma}[\![c_1\ ;\ c_2]\!] = CC_{I\sigma'}[\![c_2]\!]$ where $\sigma' = CC_{I\sigma}(c_1)$ (This is used for `begin-end` sequencing).
- $CC_{I\sigma}[\![\texttt{if}\ (e)\ \texttt{then}\ c]\!] = CC_{I'\sigma}[\![c]\!]$ where $I' = I \wedge E_\sigma(e)$.
- $CC_{I\sigma}[\![\texttt{if}\ (e)\ \texttt{then}\ c_1;\ \texttt{else}\ c_2]\!] = \text{compose}(\sigma', \sigma'')$ where

  $$I' = I \wedge E_\sigma(e), \quad I'' = I \wedge \neg E_\sigma(e), \quad \sigma' = CC_{I'\sigma}[\![c_1]\!], \quad \sigma'' = CC_{I''\sigma}[\![c_2]\!]$$

  The value of a variable v in the substitution set returned by *compose* is the simple OR of $\sigma'(v)$ and $\sigma''(v)$.
- Verilog's `case` statement is handled by pre-converting to a series of `if`s in the obvious manner.
- $CC_{I\sigma}[\![v = e]\!] = \sigma[((I)?E_\sigma(e) : \sigma(v))/v]$ where the questionmark-colon is the standard Verilog RTL conditional expression construct.
- $CC_{I\sigma}[\![v <= e]\!] = \sigma$ with the pair $(v, E_\sigma(e))$ added to a *pending update* list.

The function $E_\sigma()$ is a function which rewrites an expression where the variables are replaced with their values in the current substitution set (which are potentially complex expressions containing only input variables).

The pending update list may contain at most one entry for each v (whose type must be `reg`) and this list is also converted to RTL assignments upon completion of the block compilation.

Warnings are issued if any variable is assigned more than once, for instance via both a blocking and non-blocking assignment in the same section of code. Assignments to the same variable from separate `always` blocks results in the output net being driven by more than one gate or flip-flop, which is spotted and flagged in the final net list cross-checking phase of CSYN.

---

[1] CSYN will next be extended to support asynchronous resets to its behavioural registers. Currently flip-flops which require this must be structurally instantiated. CSYN does not yet generate transparent latches when a signal is missing from the support list, but this is no great loss, since most users of other Verilog systems run with this option disabled.

2.3 An example of CSYN input and output

Here is an input file:

```
module ADDER(clock,in1,in2,out);
  parameter size = 4;
  input clock;
  input [size-1:0] in1, in2;
  output [size:0] out;
  reg [size:0] out;
  always @(posedge clock) out <= in1 + in2;
endmodule
```

Which can be compiled to give the following output

```
// CBG CSYN Verilog hdl system.  Release 2 Beta 5. (May 95)

module ADDER(clock, in1, in2, out);
  wire u10057, u10056, u10055, u10054, u10053, u10052, u10051,
       u10050, u10049, u10048, u10047, u10046, u10045, u10044,
       u10043, u10042, u10041, u10040, u10039, u10038, u10037,
       u10036, u10035;
  output [4:0] out;
  input [3:0] in2;
  input [3:0] in1;
  input clock;
  AND2  u10055(u10055, in1[2], in2[2]);
  AND2  u10056(u10056, u10047, u10048);
  OR2  u10057(u10057, u10055, u10056);
  BUF  u10035(u10035, u10057);
  XOR2  u10053(u10053, in1[0], in2[0]);
  DFF  u10054(u10054, u10053, clock, 1, 0, 0);
  BUF  u10058(out[0], u10054);
  XOR2  u10051(u10051, u10044, u10045);
  DFF  u10052(u10052, u10051, clock, 1, 0, 0);
  BUF  u10059(out[1], u10052);
  AND2  u10043(u10043, in1[1], in2[1]);
  AND2  u10044(u10044, in1[0], in2[0]);
  XOR2  u10045(u10045, in1[1], in2[1]);
  AND2  u10046(u10046, u10044, u10045);
  OR2  u10047(u10047, u10043, u10046);
  XOR2  u10048(u10048, in1[2], in2[2]);
  XOR2  u10049(u10049, u10047, u10048);
  DFF  u10050(u10050, u10049, clock, 1, 0, 0);
  BUF  u10060(out[2], u10050);
  XOR2  u10041(u10041, u10035, u10037);
  DFF  u10042(u10042, u10041, clock, 1, 0, 0);
  BUF  u10061(out[3], u10042);
  AND2  u10036(u10036, in1[3], in2[3]);
  XOR2  u10037(u10037, in1[3], in2[3]);
  AND2  u10038(u10038, u10035, u10037);
  OR2  u10039(u10039, u10036, u10038);
  DFF  u10040(u10040, u10039, clock, 1, 0, 0);
  BUF  u10062(out[4], u10040);
endmodule
```

2.4 Two examples of non-synthesisable modules

CSYN cannot synthesise the following useful phase-frequency comparator because the registered signals are updated in an edge sensitive way by more than one 'clock'. A good implementation of this circuit uses just 12 NAND gates, but the algorithm to generate this type of circuit is a research topic.

```
module PHASEFREQ(ref, loc, faster, slower);
  output faster;  // High when local oscillator is slower than ref
  output slower;  // High when local oscillator us too fast.
  input ref;    // Reference oscillator
  input loc;    // Local oscillator
  reg faster, slower;
  wire idle = ~(faster | slower);
  always @(posedge loc)
        begin
        if (idle) slower <= 1;
        faster <= 0;
        end
  always @(posedge ref)
        begin
        if (idle) faster <= 1;
        slower <= 0;
        end
endmodule
```

In addition, CSYN cannot synthesise the following pattern generator because there is an implied thread of control. The algorithm which would generate the flip-flops required for the program counter has not been written, and is perhaps difficult.

```
reg din;
initial din = 0;
always
      begin
      @(posedge clk) din = 1;
      @(posedge clk) din = 1;
      @(posedge clk) din = 0;
      end
```

2.5 Transduction Minimisation

Philip Abbey of the Computer Laboratory has implemented the 'Transduction' technique for multi-level logic minimisation [4]. The input and output to his program are vnl files of equivalent functionality but where the output uses fewer gates (or is optimised under another metric). The transduction technique is applied heuristically to the combinatorial logic components of a circuit, leaving the flip-flop structure intact, and operates using the concept of 'permissible functions'. The permissible function at the output of a logic gate inside a multi-level logic circuit is simply a truth table indexed by the inputs to the combinatorial network and contains don't cares (X's). Heuristic modifications to the network are applied to try to increase the number of don't cares, with the effect of increasing the probability that a given gate's output can either be covered by one of its inputs or by the output of another gate in the circuit. In either case, the gate has become unnecessary and so its output is replaced with a wire to the appropriate point.

Using the Transduction method seems to reduce the CLB count for a Xilinx target by about 10 percent. This improvement is on top of the optimisation already provided by the gate-sharing algorithm described in section 2.1, which is also about 10 percent. This shows that the tools supplied with the FPGA's can sometimes be enhanced using such techniques. I will present some figures at the workshop.

Table 2.5 shows parameters for a set of five Xilinx designs, each compiled for an Xc3042 FPGA which has 144 CLBs.

Design	Lines of Verilog	VNL lines	Combinatorial gates	D-types	CLBs
AAL-10	179	2459	1858	32	80
Nasty-J	234	1850	962	70	188
Jaglink	1092	1478	976	126	124
Xcoder	838	1579	963	142	136
Cdxlx	389	1376	563	71	62

Table 2. Parameters for a few Xilinx designs.

3 ECAD Practical Classes

The CSYN compiler has been used for teaching Verilog as part of a second year course at the Computer Laboratory. The students had available the Xilinx back end tools via a batch email server. They experienced design turn around times of about 3 hours on this batch queue, and so it was rather like submitting a design for foundry fabrication. The Xilinx tools returned to them the `lca` (logic cell array) file which they could load into our Xilinx teaching cards [3]. These cards have an FPGA and switches, LEDs and other peripherals. A local program, `lcatov` will convert a `lca` file into a Verilog module containing detailed post-layout timing information for each signal. This enabled the students to perform annotated post layout simulation to see how much slower their designs will run.

The largest designs performed by the students were lift controllers which required four Xc3064 devices. A typical design was a reaction timer or other game, using 150 or so CLBs (configurable logic blocks). As a final year project, one student implemented the ARM microprocessor.

4 The CSIM simulator

To date, the design flow mostly used with CSYN is to compile everything into Cambridge HDL and then perform simulations using XSIM [5]. The XSIM simulator consists of 33000 lines of *C* code and is a robust and easy to use tool. However, recently, Mike Gordon of the Comuter Laboratory conducted a set of experiments on three commercial Verilog simulators (Verilog-XL from Cadence, ViperFree from InterHDL and Veriwell from Wellspring Solutions) in order to deduce a formal semantics for the execution model of Verilog [6]. The IEEE draft standard was also helpful [7].

The author took these semantics and implemented the CSIM simulator, based on XSIM. The performance of CSIM was compared against some other simulators for the test program in figure 4 and the resulting execution times on a SPARCstation 10 Model 30 are given in table 4. CSIM is clearly competitive in execution speed, but it is also very easy to use with its built-in interactive GUI.

Veriwell and Viperfree do not support the `vectored` Verilog range expansion mode which causes the simulator to model a bus as a single number, making the simulation run faster, but with restrictions on assignments to just parts of busses. However it is possible with all of the simulators to change the definition of `ctr` to be an integer. This reduced the simulation time by about 10 percent for Veriwell and Viperfree.

```
module CLK1MHz(o);
  output o;
  reg o;
  initial begin o = 0; forever o = #100 ~o; end
endmodule


module SIMSYS();
  wire x;
  reg [20:0] ctr;
  initial ctr = 0;
  CLK1MHz clk(x);
  always @(posedge x) ctr <= ctr + 1;
  initial #10_000_000 $finish;
endmodule
```

Fig. 1. Example program used to compare the simulators.

Simulator	Behavioural	Gate Level
Verilog XL	9	
Veriwell	8	301
Viperfree	26	290
Xsim	n/a	89
CSIM	7	180

Table 3. Execution time in seconds of the test code before and after compilation on five simulators.

5 Future Directions

Currently we are using different rules for compiling and interpreting Verilog. These rule sets cover different subsets of the (richish) Verilog language. However, we observe that once the semantics of a language are well formulated, it becomes a fairly simple matter to implement the associated compilers and interpreters, and so as semantics develop to cover a greater subset of the language, we can make our tools more powerful.

We have started the Verilog Formal Equivalence Project [8], funded by EPSRC to automate equivalence checking between different versions of a module (in VHDL these would be known as alternative *architectures*). We are especially interested in checking the equivalence of the behavioural and gate-level versions which are perhaps the input and output respectively of a synthesiser, or perhaps have come from different source files with hand recoding or isolated development. Our approach will be to define semantics for the language which can, as far as possible, be used both for compilation and simulation.

Aknowledgements are due to Olivetti Research Ltd, John Porter, Andy Harter, David Milway, Mark Hayter, Ian Pratt. Xilinx © is a trademark of the Xilinx Corporation. Verilog © is a trademark of Cadence Design Systems.

References

1. Newson, A., Milway D.: *Cambridge HDL and FDL Reference Manual* http://www.cl.cam.ac.uk/users/djg/ecadteach
2. Greaves, D.J.,: *The* CSYN *Verilog Compiler and other tools - Professional Reference Manual.* http://www.cl.cam.ac.uk/users/djg
3. Temple, S.,: *The Xilinx Teaching Board* http://www.cl.cam.ac.uk/users/djg/ecadteach
4. Muroga, S., Kambayashi, Y., Lai, H.C., Culliney, J.N.,: *The Transduction Method - Design of Logic Networks Based on Permissible Functions.* IEEE Transactions on Computers, Vol 38 No. 10 October 1989.
5. Newson, A., Milway D.: *The ORL XSIM Simulator Manual* On http://www.cl.cam.ac.uk/users/djg/ecadteach
6. Gordon, M.J.C.,: *The Semantic Challenge of Verilog HDL.* Tenth Annual IEEE Symposium on Logic in Computer Science (LICS'95), June 26-29, 1995, San Diego, California. On http://www.cl.cam.ac.uk/users/mjcg/Verilog.
7. IEEE 1364. *Section 5 - Scheduling Semantics. Draft Standard Verilog HDL.* Draft standards document. On http://www.cl.cam.ac.uk/users/mjcg/Verilog.
8. Greaves, D.J., Gordon, M.J.C,: *Checking Equivalence Between Synthesised Logic and Non-Synthesisable Behavioural Prototypes.* A three year EPSEC research project. On http://www.cl.cam.ac.uk/users/mjcg/Verilog.

A VHDL Design Methodology for FPGAs

Michael Gschwind, Valentina Salapura
{mike,vanja}@vlsivie.tuwien.ac.at

Institut für Technische Informatik
Technische Universität Wien
Treitlstraße 3-182-2
A-1040 Wien
AUSTRIA

Abstract. As synthesis becomes popular for generating FPGA designs, the design style has to be adapted to FPGAs for achieving optimal synthesis results. In this paper, we discuss a VHDL design methodology adapted to FPGA architectures. Implementation of storage elements, finite state machines, and the exploitation of features such as fast-carry logic and built-in RAM are discussed.
Using the design style described in this paper, small changes in the VHDL code can lead to dramatic improvements (a factor of 4), while optimizing key parts to the specific FPGA technology can reduce resource usage by more than a factor of 50.

1 Introduction

FPGAs are an efficient hardware target when only small series are needed, or for rapid prototyping. The FPGAs are complex enough to implement more than glue logic, including complex designs up to several thousands gates. As the logic capacity of FPGAs increases, synthesis for FPGAs is becoming more important.

To efficiently exploit increased logic capacity of FPGAs, synthesis tools and efficient synthesis methods for FPGAs targeting become necessary. One solution to designing large designs efficiently is to use VHDL [IEE88] synthesis. Several synthesis tools exist for mapping these descriptions to various FPGA families.

Using a synthesis-based approach, retargeting a design to other technologies becomes possible at little extra cost. Thus, synthesis is attractive for designing chips with small series and for rapid prototyping. When using FPGAs for rapid prototyping, synthesis can be targeted at FPGAs to exercise it for verification purposes, and later an ASIC implementation can be derived.

By using synthesis tools, the modeling, verification and implementation processes can be integrated. The major advantage of synthesis-based designs is that the same hardware description language code can be used for verification and implementation. This integrated design flow reduces the amount of code that has to be maintained and the risk of inconsistencies between different models.

Once the functional correctness of the model has been proved, the same code should be usable to generate a hardware implementation. Ideally, this process would require only recompilation with a silicon compiler to yield the final chip.

In reality, synthesis is a much longer process: the circuit description has to be evolved to a form suitable for synthesis (certain constructs are illegal for synthesis, etc.). This process is a gradual one, where components can be replaced one by one, verifying that the resulting implementation is correct.

While ideally, the synthesizable VHDL model should be the same for all target technologies, the efficiency of the resulting design is very much dependent on the description and technology used. A discussion of VHDL description efficiency for various constructs can be found in [Sel94]. These results were obtained empirically by generating various descriptions for the same semantic operation, compiling them and comparing their timing and area characteristics. In this paper, we apply this approach to FPGA synthesis.

Designing with FPGAs, one of the major differences is that logic functions of the same size cannot be traded: there is a given number of every resource, and whether it is used or not will not change chip size. On the other hand, trading a 'cheaper' (less complex) cell for a more 'expensive' (more complex) one can actually improve the device budget, if there is an ample amount of the more expensive resource available.

We discuss design strategies for generating efficient VHDL models for FPGA synthesis. These results were collected during several projects [Mau95], [Wal95], [Jau94], [SW94], [SWG94].

This paper is organized as follows: in section 2, we describe the environment used to collect the data. Section 3 and 4 discuss the implementation of storage elements and their selection, respectively, and section 5 gives the optimization of finite state machines. In section 6, we show how to optimize design to use FPGA special purpose features, such as fast-carry logic or on-chip RAM. We draw our conclusions in section 7.

2 Environment

The experiments described here were made using the Xilinx XC4000 FPGA series [Xil94a]. We have chosen this architecture mainly for tool and support availability, but also because they are a very versatile and advanced FPGA technology.

The data presented here were collected using the Synopsys VHDL design analyzer/FPGA compiler (versions 3.1a–3.3a) [Syn93], the XSI Xilinx/Synopsys interface [Xil94c] and X-BLOX as cell generator (XACT 5.1). Synopsys synthesis and XACT were targeted at a Xilinx XC4013mq240-5 FPGA. For low-level operations, we use the XACT and Viewlogic/Powerview tools for analysis and simulation [Xil94b], [Vie94a].

The code for various test circuits was written in VHDL, using the IEEE Std_Logic_1164 package [IEE93]. This package is used in most new VHDL synthesis tools and ensures code portability between tools from different vendors. The synthesis syntax for a given function block may also depend on the tool. The syntax given in this paper was tested using the Synopsys Design Analyzer.

The underlying optimizations described here are not restricted to a particular source code format. Thus, they are not restricted to VHDL, but apply equally well to other hardware description languages such as Verilog [TM91], [SST90]. In fact, many synthesis tools are independent of the source language and have front-ends for both VHDL and Verilog, as well as other special-purpose formats for lookup tables, state machines, etc.

We have also investigated the Powerview ViewSynthesis [Vie94c] VHDL synthesis system, which we had hoped would offer an integrated environment. However, the Powerview ViewSynthesis compiler has a *very* limited number of synthesizable VHDL constructs [Vie94b] and does not support the IEEE Std_Logic_1164 package.

3 Sequential Elements

In Xilinx FPGAs, each CLB contains 2 flip-flops which often go unused. Using such a flip-flop does not use any extra resources in most cases, as it will be located in the block which computes the result.

Latches have to be built using CLB function generators and require 1 CLB per bit (whereas a flip-flop normally comes for free).

Sometimes, of course, the exact functionality of either a flip-flop or a latch is required. In these cases, the appropriate type of storage element has to be used. But when the exact nature of the storage element is of minor importance, flip-flops are obviously beneficial.

The VHDL description methods for latches and flip-flops can be found in figures 2 and 1, respectively.

```
ARCHITECTURE flip_flop OF storage IS
  SIGNAL stored_value : bus;
BEGIN
PROCESS (write)
  BEGIN
  IF write'EVENT AND (write = '1') THEN
    stored_value <= value_in;
  END IF;
END process;
  value_out <= stored_value;
END flip_flop;
```

Fig. 1. VHDL description for flip-flop.

```
ARCHITECTURE latch OF storage IS
  SIGNAL stored_value : bus;
BEGIN
PROCESS (write,value_in)
  BEGIN
  IF (write = '1') THEN
    stored_value <= value_in;
  END IF;
END process;
  value_out <= stored_value;
END latch;
```

Fig. 2. VHDL description for latch.

4 Signal Selection

Selecting between two input signals is a common operation. Many conditional VHDL statement will generate a multiplexer to choose between different input sources:

```
IF (sel = '0') THEN
  out <= signal_0;
ELSE
  out <= signal_1;
END IF;
```

But multiplexers can also be introduced with other constructs, where it is less obvious. For example, choosing a particular input source with an index will normally generate a multiplexer (figure 4). These multiplexers grow with the number of input signals and signal width.

Multiplexers are expensive to implement in FPGAs, as their implementation requires many CLBs and routing resources. An alternative method of selecting an input signal from several options is to use a tri-state bus. This method is advantageous on Xilinx FPGAs, as tri-state buffers and tri-state buses (in the form of longlines) are already integrated on the chip [Wal95].

Tristate devices can be generated using the following assignment:

```
bus <= value WHEN enable ELSE (Others => 'Z');
```

Using tristate drivers, a similar signal selection can be implemented with a tristate bus (see figure 5). Depending on the FPGA part, Xilinx supports between 16 and 64 three-state busses ("longlines") per chip and between 10 and 34 tri-state buffers per longline. If these longlines are not used by any other circuitry, using them for signal selection allows to pack more functionality in a single FPGA.

```
ENTITY select IS
PORT    (
  source    : IN ARRAY (depth -1 DOWNTO 0) OF bus;
  sel       : IN std_logic_vector (log2depth -1 DOWNTO 0);
  value_out : OUT bus;
)
END select;
```

Fig. 3. Entity declaration for select.

```
ARCHITECTURE mux OF select IS
BEGIN
  value_out <= source (conv_integer(unsigned(sel)));
END mux;
```

Fig. 4. Signal selection using a multiplexer.

```
ARCHITECTURE tristate OF select IS
BEGIN
  tri_state_bus:
    FOR i IN 0 TO depth -1 GENERATE
      value_out <= source (i) WHEN (i = conv_integer(unsigned(sel)))
                   ELSE (Others => 'Z');
    END generate;
END tristate;
```

Fig. 5. Signal selection using a tri-state bus.

Tables 1 and 2 compare the FPGA resource usage for signal selection using multiplexers and tristate buffers, respectively. For narrow signals (1 or 2 bits), CLB usage is comparable. For wider signals, a selection mechanism based on tri-state functionality is preferable: the tristate implementation uses a fixed number of CLBs for generating tri-state buffer control signals, and a tri-state buffer for each signal bit, i.e. $n * w$ tri-state devices (n being the number of signals, w the width of the signal in bits). Often, these tri-state resources are unused, so this implementation increases overall FPGA resource utilization.

One caveat is the number of tri-state resources (buffers and longlines) which are available and their connectivity. Since the connectivity of tri-state buffer elements (TBUF) is fixed, and there is a limited number of longlines, there are upper bounds as to the size of the tri-state select mechanism. For example,

		bus width					
		1	2	4	8	16	32
no. of signals	2	1	1	2	4	8	32
	4	1	2	4	8	16	32
	8	3	5	14	24	44	84
	16	5	10	28	48	88	168
	32	11	26	61	103	187	355

Table 1. This table shows the number of CLBs required for selecting an output signal with multiplexing logic, as a function of the number of input signals and signal width. These results were reported by FPGA compiler 3.1a.

		bus width											
		1		2		4		8		16		32	
		CLBs	tri	CLBs	tri	CLBs	tri	CLBs	tri	CLBs	tri	CLBs	tri
no. of signals	2	1	2	1	4	1	8	1	16	1	32	1	64
	4	2	4	2	8	2	16	2	32	2	64	2	128
	8	4	8	4	16	4	32	4	62	4	128	4	256
	16	8	16	8	32	8	64	8	128	8	256	8	512
	32	19	32	19	64	19	128	19	256	19	512	19	1024

Table 2. This table shows the number of CLBs and tristate buffers required for selecting an output signal using a tristate bus, as a function of the number of input signals and signal width. These results were reported by FPGA compiler 3.1a.

the XC4010 has 40 longlines and 22 TBUFs per longline, restricting the select mechanism to a width of 40 bits (if all longlines are dedicated to a single select) and the output can be selected from a maximum of 22 signals. (In the XC4000 series, horizontal longlines can be split, so the XC4010 can also be configured as having up to 80 longlines with 11 TBUFs per longline.)

The choice of output selection mechanism has a significant influence on the size of all blocks where signals have to be selected, e.g. register files.

5 State Machines

The generation of state machines is another area where conventional ASIC synthesis and FPGA synthesis differ. When ASICs are the target technology, fully encoded representations such as binary or gray code encoding of states lead to space efficient designs, whereas the faster one-hot encoding scheme consumes more resources [Syn95b].

This is different in FPGA designs, where the state decoding logic for decoding a binary encoding would consume many CLBs, while many flip-flops on the same

	XC4000		*LSI 10K*	
FSM Encoding	time (ns)	space (CLBs)	time (ns)	space (units)
one-hot, space	18.2	8	17.0	73
one-hot, time	18.2	8	11.4	95
auto, space	56.8	7	13.9	46
auto, time	33.7	7	11.1	68
gray, space	26.6	7	18.1	58
gray, time	32.7	8	9.7	28
binary, space	52.3	10	17.2	54
binary, time	43.1	13	12.6	92

Table 3. Resource usage for FSM compilation using different encoding schemes and optimization constraints. These results were reported by FPGA compiler and design compiler, respectively (version 3.1a).

die go unused! Thus one-hot encoding is not only much faster, but also the more compact representation [AN94].

Table 3 gives the synthesis results of a simple finite state machine for the Xilinx XC4000 series and the LSI 10k ASIC library [Syn95a]. The table compares four encoding techniques available in Synopsys: one-hot encoding, a solution adapted to the particular FSM (auto), gray code encoding, and binary encoding of states.

The one-hot encoding scheme uses 6 flip-flops for state encoding, while all other implementations use 3. Although this leads to significantly larger ASIC implementations, the FPGA FSM implementation is comparable to the smallest solution. While the optimal encoding always depends on the particular state machine being used, for most state machines one-hot encoding is superior for FPGA implementations. One-hot encoding not only is the fastest encoding, but also one of the smallest representations because it exploits the availability of many flip-flops on an FPGA.

In some tools, VHDL source level encoding of the state vector may be necessary to achieve this. Synopsys supports the extraction of state machines from a design and to define an encoding to be used for the state vector. This approach is advantageous, since several different encodings can be tested and compared without having to modify the source code.

6 Usage of Xilinx primitives

It is difficult for VHDL compilers to use special purpose features which are available in FPGAs under certain conditions, such as the fast-carry logic or the builtin RAM.

Xilinx provides a partial solution to this problem by supplying a DesignWare

storage element	selection scheme	area (CLBs)	time (ns)
latches	multiplexers	570	88.2
latches	tri-state bus	402	99.8
flip flops	multiplexers	283	58.7
flip flops	tri-state bus	156	51.5
XC4000 RAM capability		8	29.2

Table 4. Occupied CLBs using five different VHDL coding strategies for a 16x16 scratch pad RAM. The test circuits were synthesized using FPGA compiler 3.2a, and routed using **ppr** (XACT 5.1). Area results as reported by **ppr**. Timing results are pad-to-pad delays as reported by **xdelay** and include the propagation delay of input and output pads. Thus, relative speed differences are more pronounced than they may appear here.

library for adders, subtracters, counters, and comparators. In Synopsys, DesignWare libraries are used to implement common, complex functional units which can be used by the design analyzer. In the X-BLOX DesignWare library, these functional units are not actually implemented using the Synopsys FPGA compiler. Instead, references to X-BLOX modules are inserted in the netlist. When the design is post-processed for final layout using XACT, X-BLOX is invoked as module generator to synthesize appropriate functional units. X-BLOX has intimate knowledge of Xilinx circuits, so it can generate logic geared to special features such as fast-carry logic.

While using the X-BLOX library helps optimize some circuits, others cannot be optimized to use these features. A solution is to include Xilinx library elements as 'components' in VHDL, and use either already available circuits in the Xilinx libraries, or to generate one's own circuits with XACT[1] or a schematic entry tool to be included in the design. These parts can then be optimized to use all features available on a particular target technology. While this approach requires more radical VHDL source code modification, the huge gains possible make this worthwhile for some resources. Isolating these features in a distinct **ENTITY** will enhance portability and restrict code changes to only a few lines of code. (These resources (RAMs, etc.) will probably also need to be adapted to a specific ASIC process to yield optimal results.)

Table 4 shows how a design can be optimized by using Xilinx XC4000 features. We compare 5 different designs for a 16x16 RAM. Depending on the VHDL description, Synopsys FPGA compiler generates either flip-flops or latches as storage elements, and uses either a MUX-based signal selection scheme or tri-state buses. These implementation specifics are orthogonal, giving four possible implementations. The synthesis results for these four different architectures show

[1] The full range of XACT tools can be used to generate these circuits: X-BLOX, **memgen**, XDE,...

```
ARCHITECTURE xilinx_ram_capability OF scratch_pad IS
  ...
  COMPONENT RAM16X1
  PORT ( D, A3, A2, A1, A0, WE : IN std_logic;
         O : OUT std_logic);
  END COMPONENT;
  ...
BEGIN
  ...
    FOR i IN 0 TO width - 1 GENERATE
      cell_ram16x1: RAM16x1 PORT MAP (
                                      D  => value_in(i),
                                      A3 => addr(3),
                                      A2 => addr(2),
                                      A1 => addr(1),
                                      A0 => addr(0),
                                      WE => write,
                                      O  => value_out(i));
    END generate;
  ...
END xilinx_ram_capability;
```

Fig. 6. VHDL code for generating a scratch pad RAM based on the Xilinx XC4000 series RAM capability. Large RAMs can be assembled using the basic RAM capability available as macros (`RAM16x1`, `RAM32x1`) in the XACT library.

how coding style in VHDL can affect resource consumption, in the example of our 16x16 scratch pad RAM yielding a 4-fold improvement.

A fifth, alternative design is based on the usage of the Xilinx XC4000 capability, where each FG function generator can be reprogrammed to act as 16x1 RAM cell. By using this customization of the RAM cell, the final design uses only 1.4% of the original design.

To generate this design, we use macros from the Xilinx XACT library, which are instantiated as VHDL components in the VHDL description (see figure 6). An alternative way to create RAMs based on the Xilinx RAM capability is to use the `memgen` tool [Xil94b] from the XACT distribution and include the resulting RAM as `COMPONENT`.

7 Conclusion

We have shown that VHDL models are highly dependent on the target technology. A slight modification in the description can cause considerable change in the implementation efficiency, especially when dealing with fixed-resource FPGAs. We have also demonstrated how VHDL descriptions can be optimized to achieve better FPGA resource utilization.

References

[AN94] Peter Alfke and Bernie New. Implementing state machines in LCA devices. In *The Programmable Logic Data Book*, pages 8–169 – 8–172. Xilinx, Inc., San Jose, CA, 2nd edition, 1994. XAPP 027.001.

[GG95] Michael Gschwind and Robert Glock. VHDL Synthesis to FPGAs Using the Synopsys Design Analyzer. Technical report, Institut für Technische Informatik, Technische Universität Wien, Vienna, Austria, 1995.

[IEE88] IEEE. *IEEE Standard VHDL Language Reference Manual.* IEEE, 1988. IEEE Standard 1076-1987.

[IEE93] IEEE. *IEEE Standard Multivalue Logic System for VHDL Model Interoperability (std_logic_1164).* IEEE, 1993. IEEE Standard 1164-1993.

[Jau94] Alexander Jaud. Implementing a Hopfield neuron with VHDL under Powerview. Personal Communication, December 1994.

[Mau95] Dietmar Maurer. Eine Implementation des MIPS R3000 Befehlssatzes in VHDL [An implementation of the MIPS R3000 instruction set architecture in VHDL]. Master's thesis, Technische Universität Wien, Vienna, Austria, 1995. (to be published).

[Sel94] Manfred Selz. *Untersuchungen zur synthesegerechten Verhaltensbeschreibung mit VHDL.* PhD thesis, Universität Erlangen-Nürnberg, Erlangen, Germany, March 1994.

[SST90] Eliezer Sternheim, Rajvir Singh, and Yatin Trivedi. *Digital Design with Verilog HDL.* Automata Publishing, Cupertino, CA, 1990.

[SW94] Valentina Salapura and Günter Waleczek. Designing from VHDL behavioral description to FPGA implementation. In *Proc. of AustroChip '94*, pages 141–146, Brunn am Gebirge, Austria, June 1994.

[SWG94] Valentina Salapura, Günter Waleczek, and Michael Gschwind. A comparison of VHDL and Statecharts-based modeling approaches. In *Proc. of ITI 94*, Pula, Croatia, June 1994.

[Syn93] Synopsys. *Design Analyzer Reference Manual.* Synopsys, Inc., 1993.

[Syn95a] Synopsys. Finite state machine tutorial source code. `${SYNOPSYS3.3a}/doc/syn/examples/fsm/proc2.vhd`, 1995.

[Syn95b] Synopsys. *Finite State Machines – Application Note.* Synopsys, Inc., 1995.

[TM91] Donald E. Thomas and Philip R. Moorby. *The Verilog Hardware Description Language.* Kluwer Academic Publishers, Boston, MA, 1991.

[Vie94a] Viewlogic Systems. *Using Powerview.* Viewlogic Systems, Inc., Marlboro, MA, 1994.

[Vie94b] Viewlogic Systems. *VHDL Reference Manual for Synthesis.* Viewlogic Systems, Inc., Marlboro, MA, 1994.

[Vie94c] Viewlogic Systems. *ViewSynthesis User's Guide.* Viewlogic Systems, Inc., Marlboro, MA, 1994.

[Wal95] Günter Waleczek. Modellierung und Synthese des MIPS/SAB R3223 in VHDL. Master's thesis, Technische Universität Wien, Vienna, Austria, 1995.

[Xil94a] Xilinx. *The Programmable Logic Data Book.* Xilinx, Inc., San Jose, CA, 1994.

[Xil94b] Xilinx. *XACT Reference Guide.* Xilinx, Inc., San Jose, CA, April 1994.

[Xil94c] Xilinx. *XACT Xilinx Synopsys Interface FPGA User Guide.* Xilinx, Inc., San Jose, CA, December 1994.

VHDL-based Rapid Hardware Prototyping Using FPGA Technology

Maziar Khosravipour, Herbert Grünbacher

Institut für Technische Informatik
Vienna University of Technology
Treitlstrasse 3 - 1822
A-1040 Vienna, Austria

email: {maziar, gruenbacher}@vlsivie.tuwien.ac.at

Abstract. This paper presents a powerful methodology for high level synthesis providing maximum flexibility at minimum cost. The concept of *rapid prototyping* is adopted for hardware designs by introducing *high level optimization* as an alternative approach to conventional methods. It is based on a combination of *VHDL* as description language, synthesis tools and *FPGA technology*.

1 Introduction

Two concepts have massively influenced the design processes of electronic devices:

1. Use of VHDL in the design process.
2. Use of field programmable gate arrays (FPGAs).

Combining VHDL for modeling, synthesis tools for technology mapping and FPGA technology for implementing builds a strong basis for rapid hardware prototyping. Unfortunately, the synthesis results heavily depend on VHDL coding primitives and style. This is especially true for FPGA technologies.

Using a hardware component for a telecommunication system as design example we show and compare the synthesis results for two functionally equivalent VHDL models. The VHDL models are functionally equivalent but the VHDL code is different. In both cases the VHDL code is technology independent and easily portable to other ASIC technologies. We used a special way of describing register files in VHDL, which reduces area in FPGA technology (i.e. number of CLBs) dramatically.

Section 2 describes the VHDL design flow for synthesis, and the design environment.

In section 3 we present the way we described the register file in VHDL and the results of technology mapping onto Xilinx FPGA Series XC4000™ and XC3000™ [1].

Section 4 summarizes our results.

2 VHDL Design Flow and „High Level Optimization"

High level synthesis is the automatic synthesis of a design structure from a behavioral specification, at levels above and including the logic level [2]. It bridges the gap between behavioral specifications and their hardware structure. The behavior is usually specified in a hardware description language. The initial structure of the synthesized logic is directly inferred from the structure of the hardware description. The quality of a design is very sensitive to the HDL coding style.

VHDL based synthesis makes it possible to synthesize a higher level description into lower level nets and map the nets to a specified technology. The description at the higher level of abstraction provides high independence of the target technology until the last steps of development. Because the description is using a computer language, software engineering methods become feasible for hardware development projects. Using different description methods for the same functionality or changing the abstraction level create different VHDL models of the same design. By creating different descriptions it is possible to „steer" the synthesis tool and find better implementations of the gate level by modifying the VHDL code at the higher level. Thus, we call this approach „High Level Optimization".

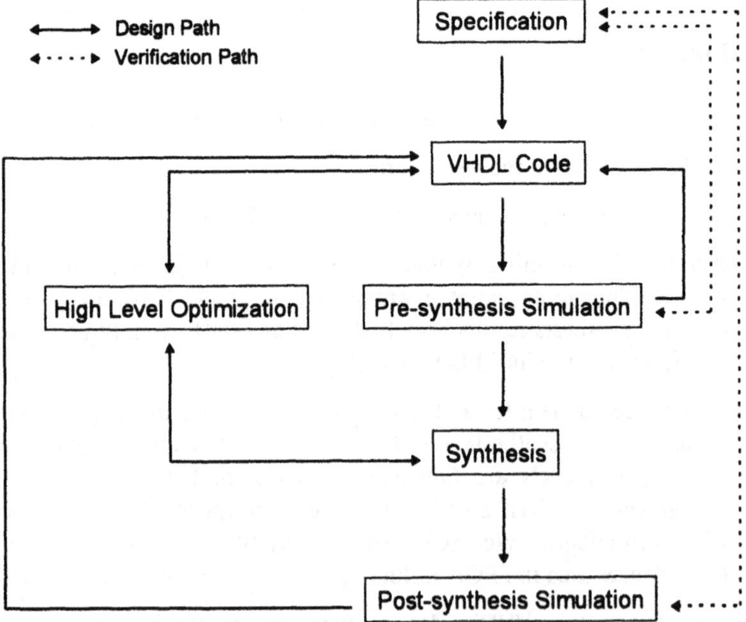

Fig. 1. VHDL Design Flow for Rapid Prototyping

The main focus of High Level Optimization is the exploration of the design space to find a good solution. This task is done by developing different VHDL models of the same functionality and comparing the results of automatic synthesis. Figure 1 shows the design flow.

In addition to this high level optimization there is still optimization done by the synthesis tool using algorithms for resource allocation and scheduling.

A powerful concept adopted from software engineering becomes applicable as the hardware design is in essence done in software: rapid prototyping. In rapid prototyping, the designer does not start with a full VHDL description of the specification but only with parts of it. Step by step, additional parts of the specification („modules" as they are called by software engineers) are described in VHDL, processed and integrated into the already developed modules. By downloading the compiled results into FPGAs, the hardware described in VHDL is easily verifiable with respect to the target environment. The success of this method highly depends on the way the specification is refined. Each module is described in VHDL and synthesized by the tools automatically.

Another alternative to optimize the implementation of a design described at a higher level of abstraction is to develop „yet another synthesis tool". In the FPGA case, optimized fitters should be used to implement logic functions into look-up tables directly.

2.1 Limitations of FPGA Technology with Respect to High Level Synthesis

FPGAs are rapidly gaining popularity due to the short design cycle time and low manufacturing cost. Since the introduction of the FPGA technology, the use of new advanced CMOS process technologies as well as architectural improvements have contributed to a large increase of FPGA capabilities (up to 25,000 gates and clock rates up to 50 MHz [1]), making this technology useful for a wide range of applications as a cost-effective alternative to other technologies.

The granularity of the FPGA devices and the complexity of the logic blocks turn to a handicap in high level synthesis. Since FPGAs consist of few configurable element types, the automatic mapping of logic functions to these elements may be very expensive in terms of resource utilization. For instance, consider the Configurable Logic Block (CLB) of a Xilinx FPGA. The combinatorial logic portion of the CLB uses a 32 by 1 look-up table to implement Boolean functions. This technique can generate two independent logic functions of up to four variables each or a single function of five variables.

In design environments with schematic entry, there is a limited number of library primitives available to the designer fitting into a single CLB. For our example, any function of up to 5 variables will fit into one CLB. For functions with 6 or more variables, the designer can either select macros (taking more than one CLB) or wire the primitives manually.

In high level synthesis, all these mapping details are not visible to the designer. Describing the hardware at a higher level of abstraction, thus, may lead to a very expensive implementation of the desired component.

Table 1 gives an overview of the Xilinx FPGA families.

Parameter	XC4013	XC3090/3190	XC2018
Number of flip-flops	1,536	928	174
Max. number of user I/O	192	144	74
Function generators per CLB	3	2	2
Number of logic inputs per CLB	9	5	4
Number of logic outputs per CLB	4	2	2

Tab. 1. Xilinx FPGA Families [1]

In the next section, we present the way we have optimized a register file description for FPGA technology. Related works for state machine synthesis are done by [3].

3 Describing Register Files for FPGAs

Our register file consists of registers of 8 bit width. The depth of the register file is between 2 and 16. The registers are addressed via address lines `ADR`, the chip is selected by the signal `CS` and the signal `RW_N` is used as Write Enable (Write is low active). The registers are written synchronously and read asynchronously. If `CS` or `RW_N` is low, the outputs of the register file are high impedance. The entity description of the register file (with 16 registers) is shown in Figure 2.

```
entity reg16 is
      port  (adr   : in std_logic_vector(3 downto 0);
             d_in  : in std_logic_vector(7 downto 0);
             cs    : in std_logic;
             rw_n  : in std_logic;
             clk   : in std_logic;
             ------------------------------------------
            d_out : out std_logic_vector(7 downto 0)
             );
end reg16;
```

Fig. 2. VHDL Entity Description of the Register File

The conventional implementation of register file uses decoders and multiplexers to select the registers to be read or written. This approach is very simple and straightforward but a significant part of the FPGA resources is used for decoding and selecting the registers as shown in Table 2.

We consider the addressing mechanism of a register file with 16 registers, a chip enable pin and a write enable input. The 6 inputs (CE, WE and 4 address lines) must be combined to select a register. The conventional implementation of the selection logic using decoders and multiplexers requires many CLBs which are not fully utilized because the flip-flops in these CLBs are not in use, nevertheless, all input and output pins of the CLB are occupied.

The architecture description for this approach is shown in Figure 3.

```
architecture mux_approach of reg16 is
   subtype reg is std_logic_vector (7 downto 0);
   type reg_file is array (0 to 15) of reg;
   signal r_rf: reg_file;
begin
REG_W : process
begin
   wait until clk'event and clk = '1';
   if (cs = '1' and rw_n = '0') then
      r_rf(conv_integer(adr)) <= d_in;
   end if;
end process REG_W;
REG_R: d_out <= r_rf(conv_integer(adr))
                         when (rw_n and cs) = '1'
                  else (Others => 'Z');
end mux_approach;
```

Fig. 3. VHDL Architecture Description of Register File (Conventional Approach)

In our approach, we select the output of the register file by switching 3-state buffers. All register outputs are connected to the data bus via these 3-state buffers. The selection of a register is performed by applying the appropriate logic level on its 3-state control line. Figure 4 shows a simple register file implementation using the 3-state approach.

In each Xilinx FPGA device (3000 or 4000 family) a pair of 3-state buffers is located adjacent to each CLB, providing access to the horizontal longlines (Figure 5). 3-state buffer control logic allows the implementation of wide multiplexing functions. Since the 3-state buffers do not use the CLB resources of the device, the area needed for the implementation of the register file is reduced dramatically.

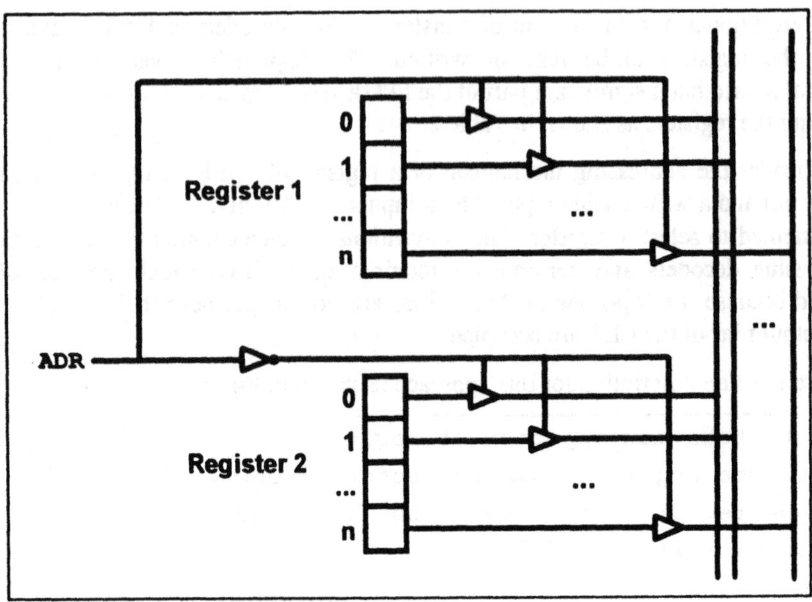

Fig. 4. 3-State Approach for Register File Implementation

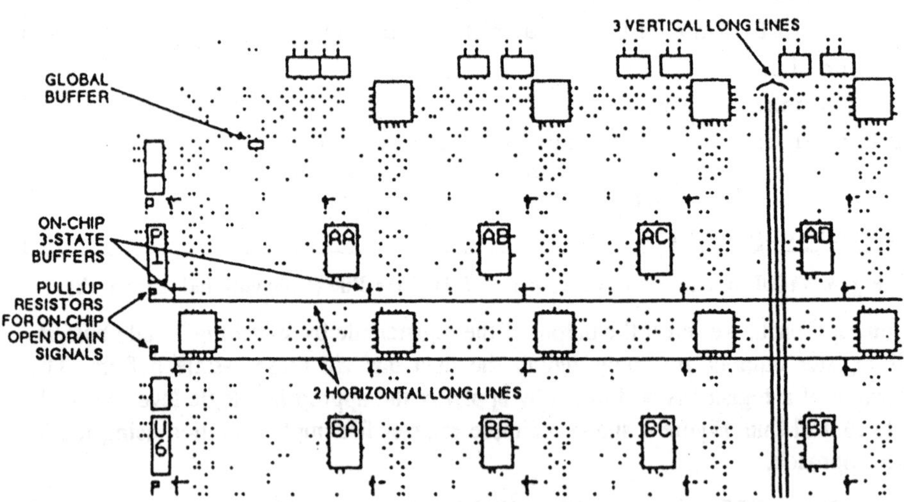

Fig. 5. 3-State Buffers and the Horizontal Longlines [1]

This approach shows an application of „High Level Optimization“, since the optimization is completely achieved by the way the register file is described in VHDL.

Figure 6 shows the VHDL architecture description of this approach.

```
architecture 3state_approach of reg16 is
   subtype reg is std_logic_vector (7 downto 0);
   type reg_file is array (0 to 15) of reg;
   signal r_rf: reg_file;
   signal d_temp : std_logic_vector (7 downto 0);
begin
REG_W : process
begin
    wait until clk'event and clk = '1';
    if (cs = '1' and rw_n = '0') then
       r_rf(conv_integer(adr)) <= d_in;
    end if;
end process REG_W;
d_temp <= r_rf(0) when adr = 0 else (Others => 'Z');
d_temp <= r_rf(1) when adr = 1 else (Others => 'Z');
d_temp <= r_rf(2) when adr = 2 else (Others => 'Z');
...
d_temp <= r_rf(14) when adr = 14 else (Others => 'Z');
d_temp <= r_rf(15) when adr = 15 else (Others => 'Z');
REG_R: d_out <= d_temp when (rw_n and cs) = '1'
                 else (Others => 'Z');
end 3state_approach;
```

Fig. 6. VHDL Architecture Description of the Register File (3-State Approach)

3.1 Results

The register file descriptions were evaluated for a depth between 2 and 16 registers. The design tools employed for this project were the Synopsys VHDL System Simulator™ [4], Synopsys VHDL Compiler™ [5], Synopsys FPGA Compiler™ [6] and Synopsys Design Compiler™ [7]. Technology mapping and resource estimation were done by using the Xilinx XACT™ Development System [8]. We examined the XC3000 as well as XC4000 family of Xilinx FPGA Device (XC3195 and XC4005). The components were clocked with 20 MHz. Table 2 shows the results of our experiments for both approaches.

Depth of Register File	Min. CLB	XC3195 Mux	XC3195 3-State	XC4005 Mux	XC4005 3-State
2	8	14	11	13	10
4	16	37	21	28	20
8	32	69	41	61	41
16	64	144	85	121	81

Tab. 2. Results of High Level Optimization

The number of CLBs used for each approach is also shown graphically in Figure 7.

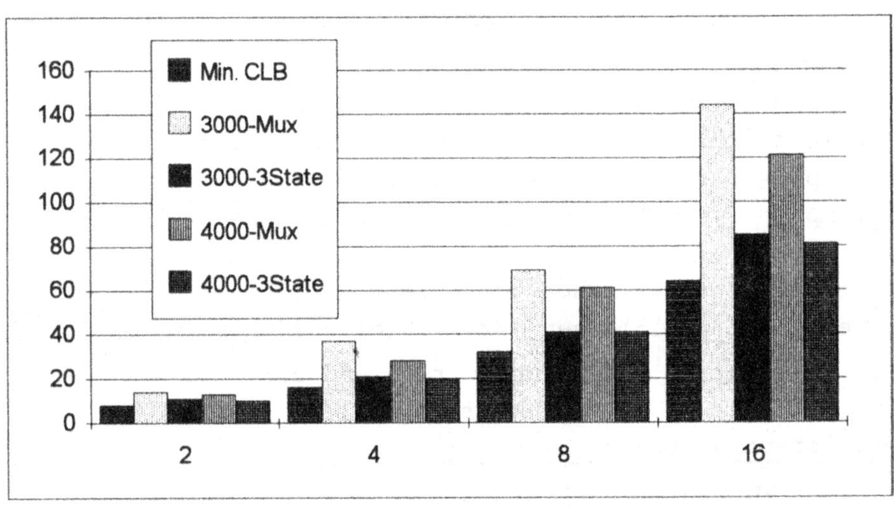

Fig. 7. CLBs used for Implementation of the Register File

4 Conclusions

FPGA technology for prototyping offers many advantages: flexibility, cost-effectiveness, design space exploration at reasonable cost.

VHDL offers good support for synthesis from higher design levels. Because of its description and abstraction capabilities VHDL supports many design methodologies. VHDL is a programming language and therefore supports software engineering methods including top-down design and rapid prototyping. The synthesis aspect is very attractive for long term projects as it becomes easy to map the design onto the latest technology. Reusable VHDL modules can form the basis for open architectures.

By combining VHDL/synthesis and FPGA technology, a powerful methodology for high level synthesis is available to designers providing maximum flexibility at minimum cost. Furthermore, the extension of conventional approaches by „High Level Optimization" opens new perspectives on high level design and development.

We have shown a design flow for rapid prototyping based on programmable gate array technology using the hardware description language VHDL. The results of our project demonstrate the feasibility of high level optimization.

Literature

[1] „The Programmable Logic Data Book", 1994, Xilinx, Inc.

[2] „A Survey of High-Level Synthesis Systems," R.A. Walker, R. Camposano (ed.), Kluwer Academic Publishers, 1991.

[3] „VHDL Code and Constraints for Synthesis," M. Selz, H. Rauch, K.D. Müller-Glaser, Proceedings of VHDL-Forum for CAD in Europe (User Paper Session), Fall'93 Meeting, Germany, 1993.

[4] „Synopsys VHDL System Simulator Reference Manual,", Version 3.3a, 1995, Synopsys, Inc.

[5] „Synopsys VHDL Compiler Reference Manual", Version 3.3a, 1995, Synopsys, Inc.

[6] „Synopsys FPGA Compiler Reference Manual", Version 3.3a, 1995, Synopsys, Inc.

[7] „Synopsys Design Compiler Reference Manual", Version 3.3a, 1995, Synopsys, Inc.

[8] „XACT Development System Reference Guide", 1994, Xilinx Inc.

Integer Programming for Partitioning in Software Oriented Codesign*

Markus Weinhardt

Universität Karlsruhe, Fakultät für Informatik
D-76128 Karlsruhe, Germany
(weinhard@ipd.info.uni-karlsruhe.de)

Abstract This paper presents a new partitioning method for software oriented hardware/software codesign. It is applied to the use of field-programmable accelerator boards. In the underlying model the dedicated hardware has no direct access to the host memory, and communication is slow. Therefore detailed data-flow information is necessary to minimize the communication overhead between host and accelerator board. The partitioning problem is formulated as an integer (linear) program which simultaneously determines which code regions should be implemented in dedicated hardware and which data has to be communicated, so that well-known optimization algorithms can be applied.

1 Introduction

The goal of software oriented hardware/software codesign is to improve system performance by moving time-critical parts of a program to hardware. This approach is attracting increasing interest since it uses simple specifications in common programming languages and allows the use of automatically designed dedicated hardware—especially field-programmable hardware—in applications beyond embedded systems. So FPGA-boards available for PCs or engineering workstations can be used to accelerate common programs.

Figure 1 shows the general design flow. A profiler determines time critical regions of the input program and average sizes of data structures used by these regions. This information is then used to determine a feasible hardware/software partitioning which maximizes the estimated speedup. The code of the regions to be implemented in hardware is transformed to an HDL specification which guides the synthesis of the corresponding circuit. The software part is processed by a conventional compiler, and the interface determines the synchronization and communication between hardware and software.

We assume as a model of computation a coprocessor board with FPGAs and local memory connected to a host computer via the system bus. Host and board can communicate through the bus, but the board cannot directly access the host memory, and communication is slow. Our goal is to select those regions

* This work has been supported by the Deutsche Forschungsgemeinschaft, Graduiertenkolleg "Beherrschbarkeit komplexer Systeme" (DFG Vo 287/5-2).

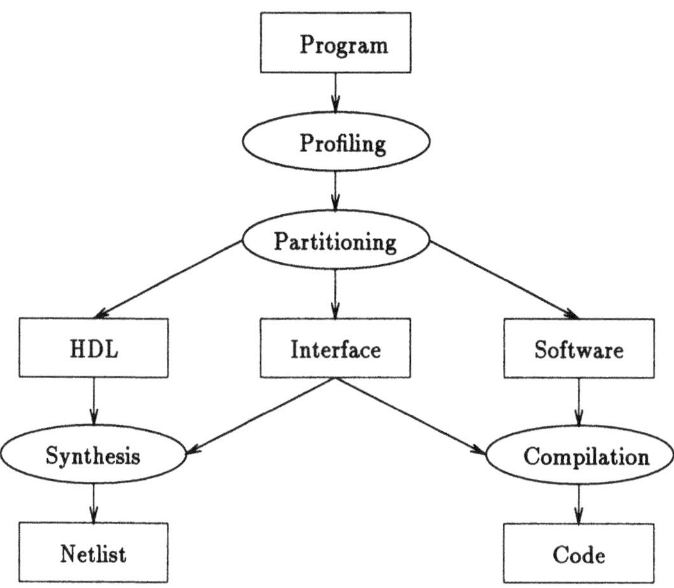

Figure1. Design flow

of the program which yield the highest speedup if moved to hardware. The computation of this partitioning should be efficient and it should consider the overhead of the required communication between host and coprocessor board. The limited host memory access places some restrictions on the regions which can be implemented in hardware, and means that data used in the hardware regions must be explicitly copied to the board. (We allocate scalar variables to FPGA registers and arrays to the local memory.) On the other hand the absence of memory access conflicts allows operating the host and coprocessor board concurrently. Some commercially available boards (e.g. [1]) implement this model of communication.

The following sections present a new partitioning method suitable for this model. We derive a linear cost function and linear constraints which define the partitioning problem formally as an *integer program*. Hence we can use the simplex and branch-and-bound optimization algorithms to compute the partitioning.

2 Preprocessing

2.1 Candidate Selection

One of the most important issues in hardware/software partitioning is the selection of adequate candidates which should be considered for hardware implementation. There is a trade-off between the granularity of the candidate and the feasibility of the optimization process. The smaller the candidate, the greater the number of different partitionings. But the large number of solutions may prevent the computation of the optimum. Larger candidates restrict the number of possible partitionings, but make it feasible to compute an exact solution.

Ernst et al. [2] treat every C statement as a candidate. So even code regions which are not likely to yield a speedup are considered. Because this results in a large number of possible solutions they use a heuristic method (simulated annealing) to determine the partitioning.

Jantsch et al. [3, 4] consider only preselected candidates which can be implemented in hardware and for which a speedup is expected. This allows finding an optimal solution of the partitioning problem using a dynamic programming technique.

We follow Jantsch's definition of candidates [3, p. 97]: A region of code is a candidate

- iff it does not include floating point operations or calls to external library and operating system functions
- and
  - if it is an inner loop or leaf function
  - or it includes only loops and calls to functions that are candidate regions themselves.

```
for (y=0; y <= maxY; y++) {
   for (x=0; x <= maxX; x++)                                  /* H1 */
      same[x] = x;                                            /* H1 */
   for (x=0; x <= maxX; x++) {
      < non-candidate code defining left, right >
      if (0 <= left && right <= maxX) {
         l = same[left];                                      /* S1 */
         < candidate inner loop using l, left, right, same;   /* H2 */
                                 defining same, left, right > /* H2 */
         same[left] = right;                                  /* S2 */
      }
   }
      < candidate inner loop using maxX, same; defining pix > /* H3 */
      < non-candidate code using pix >
}
```

Figure2. Example program structure

Figure 2 shows the structure of an example algorithm. It generates a SIRDS (Single Image Random Dot Stereogram) from a grey–map file that describes a 3D scene [5]. The regions in angle brackets are omitted for clarity. The candidate regions according to the above definition are labeled `H1`, `H2` and `H3`.

We adjust this method to the needs of our model of computation as defined in section 1 in the following way:

- We allow assignments to be added to a region if they use or define variables which also occur in the region. This can significantly reduce the amount of data which must be copied. In the example of Figure 2, additional candidates are derived by adding statement **S1** or **S2** or both to **H2**, because **same** occurs in **H2** and in both assignments.
- Because there is no direct memory access, all references in a candidate must be available at compile time to permit copying the data to the board before the execution of the candidate. This prohibits the use of pointers in candidates and means that arrays in general must be copied completely.
- Jantsch et al. consider only candidates which yield a local speedup. We also consider candidates which speed up the program only when used in combination with other regions (by reducing the communication overhead).

2.2 Dataflow Analysis

From compiler construction theory it is well known that communication needs of program parts can be determined by dataflow analysis. This method can also be used to determine the necessary communication implied by a chosen hardware/software partition. We need to compute global *definition–use chains* [6, p. 632] of the variables in the candidates. They establish a relation between the definition of a variable (i.e. an assignment to the variable) and the places where this value is used (i.e. evaluations of the variable without interleaving redefinitions). The chains lead to the following constraints for a region H implemented in hardware:

- If variable v is defined in region H and used in software, then v must be copied *to the host after* execution of H.
- If variable v is defined in software and used in H, then v must be copied *to the coprocessor board before* execution of H.

3 An Integer Program for Partitioning

In the following we assume that the candidate–regions and definition–use chains have been computed as defined in section 2. Now we intend to solve this partitioning problem:

> Find the subset of candidate regions to be implemented in hardware which yields the greatest estimated overall program speedup. The estimate must consider the hardware speedup, the implied communication overhead and the limited amount of hardware available.

Since a definition–use relation may exist between all candidates, the decision to implement a region in hardware can influence the necessary copy operations of all other candidates. Hence this problem is not decomposable. So we have to find an overall optimum considering all candidates simultaneously. Because the number of candidates is restricted, it is feasible to find an exact solution by representing the problem as a *0, 1 integer program* [7, ch. 4.6] and solving the program with standard methods. A 0, 1 integer program is defined by a real $m \times n$ matrix A, a real m-vector b, and a real n-vector c. Solutions are all vectors $x \in \{0,1\}^n$ that satisfy the linear constraints $A \cdot x \leq b$ and $x \geq 0$ and that minimize the linear cost function $c^T \cdot x$.

3.1 Notation

I	set of candidate indexes
$\{H_i\}_{i \in I}$	set of candidate regions
$I_0 = I \cup \{0\}$	extended set of candidate indexes
$\{H_i\}_{i \in I_0}$	set of candidate regions including pseudo-region
H_0	pseudo-region representing code always implemented in software
$call(i)$	estimated number of calls of H_i
$HWtime(i)$	estimated execution time of H_i if implemented in hardware
$SWtime(i)$	estimated run time of H_i if implemented in software
$HWarea(i)$	estimated hardware resources (number of logic blocks) needed for H_i
$HWres$	available hardware resources (number of logic blocks)
V	set of variables used in candidates
$size(v)$	size of $v \in V$ in bytes (for arrays their estimated average size is used)
$Ttime$	time needed to copy a byte between host and board
$v \triangleleft H_i$	$v \in V$ occurs in H_i

Table1. Notation

Table 1 defines the required notation. The definition–use chains are represented by a predicate $DU(i,j,v)$ for $i,j \in I_0$, $v \in V$:

$$DU(i,j,v) \Leftrightarrow v \text{ is defined in } H_i \text{ and used in } H_j$$

In order to treat all dependencies uniformly, a pseudo-region H_0 represents the parts of the program always implemented in software. For regions which are not disjoint we must remove intra-region dependencies from DU.

The partitioning problem is represented by these binary variables:

$$\forall i \in I_0 : x_i = \begin{cases} 1, & \text{if } H_i \text{ is implemented in hardware} \\ 0, & \text{otherwise} \end{cases}$$

$$\forall i \in I, v \in V, v \triangleleft H_i : in_{i,v} = \begin{cases} 1, & \text{if } v \text{ must be copied to the board before} \\ & \text{the execution of } H_i \\ 0, & \text{otherwise} \end{cases}$$

$$\forall i \in I, v \in V, v \triangleleft H_i : out_{i,v} = \begin{cases} 1, & \text{if } v \text{ must be copied to the host after} \\ & \text{the execution of } H_i \\ 0, & \text{otherwise} \end{cases}$$

3.2 Cost function

The solution of the integer program minimizes the overall runtime of the program, represented by the following *cost function* $C = C_1 + C_2$, where

$$C_1 = \sum_{i \in I} [HWtime(i) - SWtime(i)] \cdot call(i) \cdot x_i \tag{1}$$

$$C_2 = \sum_{i \in I, v \in V, v \triangleleft H_i} [in_{i,v} + out_{i,v}] \cdot Cnum(i,v) \cdot size(v) \cdot Ttime \tag{2}$$

C_1 represents the execution time of the candidates. Note that we do not count the execution time of any regions implemented in software. This is because we have to find the partition which minimizes the execution time *relative* to the pure software solution (for which $C_1 = 0$). C_2 represents the communication overhead. Because constants only have to be copied once, we define $Cnum(i,v) = 1$ for constants, and $Cnum(i,v) = call(i)$ otherwise. For the pure software solution, $C_2 = 0$.

3.3 Constraints

The following constraints are needed to define the admissible solutions:

The software part can never be moved to hardware:

$$x_0 = 0 \tag{3}$$

This constraint for pseudo region H_0 is needed for the technical reasons mentioned above.

Only one of a subset of overlapping candidates can be implemented in hardware:

$$\forall i \in I : \sum_{j \in I, H_i \cap H_j \neq \emptyset} x_j \leq 1 \tag{4}$$

In Figure 2 this constraint applies to **H2** and its extensions.

The following two sets of constraints represent the dependency conditions given in section 2.2.

If $DU(i,j,v)$, H_i is in hardware and H_j is in software, then v must be copied to the host after execution of H_i:

$$\forall i,j \in I_0, v \in V, DU(i,j,v), H_i \not\supseteq H_j : x_i - \sum_{k \in I_0, H_k \supseteq H_j} x_k \leq out_{i,v} \tag{5}$$

If $DU(i,j,v)$, H_j is in hardware and H_i is in software, then v must be copied to the coprocessor board before execution of H_j:

$$\forall i,j \in I_0, v \in V, DU(i,j,v), H_j \not\supseteq H_i : x_j - \sum_{k \in I_0, H_k \supseteq H_i} x_k \leq in_{j,v} \qquad (6)$$

Note that a candidate is implemented in hardware if it or any region containing it is in hardware. (Hence the sums in the above inequalities.)

Finally, the selected candidates must fit in the available hardware resources:

$$\sum_{i \in I} HWarea(i) \cdot x_i \leq HWres \qquad (7)$$

4 Results

We have applied our method to an example program containing the algorithm of Figure 2. To solve the integer program we used the mixed IP–solver [8]. The method determined the regions `H1`, `H2` (extended by `S1` and `S2`) and `H3` to be implemented in hardware. In the resulting system, the array `same` (containing intermediate results) is only used in the hardware region and therefore never has to be copied—it does not have to be allocated in the host memory at all. Copy operations were inserted only for `maxX` (to `H1` and `H3`), `left`, `right` (to extension of `H2`) and `pix` (from `H3`). This example shows that considering exact dataflow information for partitioning has a significant impact on the overall quality of the resulting hardware/software solution.

5 Conclusions

We presented an IP–model for hardware/software partitioning in software oriented codesign which for the first time considers detailed data–flow information. It allows efficient computation of regions to be implemented in hardware, along with the implied copy operations. We have demonstrated the impact of the method on the selection of hardware regions. The method can also be used for large programs, because the number of candidates is limited in most cases. Although integer programs are NP–hard in general, practical problems of moderate size can be solved efficiently with the standard algorithms.

Future work should improve the method by taking into account the limited size of local on–board memory. Additionally, a more precise cost function could eliminate the consideration of some unnecessary communication.

References

1. M. Thornburg, S. Casselman, and J. Schewel. *Engineers' Virtual Computer Users Guide — EVC1s.* Virtual Computer Corporation, 1994.

2. R. Ernst, J. Henkel, and Th. Benner. Hardware–Software Cosynthesis for Microcontrollers. *IEEE Design & Test of Computers*, pages 64–75, December 1993.
3. A. Jantsch, P. Ellervee, J. Öberg, A. Hemani, and H. Tenhunen. A software oriented approach to hardware/software codesign. In *Proc. of the Poster Session of CC'94 - Internat. Conf. on Compiler Construction*, Technical Report, Dept. of Comp. and Inform. Science, Linköping Univ., Sweden, April 1994.
4. A. Jantsch, P. Ellervee, J. Öberg, A. Hemani, and H. Tenhunen. Hardware/software partitioning and minimizing memory interface traffic. In *Proc. of European Design Automation Conf. '94*. IEEE Computer Society Press, September 1994.
5. H. W. Thimbleby, S. Inglis, and I. H. Witten. Displaying 3D images: Algorithms for single–image random–dot stereograms. *Computer*, pages 38–48, October 1994.
6. A.V. Aho, R. Sethi, and J.D. Ullman. *Compilers — Principles, Techniques, and Tools*. Addison-Wesley, 1986.
7. Th. Lengauer. *Combinatorial Algorithms for Integrated Circuit Layout*. Teubner/John Wiley & Sons, 1990.
8. M. Berkelaar. Unix manual page of `lp_solve`. Eindhoven University of Technology, Design Automation Section, 1992.

Test Standard Serves Dual Role as On-board Programming Solution

Kristin Ahrens, Advanced Micro Devices

Programmable logic device (PLD) manufacturers and users have found that the Joint Test Action Group (JTAG) boundary-scan standard (IEEE 1149.1), originally developed to test PCB connections, can also serve as a standard for on-board programming. Several, but not all, of the large CPLD and FPGA vendors have embraced on-board programming for non-volatile devices. The acceptance of on-board programming among high-density CPLD users continues to grow due to such benefits as lower design cost and faster time to market. On-board programming methods refer to programming algorithms which do not require voltages greater than five volts. This is also known as in-system programming (ISP). Today there are three PLD vendors with on-board programming, and two of these use JTAG circuitry for programming: AMD's MACH 3/4 and Altera's MAX9000. Lattice ispLSI devices use a proprietary scheme.

JTAG, On-board Programming Features

The advantages of JTAG, on-board programming are easier prototyping and board debug, and easier field update and maintenance. JTAG testability is an excellent complement to on-board programming. PCB connections can be checked before programming, to ensure that there are no shorts which could damage the device once it has been programmed. JTAG, on-board programming can supply features for device/design tracking, output control during programming, pattern security, and allow test tools to be used for programming.

When programming is done off-board, devices are marked to indicate pattern and placed in inventory before being placed on a board. There may be several PLDs of the same type that are programmed with different patterns. Once a device is programmed with a new pattern, it is assigned a new inventory number because it then must be assigned to a particular place on the board. Devices programmed on-board are soldered on a board while they are still blank. Only one inventory number is needed for blank devices, but once they are on the board, the optional JTAG, 32 bit programmable register, USERID, can be used to differentiate PLDs. This register is often used for revision control during prototype. USERID can also be used in field updates as a check for the rev. of the last field update. This information can be accessed using any JTAG test tools and can be retrieved from any JTAG chain.

Many on-board programmable devices tri-state outputs while programming. This is satisfactory for most cases, but there are instances where the ability to define the state of the outputs while programming is useful. The JTAG boundary scan cells can drive any combination of values on the outputs and can be used during programming. In the MACH devices the state of an output during program is dictated by its boundary scan cells (output and output enable). This gives the user ultimate control of the outputs during programming.

JTAG, on-board programming algorithms are easily understood and available to designers as well as those who would like to steal designs. Design security features become more imperative. Security for MACH devices is implemented by inhibiting the JTAG instructions for programming and verifying. If the security fuse is programmed, then the device will not recognize these instructions and will go into bypass mode every time they are entered. The security fuse is only erased when the entire programmable array is erased.

Programming algorithms which work within the JTAG standard for on-board programming allow designers to easily understand algorithms and debug programming problems-- such as noise inserting bits into the data stream. The increasing popularity of JTAG makes it an obvious choice for an on-board programming standard. The JTAG standard is a flexible one that allows for manufacturer-defined instructions and manufacturer-defined JTAG registers. Devices can be programmed in any JTAG chain using the four mandatory JTAG pins.

Example Algorithm

JTAG compliant programming algorithms may vary between CPLD vendors, and possibly between CPLD families from the same vendor. All compliant algorithms, however, will be similar to this model. There are currently two CPLD vendors with JTAG compliant algorithms and these algorithms differ mainly in number of instructions and special feature instructions for test and security.

All of these algorithms are implemented using manufacturer-defined JTAG registers and manufacturer defined instructions which allow access (verify) and control (program) of internal nodes. In this simplified example, two manufacturer-defined registers and five manufacturer-defined instructions. The two manufacturer-defined registers are the ADDRESS register which selects one of N rows in the programmable array. Each row has M bits of data. The DATA register provides access and control to the M bits in the selected row (Figure 1).

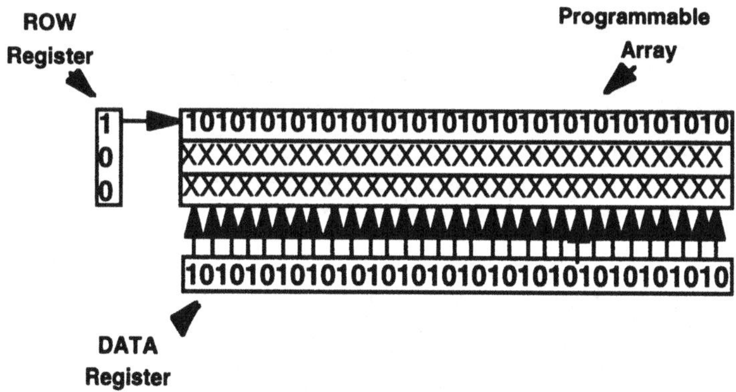

Figure 1. Array Accessing Scheme

There is a minimum set of five programming instructions required. These are
(1) ERASE which erases the entire array or sets all bits in the programmable array to 0. This is necessary because program can only
(2) ROW which places the ADDRESS register between the serial in and serial out pins. After this instruction is placed in the JTAG instruction register, a row address N bits long is loaded into the ADDRESS register.

(3) DATA which places the DATA register between the serial in and serial out pins. When this register is selected, its contents can be read and changed by serially shifting in M bits of new data. The data register is associated with the row selected by the ADDRESS register.
(4) PROGRAM transfers the data in the DATA register to the row selected by the ADDRESS register.
(5) VERIFY transfers the data in the selected row into the DATA register. Once data has been transferred into the DATA register, it can be serially shifted out and checked against a JEDEC file.

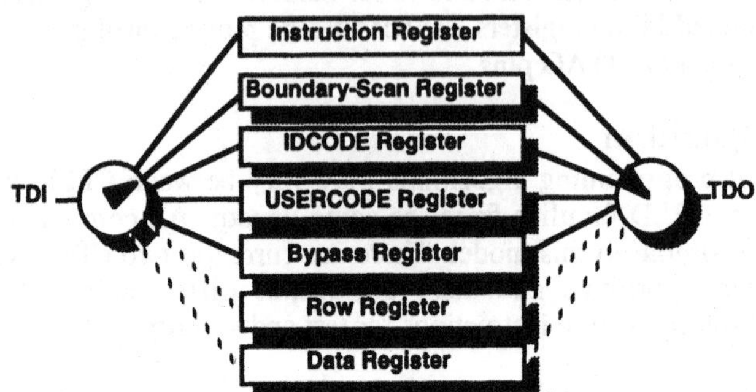

FIGURE 2. Programming registers ccan be accessed through the serial data JTAG pins

Programming software follows these steps. First shift in the ERASE instruction. This erases the entire programmable array. Programming a device without erasing first is much like reprogramming a bipolar device--the new pattern will be programmed over the old one.

To program the first row, first enter the ADDRESS instruction (this places the ADDRESS register between TDI and TDO as in Figure 2) and then serially shift in the address of the first row. The next step is to shift in the ROW instruction (this places the DATA register between TDI and TDO) and then the M bits of data to be shifted into ROW 1. Finally, the program instruction is shifted in. After this instruction is in the instruction register for the specified wait time, the first row is programmed. This procedure is repeated for all N rows.

Verifying each row is a similar procedure. First enter the ADDRESS instruction and shift in the address of the first row. Then shift in the VERIFY instruction. This instruction transfers the data in the first row to the DATA register. The DATA register can then be selected and the data can be shifted out and compared to the original JEDEC file. This procedure is repeated for each row.

Software shifts all of the necessary instructions and data, it is helpful however to know a few things about JTAG when programming. The software will serialize the JEDEC data and compare data read from a device with the JEDEC file. For JTAG programming, the programming software must also be able to put any device in bypass mode given only the length of its instruction register.

The programming method described requires two types of software files: JEDEC (.JED) files and Chain (.CHN) files. The chain file describes the JTAG chain in which the device(s) to be programmed is (are) located. To be able to place non-programmable devices in bypass, the length of each device's JTAG instruction register must be specified. The chain file is also where special JTAG instructions such as security and state of the outputs while programming can be specified.

In addition to the programming instructions, the software should support several JTAG features to take advantage of JTAG on-board programming. First, the software should read the IDCODE of each programmable device. This ensures that the JTAG chain is intact and in the right order and will compare the device specified in the JEDEC file with the device specified in the chain file. The software should also allow a user to access the USERID (user programmable) register if available. Before programming, the software will use the SAMPLE/PRELOAD JTAG instruction to set the outputs to a known state during programming.

```
; Fields: 'Part_ref' Part_type action IR jed_file /:[-s1][-s2] -f output
-o {0/1/Z/X};
; s1 and s2 refer to levels of security
; -f where the user specifies the output file, in this case output.out
;-o allows the user to specify the state of the outputs while
programming

'U1'  MACH445 p 6  X.JED/:  -s1 -s2  -f X.OUT  -o Z
'U2'  NONPLD  n 3
'U3'  NONPLD  n 5
```

Figure 3. MACHpro configuration File

Universal JTAG programming tools are beginning to appear. The first of which is the Corelis/JTAG Technologies JTAGPROG. This is an upgrade to the proprietary software offered by silicon vendors. This software allows users to test infrastructure (JTAG chain integrity) before programming. Silicon vendor software typically does not support non-programming test functions.

JTAG On-board Programming Issues

The JTAG algorithm can easily be implemented in test environments. A JTAG compliant algorithm can be expressed in the Serial Vector Format (SVF) often used in test equipment (e.g., Corelis, Genrad, JTAG Technologies). This SVF format is a proposed amendment to the current JTAG standard and is expected to be approved in 1995. An SVF file can contain all information required for programming. The growing popularity of the SVF format has prompted other test equipment vendors to create SVF translators so that their respective products can accept the SVF format. SVF can be translated into other formats for other test tools. HP offers an SVF to PAT converter for use with their HP3070 system. The SVF file is usually created by the vendors development or programming software.

```
SIR  12  TDI (0FF);        ! Shift in instruction  3 (row)
SDR  81  TDI (100000000000000000000);          ! Load 0s except last 1 for row
SIR  12  TDI (13F);        ! Shift in instruction  4 (data)
S D R                          7 9 3                                  T D I
(07FEFFDFFBFF7FEFFDFFBFF7FD0841082104208410821042FF
          9FFBFF7C6F8DF1BE30C618DBF3FF7FEFFDFFBFF7FEFFDFFBFE
           7FFFFFFFFFFFFFFFFE73CE7FFCFFDFFBFF7FEFFDFFBFF7FEFF9
           FFFFFFDFFFEFFFE7DDF7FFFF3FF7FEFFDFFBFF74208C33FFE);
! Programming row   0
SIR  12  TDI (1BF);        ! Shift in instruction  6 (program)
```

Figure 4. SVF excerpt for programming one row of the array

Because on-board programming time may be added to test time, minimum program times are also an issue for those who wish to program as a part of their test flow. Current JTAG, programming algorithms program a 256 (MACH465) macrocell device in a minimum of 4.5 seconds. Only .5 seconds of this is data shifting time, the rest is a programming wait time. If devices are programmed in parallel, then this wait time can be shared between devices in a chain. For example, programming two MACH465s (with different patterns) in a serial chain would take 5 s. While there are no hardware limitations for this type of programming, the software would be complex and does not yet exist. As JTAG programming gains popularity, this type of programming will likely become available.

On-board Programming: Proprietary vs. Standard

On-board programming methods that do not use JTAG exist, but these algorithms are not compatible with those of other vendors, and these devices cannot be programmed in JTAG chains. The JTAG standard operates on the premise that all devices contain the same control state machine. If a device is placed in a chain and does not use the same state machine, there will be chaos (unknown states) in the JTAG state machine. It would be possible that boundary-scan cells could be driven into contention.

Not having JTAG on on-board programmable devices is a disadvantage because boundary scan allows a user to test for opens and shorts on a board before programming while the I/Os are tri-stated, avoiding damage to the device. If a designer is not interested in boundary-scan test, however and uses only one PLD vendor, then a proprietary method may suffice. If either condition is not met, then a standard approach is preferable. On-board programming requires board designers and manufacturers to be knowledgeable about the programming algorithm. Previously, the off-board programmers shielded the user from such details. Industry is not likely to accept several proprietary algorithms, especially when those based on JTAG offer a more standardized solution.

Conclusion

Currently, there are both proprietary and JTAG compliant on-board programming schemes, resulting in confusion and frustration among PLD users. This situation is reminiscent of a similar issue which confronted the industry in its early years: PLD vendors used different file formats to describe fuse connectivity in PLDs. Then, as now, multiple schemes complicated program and verification procedures. Finally a standard emerged from the Joint Electronic Device Engineering Council known as JEDEC. Once the standard became widely accepted, this confusion was greatly reduced. The use of JTAG for on-board programming is an emerging standard and simplifies on-board programming procedures. Any designer familiar with JTAG can easily understand the algorithm and use test equipment to program as well as test.

Advanced Method for Industry Related Education with an FPGA Design Self-Learning Kit

U. Zahm, T. Hollstein, H.-J. Herpel, N. Wehn, M. Glesner
Institute of Microelectronic Systems, Darmstadt University of Technology,
Karlstr. 15, 64283 Darmstadt, Germany

Abstract—In this contribution a self-learning kit for FPGA design is described, which is targeted to staff training in small and medium enterprises. The kit consists of a HyperCard-based learning software and a hardware tester board, which is controlled by a convenient MS-Windows based control software.

1. INTRODUCTION

Field programmable gate arrays (FPGAs) combine the flexibility of mask programmable gate arrays (MPGAs) and "time to market" advantages of programmable logic devices (PLDs) by relatively small fixed costs. Reducing the number of circuits on board, they increase the reliability of the whole board. Depending on the required performance, they can replace ASICs (Application Specific Integrated Circuits) for small series. For these reasons, they are well accepted for rapid prototyping purposes. Their part in the programmable logic market actually runs up to more then 20% and it is still increasing.

New products need advanced technologies and competitive manufacturing techniques. Introducing new technologies involves education and training of the staff. Therefore, large enterprises maintain specialised training centers to provide education and training possibilities to their staff. In small and medium sized enterprises (SMEs) we have a different situation. They cannot afford their own training departments, but also participation at external courses often exceeds their financial frame. They often can send only one participant who has to transfer his knowledge to colleagues after visiting a one- or two-day course. Since the market demands innovation a more efficient education possibility has to be found. A computer-based self-learning kit provides especially for SMEs considerable advantages:

- Adaptive learning rate: Due to his personal state of knowledge the learning person can decide how fast he can proceed and what fields he has to learn in detail and what topics he can skip.
- Flexible learning times: There is not a special time for training. The user can learn when his work allows it. He can freely select the amount of material to evaluate per session.
- Learning on the field: The learning person can integrate the self-learning kit with theory, tutorials, and practicals in his normal working environment.
- Quick introduction into topic: Without a teaching support a beginner of new tools has to study meters of manuals before being able to do the first steps. Due to the

didactic structure of the kit, the learning person can have an early success without an intensive study of manuals.

This paper is organized as follows: Section 2 describes how our self-learning kit is integrated into the COMETT project AMES and section 3 gives an overview over the main structure of the kit. The learning software is described in section 4 and the hardware with its control software in section 5. A short summary concludes the paper.

2. Project Integration

The system described in this contribution has been developed in the frame of an EU funded COMETT project, which is called AMES (Advanced Microelectronics Educational Service) [1]. It is part of a so-called 'Ensemble Didactique Digital Design' and will be integrated together with kits, developed by other project partners. The overall structure of the project (partners and topic of the contributions) can be seen from figure 1.

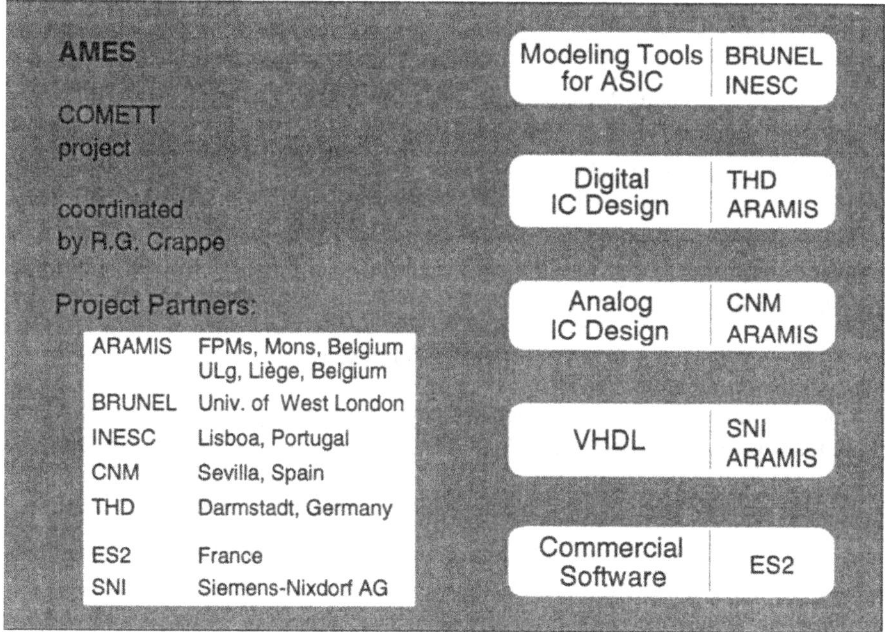

Figure 1: AMES Project Overview: Partners and Tasks

3. Structure of the Self-Learning Kit 'FPGA Design'

The entire self-learning kit, shown in figure 2, comprises two components. The Apple Macintosh-based teaching software module is described in section 4 and the hardware tester module, consisting of a tester board and the related control software for the board, in section 5.

The hardware tester board can be connected to a PC via RS-232 interface. It is controlled by a comfortable MS-Windows[1] based operating software, running on an PC. The Xilinx design tools are also installed on a PC.

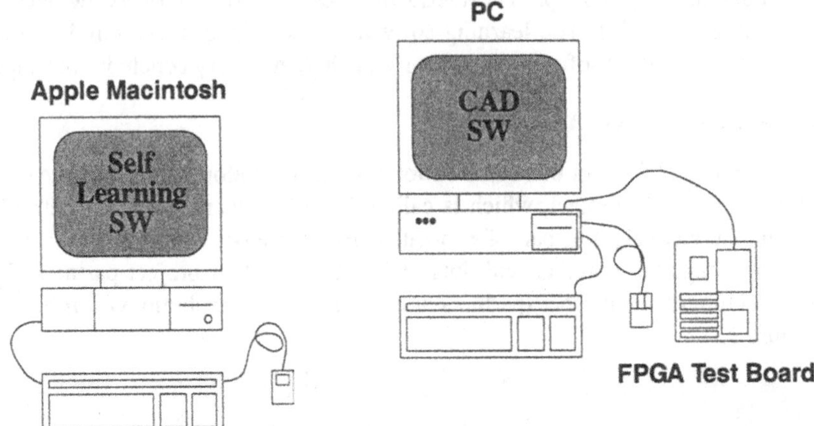

Figure 2: Global Structure of the Self-Learning Kit

By introducing the PowerPC[2] processor we are given the possibility to integrate the whole system on one single computer. A PowerPC is able to run the Apple System 7.x[3], MS-DOS[4], and MS-Windows at the same time. This integration allows the user a parallel access to the teaching material, and to the practical circuit design of the exercise (the traffic light controller) by simultaneously using the CAD tools and the belonging tutorials.

4. MACINTOSH-BASED LEARNING SOFTWARE

The learning software of our education package is implemented on an Apple Macintosh using the HyperCard[5] Software. HyperCard provides convenient development features and HyperTalk, an easy to learn scripting language. The user surface which can be created is very comfortable and may contain a lot of animation.

As can be seen in figure 3, the whole software is implemented by four HyperCard stacks. The main stack, which is normally first proceeded by the user, contains the theory about commercially available FPGAs with emphasis on Xilinx LCAs and the design cycle of look-up table based FPGAs [2, 3, 4]. Linked to the theory stack, the index stack can be found containing all main technical terms of the theory stack.

---

[1] Trademark of Microsoft Corp.

[2] by Apple Computer Inc., IBM Corp., and Motorola Corp.

[3] Trademark of Apple Computer Inc.

[4] Trademark of Microsoft Corp.

[5] Trademark of Claris Corp.

From the theory stack the user can switch to the exercise stack. Here, the learning person has to complete a Xilinx XC4000 design of a traffic light controller. Manuals of the tools needed for the exercise are provided in the tutorial stack. Links exist to the main card of the tutorial as well as to the description of the tools [5, 6, 7] to be used at the current stage of the exercise.

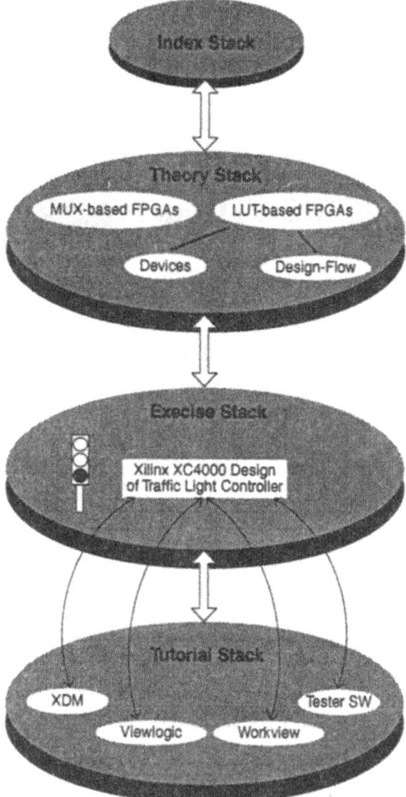

Figure 3: Structure of the Macintosh-based Learning Software

A. The Theory Stack

On the main topic card, shown in figure 4, the user can choose, if he likes to learn about available multiplexer based FPGAs or look-up table based ones. The multiplexer based FPGAs are divided in Actel, QuickLogic, Algotronix, and Concurrent Logic. For every FPGA the type of the logic block architecture and the routing resources are described.

Concerning look-up table based FPGAs, information on Xilinx devices can be selected. The user has the choice between learning about hardware or about software. In the hardware part, the Xilinx LCAs of the XC2000, XC3000, and XC4000 families are described. In the software part, the design flow can be found: design entry, netlist generation, logic optimisation, technology mapping, placement, and routing. For every step an existing tool is described, for example the chortle-cf technology mapper.

A lot of animation is implemented and various buttons to browse the stack. An index stack is linked to this theory stack. Clicking to the index button the user comes to the index main card. To find the term FPGA, the user clicks on the register tab F to get the page with F terms, clicks on the term FPGA to obtain a selection of page number where the term FPGA appears. Choosing a page number with a mouse click brings him back to the theory stack to the chosen page.

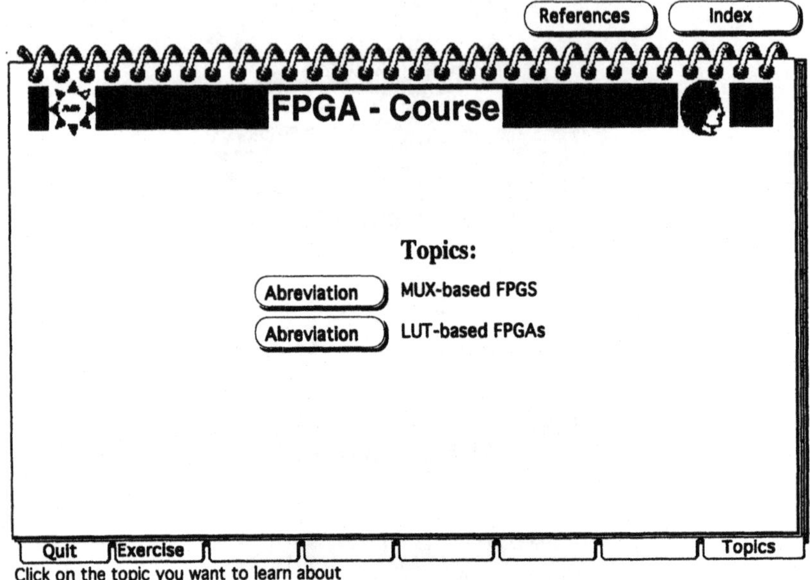

Figure 4: Main Topic Card of the Theory Stack

A second link exists to the exercise stack. The first click on the exercise button brings the user to the design task definition card of the exercise stack, depicted in figure 5. Afterwards, flipping between the two stacks means that the user returns to the same page of the stack as he was at leaving.

B. The Exercise Stack

The Exercise stack starts with the description of the problem. A traffic light controller has to be implemented on a Xilinx XC4000 LCA. The design is partially given and should be completed by the learning person. The design is shown in hierarchical manner. Afterwards, the step by step exercise is described. First the finite state machine has to be designed with a schematic editor. We recommend VIEW*logic*'s Workview[6]. At this card there are specific links to the tutorial stack: to the Xilinx

---

[6] by VIEW*logic* Systems Inc.

Design Manager[7] overview, to the VIEW*logic* overview, and to the Workview. Such links we have implemented on any page, were working with design tools is described.

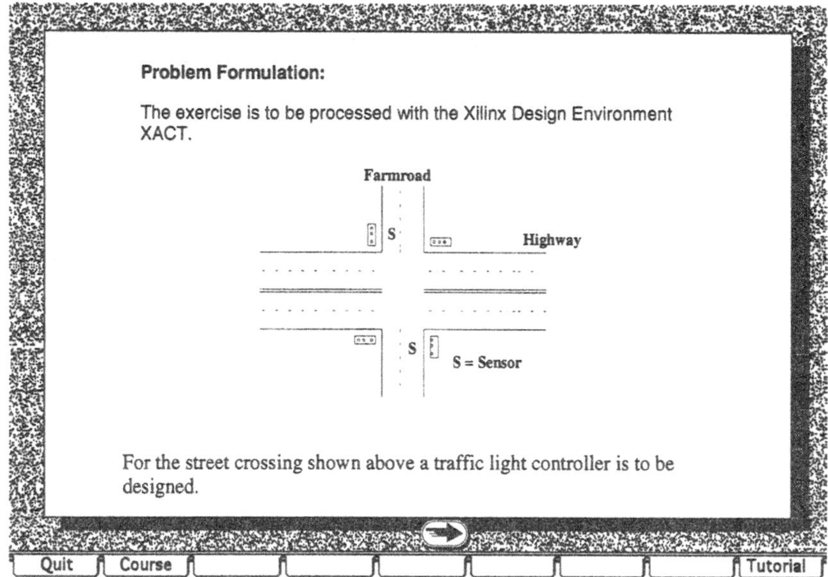

Figure 5: Problem Formulation Card of the Exercise Stack

After having designed the finite state machine and fitting it into the given design, the user has to generate the netlist and perform a simulation with unit delay. The stimuli file for the simulation is given and can be viewed by clicking the Stimuli button. To compare the obtained results, the user can click the Results button to see our simulation results.

After delay simulation the PPR (partitioning, placement, and routing) algorithm has to be applied to the netlist and then a timing simulation can be performed afterwards. The results of the timing simulation can be compared with our results in the same way as the delay simulation results.

The last step of the exercise is programming an LCA and run a hardware validation. The learning person has to generate a bitstream, to download it to a XC4000 LCA on the tester board, and to validate the functionality by a hardware test.

C. The Tutorial Stack

As it can be seen on the main topic card of the tutorial stack, depicted in figure 6, the tutorial stack contains manuals of three systems: Xilinx Design Manager, VIEW*logic*, and the control software of our hardware test board. If the user chooses XDM, VIEW*logic*, or Testing Overview, the PC program surface of the correspon-

---

[7] by Xilinx Inc.

ding system appears. The HyperCard picture can be used in the same way as it would be done with the real surface. However, instead of invoking a program, the user flips to the description card of the chosen program.

The description card of special programs can directly be chosen from the main topic card. The description pages exist of one or more very complex cards with a lot of browsing possibilities. To lead the user to the browsing possibilities various help buttons are implemented.

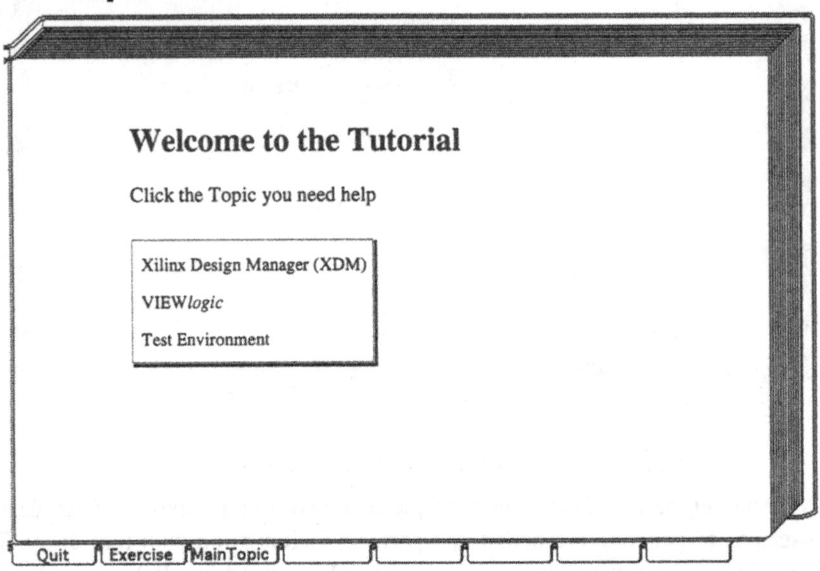

Figure 6: Main Topic Card of the Tutorial Stack

When returning to the exercise stack it will be re-entered at the same card where this stack had been left before. Which card of the tutorial stack is reached depends on the button the user clicks. The Tutorial button leads to the main topic card, the buttons XDM, VIEW*logic*, and HWTest let appear the PC surface of the Xilinx Design Manager, VIEW*logic*, or the control software of the hardware test board, respectively. The buttons, called like special programs, lead directly to the description card of that program.

5. FPGA TESTER HARDWARE

In the exercise stack, described above, the user is encouraged to complete the design of a traffic light controller. To implement his design on a real FPGA, we developed the hardware tester board, depicted in figure 7. The board is connected to the PC via serial RS-232 interface.

The stimuli generated for simulation of the design on the different levels can be applied as test vectors for a hardware test on the FPGA mounted on the tester board. The output signals are sampled with a clock rate which is eight times the input clock rate for the test vectors. Sampled output signals are transferred back to the PC where they are visualised as waveforms together with the waveforms of the input signals.

Figure 7: The Hardware Tester Board

The structure of the board is depicted in figure 8. The board is composed by the following modules:

- Data from the PC is sent via serial RS-232 interface to the microcontroller module which controls the board functions.

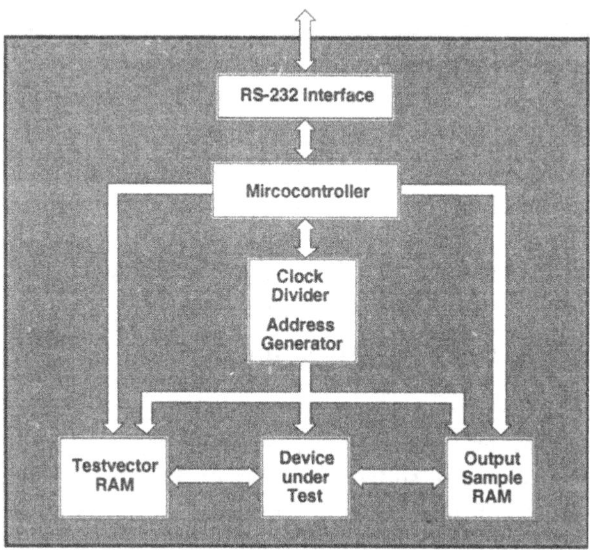

Figure 8: Structure of the Tester Board

- The clock divider and address generator is implemented on a Xilinx XC4003 PG120 LCA.
- The test vectors are stored in a 16 k input RAM and the sampled output vectors in a 128 k output RAM. Analogous to the sample clock frequency, also the size of the output vector RAM is eight times the input vector RAM size.
- The FPGA under test is mounted on a special test adapter (device under test in figure 8). The test adapter can be replaced by adapters for other devices. However, only Xilinx LCA devices can be directly configured from the PC via serial RS-232 interface on board without being dismounted from the adapter.

The manufacturing costs of the tester board is actually about 500 ECU. The table in figure 9 summarises the main tester board data:

Test Device Input Vector Size	max. 32 Bit
Test Device Output Vector Size	max. 32 Bit
Test Vector RAM Size	16 k
Output Vector RAM Size	128 k
Test Clock Frequency	0.8, 4, 8, or 16 MHz
Serial Interface	4800 .. 38400 Baud

Figure 9: Main Tester Board Data

The interaction between the tester board and the PC is done by the related control software running under MS-Windows. The communication parameters for the interaction between the board and the PC (serial port, the baud rate, and waiting period) can also be configured by this software as well as best board parameters (input/output frequency, number of transferred values, and scale factor for the waveform display). After configuration of the board, the bitstream of the design can be downloaded in order to (re-)configure the FPGA.

Test vectors can be applied via menu. Waveforms of the input signals (clock, reset, and signal) are displayed in red colour and the output signals in blue colour. The input waveform can be modified by mouse and new waveform channels (input or output) can added or deleted on the screen. Zoom in and zoom out functions and an undo button are provided as special user convenience.

6. Conclusion and Future Work

We introduced a self-learning kit for FPGA design with emphasis to Xilinx LCAs. The kit consists of a Apple Macintosh-based learning software and a dedicated hardware with a related control software running on PC. The whole system can be installed on a computer with a PowerPC Processor.

The kit was designed for integrated learning and practical exercises. The target group are small and medium sized enterprises where advanced education is necessary but external courses too expensive.

We are currently working on enhancing the control software of the tester board. We intend to give the possibility of direct waveform comparison of simulation and tester results on one screen.

7. REFERENCES

[1] C. Jeffries, B. Merison, S. Sternberg, C. Schaessens: The COMETT Guide to Developing a House Guide; Cambridge Learning Systems Limited, 1992.

[2] S. D. Brown, R. J. Francis, J. Rose, Z. G. Vranesic: Field Programmable Gate Arrays; Kluwer Academic Publishers, 1992.

[3] P. K. Lala: Digital System Design using Programmable Logic Devices; Prentice Hall, 1990.

[4] H. J. Herpel, T. Hollstein, M. Glesner: FPGAs - An Easy Entry in Microelectronics for SMIs; Comett Course, Liège, November 1992.

[5] VIEWlogic Manual for Use with Workview; Viewlogic Systems, Inc., Massachusetts, May 1991.

[6] XACT Reference Guide, Volume 1, 2, 3; Xilinx, Inc., San Jose, April 1994.

[7] XACT User Guide; Xilinx, Inc., San Jose, April 1994.

FPGA Implementation of a Rational Adder

Tudor Jebelean

RISC-Linz, A-4040 Linz, Austria

Abstract. A systolic coprocessor for the addition of signed normalized rational numbers is implemented using field programmable logic from Atmel. The circuit is structured as a sandwich of systolic arrays implementing the necessary subtasks: integer GCD, exact division, multiplication and addition/subtraction. In particular, the implementation of GCD and of exact division improve significantly (2 to 4 times) previously known solutions. In contrast to the traditional approach, all operations are performed least-significant digits first, which allows bit-pipelining between partial operations at reduced area-cost. The actual implementation for 8-bit operands consumes 730 cells (3,500 equivalent gates) and runs at 25 MHz (5 MHz after layout).

1 Introduction

During the last decades, research in computer arithmetic was focused on fixed precision (floating point) computations, which are used in numerical calculations. However, we are currently witnessing a fast increase of interest in exact computations in various areas of mathematics and engineering [4, 3, 5]. Developing coprocessors for arbitrary rational arithmetic is therefore important both from the practical and the theoretical points of view.

Although parallel [systolic] algorithms for the various integer operations have been studied for a long time, only recently some attempts at building specialized hardware for rational arithmetic have been performed [8, 7, 15].

Addition of rational numbers contains all the integer operations: greatest common divisor (GCD), exact division, multiplication and addition/ subtraction. We consider normalized inputs (the numerator and the denominator are relatively prime) and signed. This contrasts with the approach in [12], were the inputs are not considered normalized.

The approach is bit serial, least-significant digits first (LSF) and systolic. We employ purely systolic algorithms, that is all the communications are local. Moreover, communication with the host takes place only at one end of the systolic array.

The mathematical background and the rationale of the design are presented in detail in [13]. In the present paper we concentrate on the FPGA implementation details. The overall structure of the adder is a "sandwich" of systolic arrays (see fig.1) and it corresponds to the algorithm for addition of rational numbers given in [9]. First the denominators are fed LSF serially into the *GCD computation* array GCD. Here the trailing null bits are shifted out and the GCD is computed, which together with the shifted denominators is sent in parallel to

* Supported by Austrian FWF grant P10002-PHY.

a double array for *exact division* DIV. The GCD is also used in one multiplication. The quotients from the exact division are sent serially to the multiplication arrays MUL, where they interact with the LSF serially fed numerators. Finally, the results of the above multiplications is piped through a carry-feedback adder / subtracter, giving the numerator of the result in LSF serial fashion. The denominator is produced, also serially, by the last multiplication array.

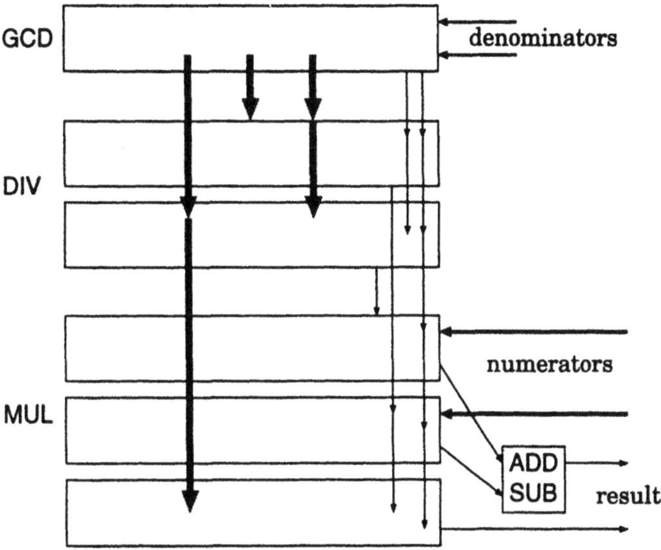

Fig. 1. Structure of the rational adder.

2 GCD computation

The GCD array is based on the *plus-minus* algorithm of [2], simplified as shown in [11]. In contrast to previous approaches [2, 10], the array is composed of two separate units (see fig.2):

- a *feeder* of constant time an area performs the preprocessing of the operands;
- an *array of systolic cells* computes the shifted GCD.

The **feeder** is a pipeline of 4 blocks as shown in fig.3. The block F0 receives the operands A and B, least-significant bits first, 4 bits at a time. Additionally the host has to indicate the end of each operand by setting the input END. Using the input signal END, the feeder also generates a tag bit TAG which indicates the variable length of the operands in the GCD array. F0 skips the trailing 4-bit blocks which are null at both operands, by computing an 8-input OR signal START which activates the next functional blocks in cascading mode. Also, F0 passes on a 4-input OR signal S2 which activates F1. If S2 is set, then F1

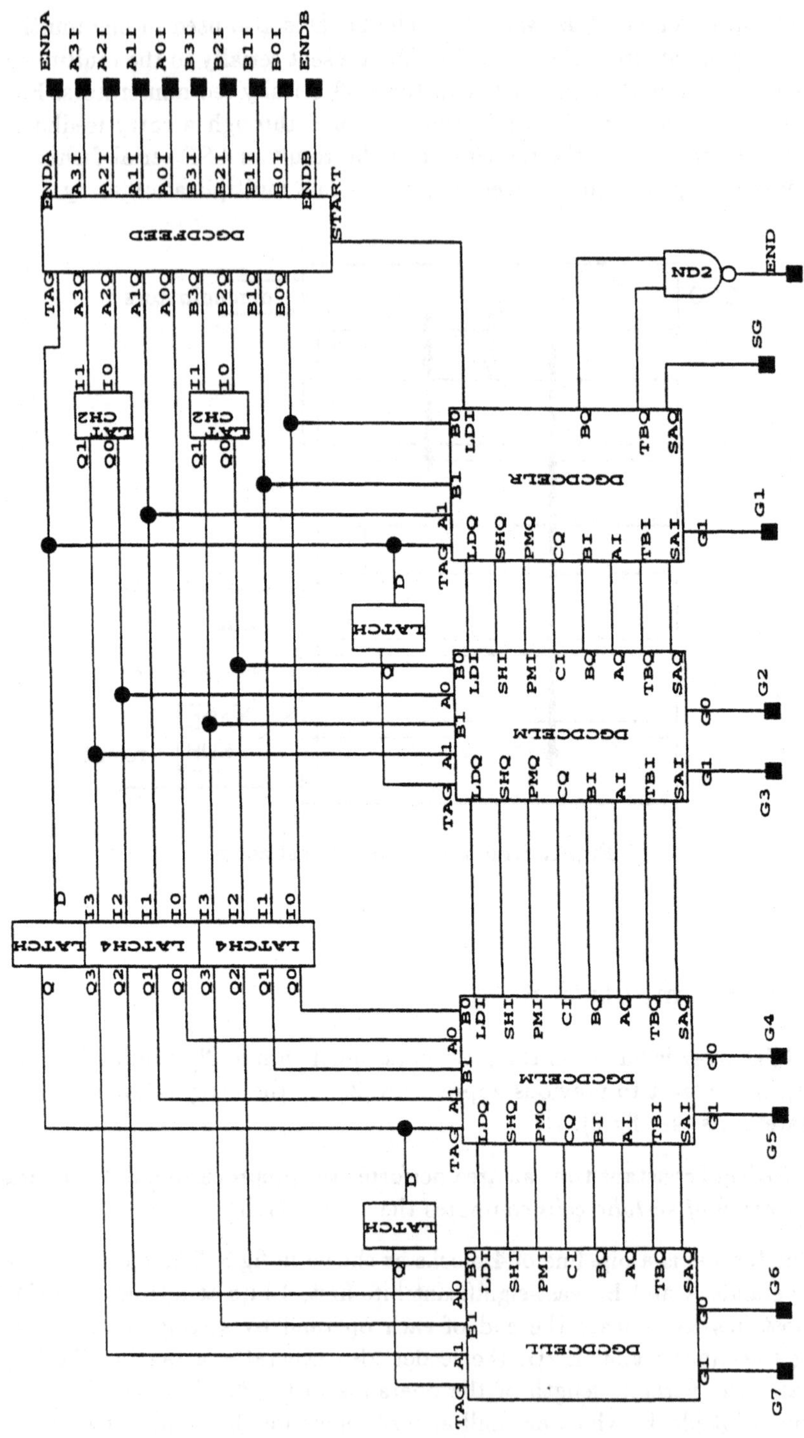

Fig. 2. Structure of the GCD unit.

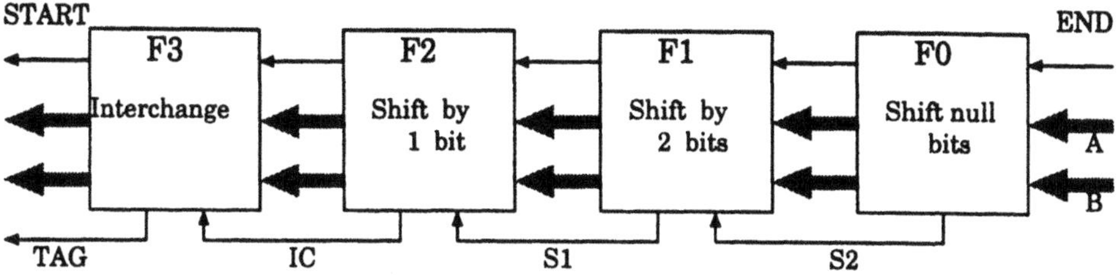

Fig. 3. Feeder of the GCD array.

will shift by 2 bits each of the operands. This is done using an array of latches and of multiplexors. Additionally, an 2-input OR will generate the activating signal S1 for F2. Similarly, if S1 is set, then F2 will shift by 1 bit each operand. Additionally, F2 sets IC if the operand A is even. If IC is set, then F3 will interchange the two operands. The feeder consumes an area of 97 Atmel cells (469 equivalent gates), which does not depend on the lengths of the inputs. The delay through the feeder is also constant: 3 clocks.

An area less-expensive implementation could be realized by performing all the operations on 1-bit streams and using serial to parallel converters. However, such an implementation would need an additional clock and would also require a very fast operation from the host.

The **array of systolic cells** for GCD computation (fig.2) works in the following manner:

- The operands are piped right-to-left at double speed, using latches as shown in fig.2 and are loaded into the appropriate cells when the signal LD is set.
- The rightmost cell (fig.4) examines the least-significant bits a_0, a_1, b_0, b_1 of the operands and depending on them generates a 2-bit code (PM and SH) of the operation to be performed (SHift B, Plus, or Minus).
- The op-code is propagated systolically from right to left, activating the other cells.
- In order to avoid idle steps, cells are clustered 2 by 2, both in time and space (see also [13, 6]), as shown in fig.5. There are two kinds of functional units: the upper units perform the shift/add/subtract (fig.6), while thee other operate upon the tags (fig.7).
- The rightmost cell (fig.4) decides upon the termination of the computation by inspecting the bit b_0 and the least significant tag of B. In this moment, the GCD of the shifted operands is available as G, in parallel in all the cells. A signal SG indicates the sign of the result. Likewise, the array will also store the shifted values of the input operands, which are needed for the subsequent exact division.

In the Atmel FPGA implementation, a GCD array for 8-bit inputs consumes 224 cells (1,129 equivalent gates) and has a clock cycle of 39.6 ns. Together with the feeder it takes 333 cells (1,598.5 equivalent gates). This is 72% of the area of

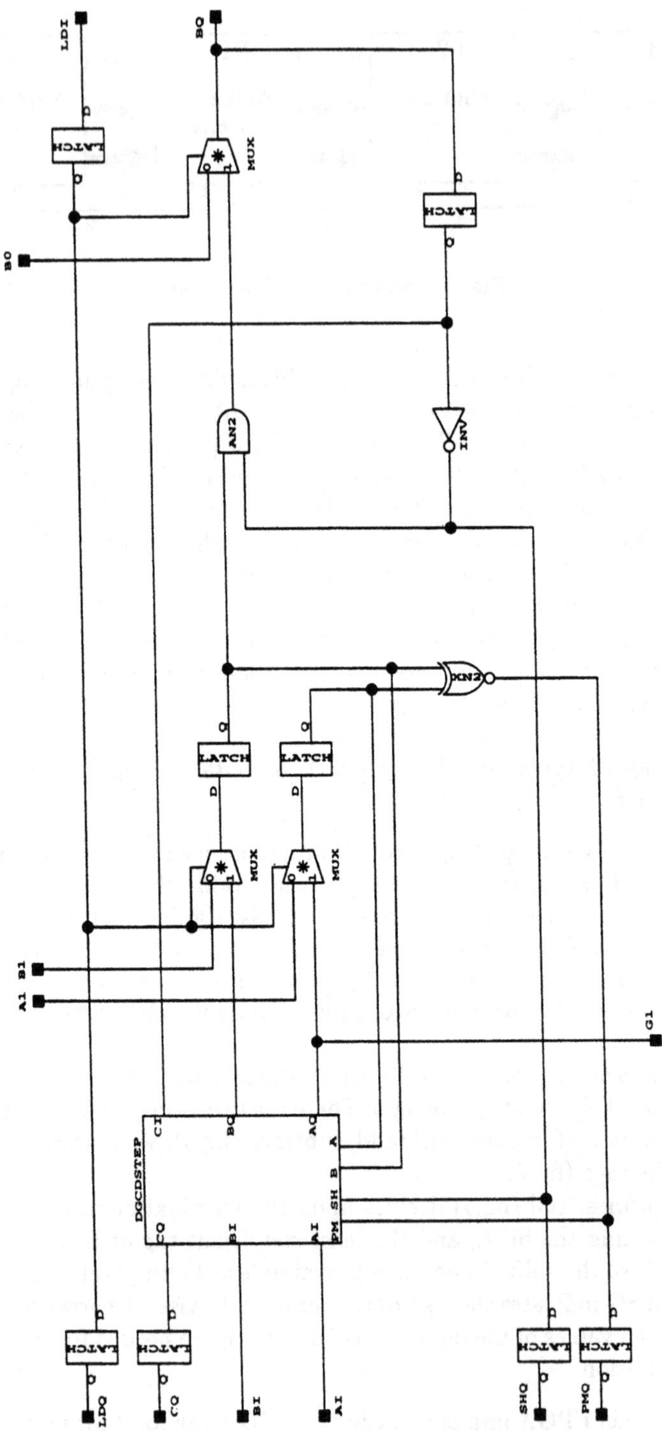

Fig. 4. The rightmost cell of the GCD array.

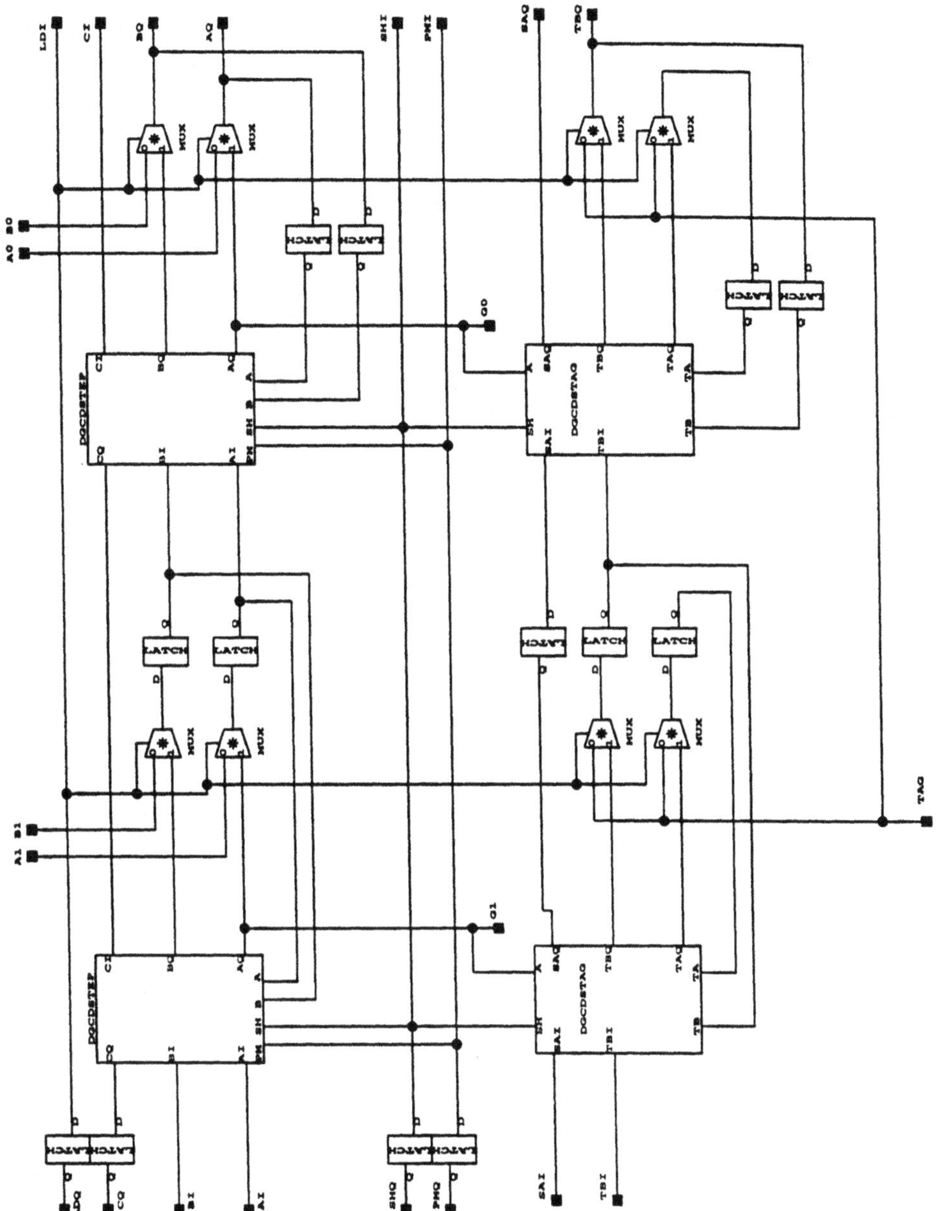

Fig. 5. Structure of the GCD cell.

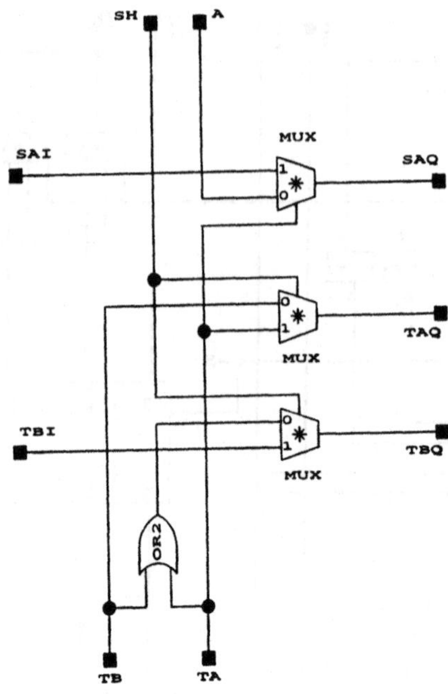

Fig. 6. Functional unit in GCD cell: operands.

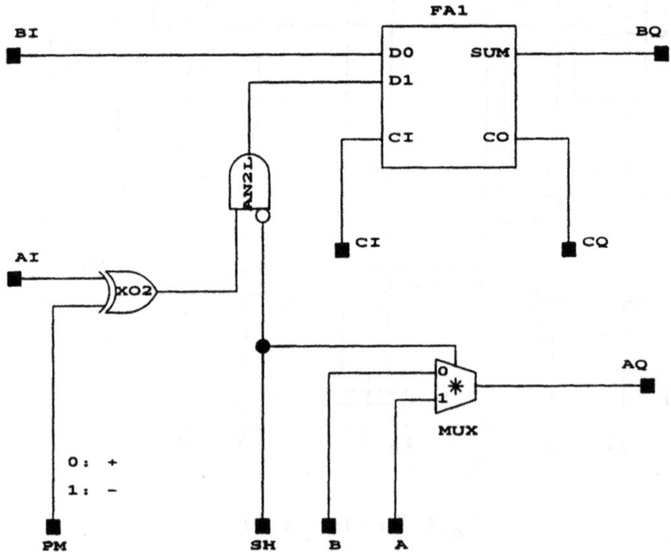

Fig. 7. Functional unit in GCD cell: tags.

the design in [11] and 62% of the clock time. Moreover, the present array needs two times less steps. After layout, the cell count is 953 (60% of the old design) and the clock time is 205 ns (47%), hence the speed of new array is 4 times higher.

3 Exact division

The design of the array for *exact division* [13] is based on the systolic algorithm presented in [10] and has a similar structure to the one presented in [14] - except the fact that our array is parallel-parallel. The middle cell and the rightmost cell of the array are shown in fig.8 and fig.9. There, G0 and G1 represent two successive bits of the divisor, while SG is the sign bit, generated by the GCD array. A0, A1 are two successive bits of the dividend, their loading is regulated by the signal TR, which is generated by the output END of the GCD array. The quotient X is generated in LSF serial manner, and it is also pipelined leftward through the array.

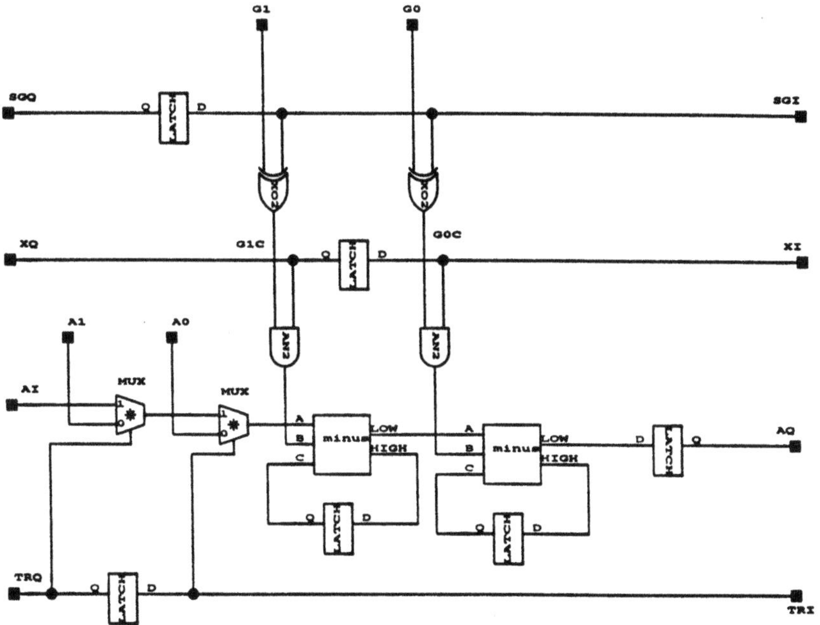

Fig. 8. The middle cell in exact division.

The implementation of an 8-bit array for 2 divisions consumes 160 cells (534 equivalent gates). This is 24% of the array presented in [10] (535 cells, 2,183.5 equivalent gates) - hence a 4-time area improvement.

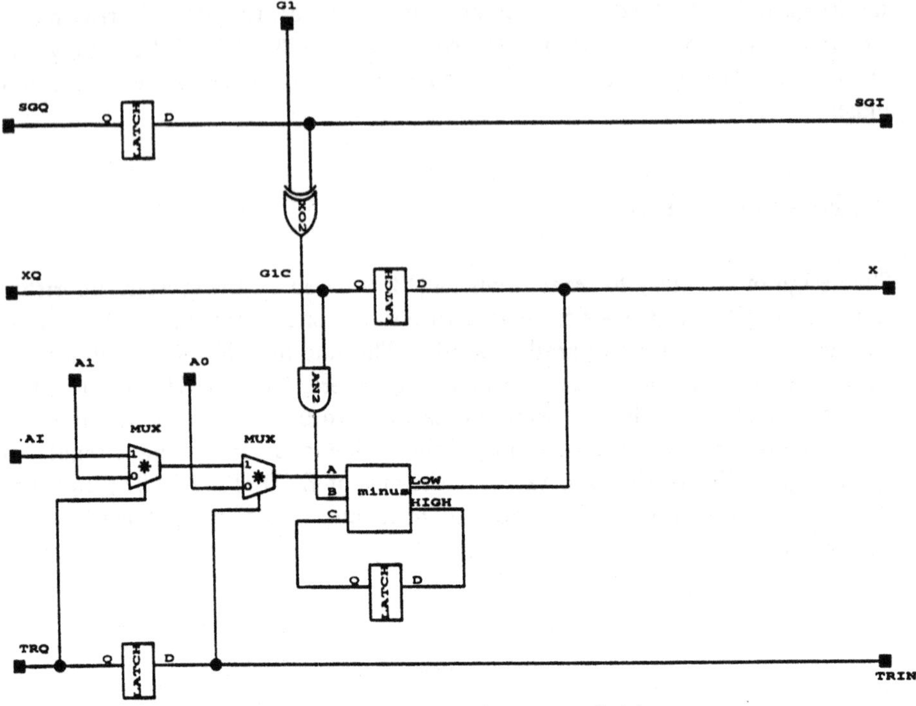

Fig. 9. The rightmost cell in exact division.

4 Conclusions

We do not describe in detail the *multiplier* array, which is a straightforward implementation of the design introduced in [14]. However, the upper two arrays (fig.1) also use a double-speed 4-bit wide bus for pipelining the inputs (the numerators of the rational operands). These buses are similar to the one used for the GCD input (fig.2). Thus, these multipliers are fully serial-serial, as Atrubin's [1], but with a lower area consumption.

The implementation of one array for 8-bit operands consumes 82 cells (273 equivalent gates). Three such arrays are needed in the arithmetic unit - the last one, however, does not need the pipeline because the input is taken in parallel from the GCD array.

Due to the LSF manner in which the products are generated, the final addition / subtraction can be performed by a simple carry-feedback unit (fig.1).

The area consumption of the complete arithmetic unit, in the implementation for 8-bit operands is 729 Atmel cells (3,428 equivalent gates). The clock-time is 40 ns (like the GCD array), hence a speed of 25 MHz is possible (before layout). Automatic layout was not successful on a 6005 chip, but most probably the new 6010 chip will be able to accommodate an 8-bit implementation. (Due to the fully systolic characteristics of the algorithms, longer operands can be treated by simply tiling several chips.) A slowdown of 5 is noted after layout, hence a speed of 5 MHz. Since $3n$ steps are necessary (in average) for adding n-bit operands, 32-bit operands would need about 100 steps by 200 ns, that is 20 μs. An optimized C program on a DECstation 5000/240 solves the problem in 14.3 μs, hence the timings are in the same range. However, for longer operands the speed-up will increase linearly, because the time complexity of the systolic array is linear, while the timing of the software algorithms increases quadratically.

References

1. A. J. Atrubin. A one-dimensional iterative multiplier. *IEEE Trans. on Computers*, C-14:394–399, 1965.
2. R. P. Brent and H. T. Kung. A systolic algorithm for integer GCD computation. In K. Hwang, editor, *Procs. of the 7th Symp. on Computer Arithmetic*, pages 118–125. IEEE Computer Society, June 1985.
3. M. Bronstein, editor. *ISSAC'93: International Symposium on Symbolic and Algebraic Computation*, Kiev, Ukraine, July 1993. ACM Press.
4. B. Buchberger, G. E. Collins, and R. G. K. Loos (eds). *Computer Algebra, Symbolic and Algebraic Computation.* Springer Verlag, Wien-New York, 1982.
5. A. M. Cohen and L. J. van Gastel, editors. *SCAFI'92: Studies in Computer Algebra for Industry II*, Amsterdam, 1992. Report Series of the Computer Algebra Netherlands Expertise Center.
6. Sh. Even and A. Litman. On the capabilities of systolic systems. *Math. Sys. Theory*, 27:3–28, 1994.
7. A. Guyot and Y. Kusumaputri. OCAPI: A prototype for high precision arithmetic. In A. Halaas and P. B. Denyer, editors, *VLSI'91*, pages 11–18. IFIP, North Holland, 1991.
8. A. Guyot, Y. Herreros, and J.-M. Muller. JANUS, an on-line multiplier/divider for manipulating large numbers. In M. J. Irwin and R. Stefanelli, editors, *ARITH-8: 8th IEEE Symposium on Computer Arithmetic*, pages 106–111, Como, Italy, May 1987. IEEE Computer Society Press.
9. P. Henrici. A subroutine for computations with rational numbers. *Journal of the ACM*, 3:6–9, 1956.
10. T. Jebelean. Systolic normalization of rational numbers. In L. Dadda and B. Wah, editors, *ASAP'93: International Conference on Application-Specific Array Processors*, pages 502–513. IEEE Computer Society Press.
11. T. Jebelean. Designing systolic arrays for integer GCD computation. In P. Capello, R. M. Owens, E. E. Swartzlander, and B. W. Wah, editors, *Proceedings of ASAP '94*, pages 295–301. IEEE Computer Society Press.
12. T. Jebelean. Rational arithmetic using FPGA. In W. Luk and W. Moore, editors, *More FPGAs*, pages 262–273. Abingdon EE&CS Books, Oxford, 1994. Proceedings of FPLA'93: International Workshop on Field Programmable Logic and Applications, Oxford, UK, September 1993.
13. T. Jebelean. Design of a systolic coprocessor for rational addition. In *Proceedings of ASAP'95*. IEEE Computer Society Press, in print.
14. P. Kornerup. A systolic, linear-array multiplier for a class of right-shift algorithms. *IEEE Trans. on Computers*, 43:892–898, 1994.
15. C. Riem, J. König, and L. Thiele. A Case Study in Algorithm–Architecture Codesign: Hardware Accelerator for Long Integer Arithmetic. In *Proc 3rd International Workshop on Algorithms and Parallel VLSI Architectures*, Katholieke Universiteit Leuven, Belgium, 1994.

FPLD-Implementation of Computations over Finite Fields $GF(2^m)$ with Applications to Error Control Coding

André Klindworth

University of Hamburg, Dept. of Computer Science
Vogt-Koelln-Str. 30, 22527 Hamburg, Germany
E-mail: klindwor@informatik.uni-hamburg.de

Abstract. This paper investigates the implementation of computations over finite fields $GF(2^m)$ using field-programmable logic devices (FPLDs). Implementation details for addition/subtraction, multiplication, square, inversion, and division are given with mapping results for Xilinx LCAs, Altera CPLDs and Actel ACT FPGAs. As an application example, mapping results for complete encoders for error-correcting codes are also presented. Finally, new opportunities emerging from FPLD technology for data transmission systems with dynamic code adaption are discussed.

1 Introduction

Finite fields have seen an ongoing interest of the scientific community for more than three decades. This is due to the fact that computations over finite fields play a key role in many important applications such as cryptography [12] and error-control coding [3], [4], [9]. Efficient implementations of algorithms for encoding and decoding usually make use of a considerable amount of computations in finite fields $GF(2^m)$. In many cases, a hardware implementation is mandatory to meet strong performance requirements.

2 Construction of finite fields

This section gives a brief introduction to the construction of finite fields and their main properties. For details the reader may refer to [8], [2], [9], [6], or [4].

The finite field over a set F with q elements is denoted by $GF(q)$. In such a field, two operations are defined which are called *addition* (denoted by "+") and *multiplication* (denoted by "·"), respectively. The smallest finite field is the field $GF(2)$ over the set $\{0, 1\}$. In this field, addition is defined as integer addition modulo-2 (logical XOR) and multiplication is defined as integer multiplication modulo-2 (logical AND).

With respect to hardware implementation, the most important class of finite fields are the fields $GF(2^m)$ for small m ($m \leq 16$). These fields consist of all polynomials over $GF(2)$ with degree smaller than m. To form a field, addition is

Element Number	Polynomial Representation	Polynomial Basis Rep.
0	0	0000
1	1	0001
2	x	0010
3	x^2	0100
4	x^3	1000
5	x^3+1	1001
6	x^3+x+1	1011
7	x^3+x^2+x+1	1111

Element Number	Polynomial Representation	Polynomial Basis Rep.
8	x^2+x+1	0111
9	x^3+x^2+x	1000
10	x^2+1	0101
11	x^3+x	1010
12	x^3+x^2+1	1101
13	$x+1$	0011
14	x^2+x	0101
15	x^3+x^2	1100

Table 1. Elements of $GF(16)$, constructed with $p(x) = x^4 + x^3 + 1$.

defined as polynomial addition and multiplication is defined as polynomial multiplication modulo an irreducible polynomial $p(x)$ with degree m and coefficients from $GF(2)$. $p(x)$ is called the *generator polynomial* of the field.

As an example, table 1 lists all elements of the field $GF(16)$ constructed with $p(x) = x^4 + x^3 + 1$.

The m coefficients of the polynomial representation of a field element form a vector of the m-dimensional vectorspace over $GF(2)$. An element of $GF(2^m)$ usually is represented with respect to a particular basis of this vectorspace. The second column in table 1 shows the vector representation of each element with respect to the basis $\{\alpha^3, \alpha^2, \alpha, 1\}$ with $\alpha = x$. This vector is called the *polynomial basis* (PB) representation of a field element.

3 Computations over $GF(2^m)$

3.1 Addition/Subtraction

In a finite field with characteristic 2, the additive inverse of β is β itself, that is, addition and subtraction are identical operations. Let $\beta, \gamma \in GF(2^m)$ be given in vector representation with respect to the basis $\{\alpha_0, \ldots, \alpha_{m-1}\}$. Then adding β and γ is done by modulo-2 addition of corresponding components.

Obviously, a parallel hardware realisation will use m XOR-gates while a single XOR-gate will be sufficient for bit-serial addition.

3.2 Multiplication

Multiplication in $GF(2^m)$ is much more complicated than addition. Several different hardware implementations have been proposed in the literature (e.g. [1], [13]). Their cost depends on the basis and generator polynomial chosen for the construction of the field.

For the hardware realisation of parallel multipliers, the coefficients of the product each have to be generated by a considerable amount of combinatorial logic. Fig.1 shows as an example a parallel multiplier circuit for the field $GF(16)$.

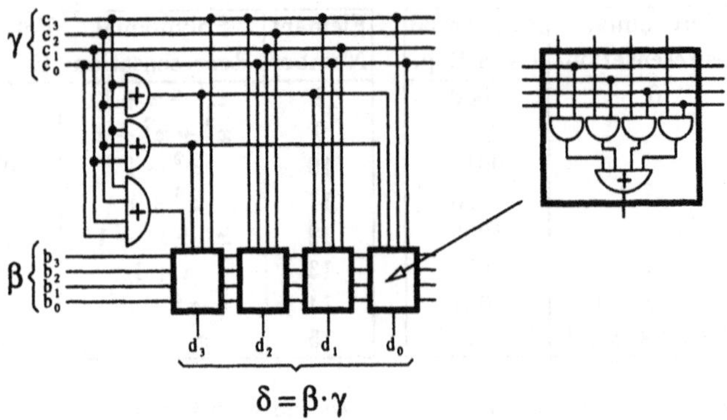

Fig. 1. A PB parallel multiplier over $GF(16)$ with $p(x) = x^4 + x^3 + 1$

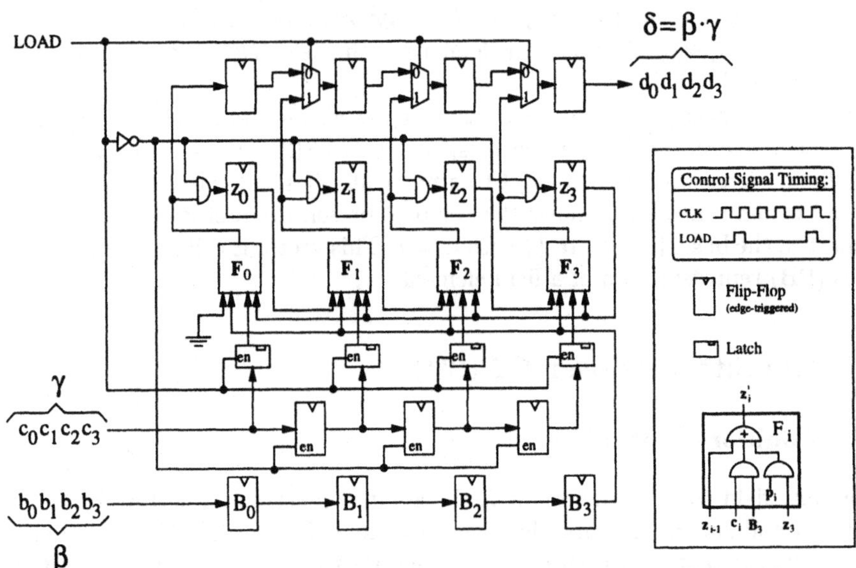

Fig. 2. A classical shift register multiplier over $GF(16)$.

This multiplier uses the PB representation and may be designed according to the procedure described in [7].

Mastrovito has shown that the classical shift register multiplier of Bartee and Schneider [1] indeed is an area- and time-efficient implementation of a bit-serial multiplier [7]. The remainder of this section briefly describes this kind of multiplier and presents its hardware implementation.

With PB representation, multiplying a field element β by α can be done by shifting the components of *beta* to the left and conditionally adding the coefficents of $\alpha^m = \sum_{i=0}^{m-1} a_i\, \alpha^i$:

$$\begin{aligned}
\alpha \cdot \beta &= \alpha \cdot \sum_{i=0}^{m-1} b_i\, \alpha^i = b_{m-1}\, \alpha^m + \sum_{i=1}^{m-1} b_{i-1}\, \alpha^i \\
&= \sum_{i=0}^{m-1} b_{m-1}\, a_i\, \alpha^i + \sum_{i=1}^{m-1} b_{i-1}\, \alpha^i \\
&= \sum_{i=1}^{m-1} (b_{m-1}\, a_i + b_{i-1})\, \alpha^i + b_{m-1}\, a_0 \quad .
\end{aligned}$$

Bit-serial multiplication of two arbitrary field elements β and γ processes each component of β sequentially. If the current component is 1, γ is added in parallel to an accumulator, which is initially set to zero. After that, the accumulator is multiplied by α and the next component of β is processed. After the m components of β have been encountered, the accumulator contains the product $\beta \cdot \gamma$.

Fig. 2 shows a circuit implementing a bit-serial multiplier for $GF(16)$. Besides the accumulator described above, the multiplier uses three buffer registers which realize a serial-in serial-out data interface.

In many applications where multiplication in $GF(2^m)$ is needed, one of the factors is a constant and thus the circuits presented in this subsection can be simplified.

3.3 Square

A circuit for the multiplication of field element β by itself is comparatively simple. A parallel implementation only has to permute some of the bits b_i and determine the remaining bits by a few (less than m) XOR-operations [7]. As an example, fig. 3 shows a parallel squarer for $GF(2^m)$.

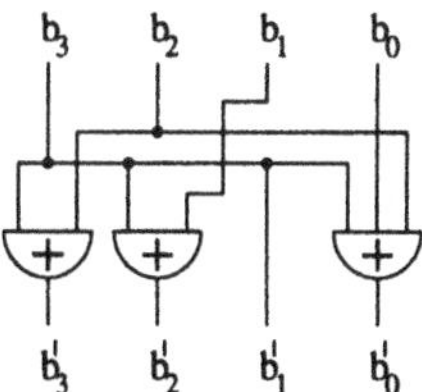

Fig. 3. A parallel squaring circuit for $GF(16)$.

3.4 Inversion/Division

To compute the multiplicative inverse of an element is much more complicated than multiplication. In many hardware implementations, simply tabulating all inverses in a ROM proved to be the most effective solution (e.g. in [11]).

As an alternative to the look-up table approach, bit-serial inversion circuits have been proposed in the literature ([13],[7],[1]) which use serial-to-parallel operand conversion and a parallel multiplier. Such an implementation makes use of the equation $\forall\beta \in GF(2^m): \quad \beta^{2^m} = \beta$

from which follows $\beta^{-1} = \beta^{2^m-2} = \beta^2 \cdot \beta^4 \cdot \beta^8 \dots \beta^{2^{m-1}}$.

From this relationship, an iterative algorithm for calculating the multiplicative inverse of β can be derived and is shown along with its hardware implementation in fig. 4. The circuit uses two buffer registers for serial-to-parallel and parallel-to-serial conversion, a parallel multiplier, and a parallel squaring circuit.

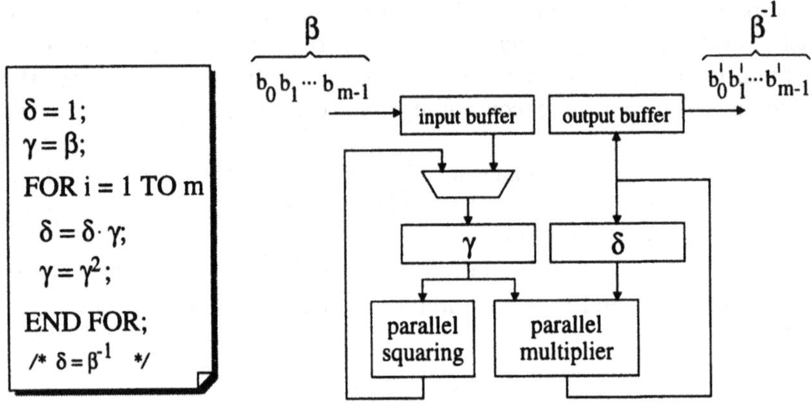

Fig. 4. Bit-serial inversion over $GF(2^m)$ by iterative calculation.

In the case of division, multiplication of the inverted divisor by the dividend can be easily integrated in the circuit of fig. 4 by initializing δ with the dividend instead of 1.

4 Implementation results

4.1 FPLD Architectures

This section presents mapping results for three major FPLD architectures, namely for the PAL-like Altera MAX7000 devices, for the multiplexor-based architecture of the Actel ACT-3 series, and for two representatives of look-up table (LUT) FPLDs: the Xilinx XC4000 LCAs (mixed 3- and 4-input LUTs) and Alteras FLEX8000 series (4-input LUT only). Table 2 gives the number of blocks available in the largest device of each architecture. This allows a normalized comparison of the architectures' respective capacities in terms of *utilization rate*, which is defined as the number of functional blocks used for the implementation of a given function, divided by the number of available blocks in the largest device of the device series.

Device Series	Largest Device	number of functional blocks
MAX7000	EPM7256	256 LCs
FLEX8000	EPF81500	1296 LEs
XC4000	XC4013	576 CLBs
ACT-3	A14100A	1377 LMs

Table 2. Device capacities

4.2 Design method

As has been shown in section 3, multiplication plays a key-role in the implementation of computations over $GF(2^m)$. Hence, the presentation of mapping results focusses on the implementation of multipliers. Delays of mapped circuits are always given for the fastest available device into which the respective design completely fits.

For the Altera devices, mapping results have been produced by automatic synthesis using the vendor's MAX+plusII tool, version 5.2. Mapping for the PAL-based MAX7000 CPLDs had to be guided by manual intervention to find an acceptable realization. The Xilinx FPGAs have been mapped using the Synopsys FPGA Compiler, version 3.2 and Xilinx XACT 5.1. Mapping to Actel's ACT-3 FPGAs has been done manually, since neither libraries for the Synopsys FPGA Compiler nor Actel's ALS development tool had been available at our institute. As all circuits exhibit a highly regular structure, hand-mapping supported by a general-purpose logic synthesis tool (`MIS-II` [5]) proved feasible and results should be comparable to those for the other devices. Reported delays are for the physically mapped circuits, except for ACT-3 devices for which delays have been estimated according to the vendor's rules-of-thumb considering logic depth and signal fanouts.

4.3 Mapping Results

Table 3 gives the mapping results for parallel multipliers for the fields $GF(2^m)$ with $m \in \{4, 6, 7, 8, 9, 10\}$. For each architecture and each field, the number of functional blocks used, the resulting device utilization, the number of block levels in the critical path, and the maximum clock frequency for a register-to-register multiplication are given. For comparative reasons, the rightmost column of table 3 lists the data for a mapping on 2-input NAND/NOR gates. This data has been obtained using `MIS-II` [5].

From this table it can be seen that the PAL-like architecture of the MAX7000 series falls well behind the other devices. Its macrocell structure does not match the XOR-rich logic functions that have to be generated and hence much of the logic resources are wasted. When compared to the respective maximum device capacity, it implements the multiplier five times less efficiently than the FLEX devices, while it is at most 30 % faster. Difficulties in implementing XOR-rich logic functions proved to be a drawback for the MUX-based devices too. Although they give significantly better utilization than the PAL-like architecture,

m	Altera MAX7000	Altera FLEX8000	Xilinx XC4000	Actel ACT-3	2-input NAND/NOR
4	15 LCs 5.8 % 3 levels 42 MHz	14 LEs 1.0 % 2 levels 65 MHz	5 CLBs 0.8 % 2 levels 43 MHz	21 LMs 1.7 % 4 levels ca. 52 MHz	97 gates - 10 levels -
6	27 LCs 10.5 % 3 levels 42 MHz	27 LEs 2.0 % 3 levels 40 MHz	11 CLBs 1.9 % 1 level 52 MHz	45 LMs 3.3 % 5 levels ca. 48 MHz	227 gates - 17 levels -
7	41 LCs 16.0 % 3 levels 42 MHz	40 LEs 3.1 % 3 levels 37 MHz	17 CLBs 3.0 % 2 levels 42 MHz	64 LMs 4.7 % 5 levels ca. 45 MHz	381 gates - 20 levels -
8	55 LCs 21.5 % 3 levels 42 MHz	55 LEs 4.2 % 3 levels 33 MHz	28 CLBs 4.9 % 3 levels 31 MHz	84 LMs 6.1 % 6 levels ca. 38 MHz	569 gates - 16 levels -
9	62 LCs 24.2 % 4 levels 35 MHz	61 LEs 4.7 % 4 levels 28 MHz	27 CLBs 4.7 % 3 levels 34 MHz	105 LMs 7.6 % 6 levels ca. 38 MHz	602 gates - 16 levels -
10	76 LCs 29.7 % 4 levels 35 MHz	76 LEs 5.9 % 4 levels 28 MHz	34 CLBs 5.9 % 3 levels 30 MHz	132 LMs 9.6 % 6 levels ca. 35 MHz	773 gates - 16 levels -

Table 3. Mapping results for parallel multipliers over $GF(2^m)$

Device	resource usage	levels	max. freq.
MAX7000	$5\,m$ MCs	1	67 MHz
FLEX8000	$6\,m$ LCs	2	49 MHz
XC4000	$3\,m$ CLBs	1	58 MHz
ACT-3	$6\,m$ LMs	2	ca. 70 MHz

Table 4. Mapping results for bit-serial multiplier over $GF(2^m)$

MUX-based FPLDs do not reach the capacity of LUT-based architectures. But ACT-3 FPGAs tend to be faster with increasing field dimension m.

Table 3 also shows that the additional 3-input LUT in the XC4000 can be used effectively to reduce the number of CLBs and/or the logic depth in comparison with the pure 4-input LUT structure of the FLEX8000 devices. On average, both device series exhibit the same utilization and speed. But in contrast to Altera's FLEX8000 devices, unused registers of CLBs that implement combinatorial logic functions may still be used for other circuitry in a XC4000 device, e.g. for serial-to-parallel and parallel-to-serial conversion in the bit-serial inversion circuit of fig. 4. This feature would increase the capacity of XC4000 devices when compared to FLEX8000. A similar feature in Altera's new MAX9000 devices will improve the utilization of the MAX7000 series in the same way.

Mapping results for bit-serial multipliers are given in table 4. The complexity of the circuit is proportional to m for all architectures. The PAL-like architecture is almost as fast as the MUX-based architecture, but the latter has a much lower utilization rate. When speed is the main concern, both devices outperform the LUT-FPLDs which again exhibit the highest capacity. Here, the H-block of the Xilinx XC4000 CLB can be used effectively to reduce the logic depth to a single CLB. In addition, since buffering can be overlayed with combinatorial logic in the Xilinx device, XC4000 LCAs also implement serial multiplication more area-efficient than FLEX8000 CPLDs. The latter use approximately one half of their logic resources for buffering purposes.

5 Error Control Coding

5.1 Error-correcting Block Codes

One of the most important application areas of computations in finite fields $GF(2^m)$ is error control coding, namely the implementation of algorithms for encoding and decoding of linear block codes. The general idea of error-correcting block codes (ECBC) is to protect a sequence of k m-bit symbols by $(n-k)$ *parity* or *check symbols* so that the Hamming distance between any two distinct codewords of length n is not less than a selected lower bound d. Then a decoder may reconstruct the original codeword even in the presence of up to $\lfloor(d-1)/2\rfloor$ errornous symbols. *Cyclic linear block codes* over $GF(2^m)$ proved to be especially well suited for hardware implementation of encoders and decoders. Within this class, the check bits are chosen so that the resulting codeword $(c_{n-1}, c_{n-2}, \ldots c_1, c_0)$, when interpreted as a polynomial $c(x) = \sum_{i=0}^{n-1} c_i x^i$, is divisable by a polynomial $g(x)$ of degree $(n-k)$. $g(x)$ is called the generator polynomial of the code.

5.2 Cyclic Linear Block Encoders

In order to guarantee that $c(x)$ generated from an information polynomial $i(x)$ is a multiple of $g(x)$, a systematic encoder computes $c(x) = i(x)x^{n-k} - r(x)$, where $r(x) = (i(x)x^{n-k}) \bmod g(x)$ [3]. Such an encoder is sketched in fig. 5. It can be seen from this figure that one should search for a generator polynomial $g(x)$ with the least number of distinct non-zero coefficients since as many $GF(2^m)$-multipliers are needed.

Table 5 lists mapping results for the implementation of encoders for a popular class of ECBCs: the Reed-Solomon codes. For each of the three codes, results for an encoder which uses either a parallel or a serial multiplier circuit are given.

Table 5 also summarizes the main parameters of the codes, namely the type of the code, the code's blocklength n, the number of information symbols per block

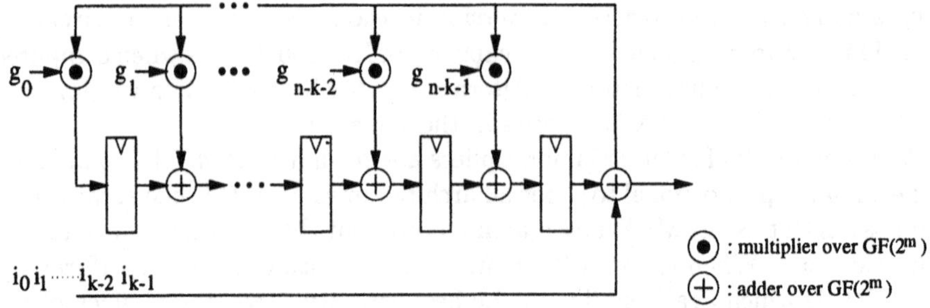

Fig. 5. Principle structure of an encoder for a cyclic linear block code

k, the number of distinct non-zero coefficients of $g(x)$, g, the minimum distance d, and the number of correctable symbol errors per block t. According to Berlekamp [3], the first type of code named *rs31* has been adopted by the Joint Tactical Information Distribution System (JTIDS) of the U.S. Armed Forces. The type of code named *rs255* has widely been used in space communication and therefore is sometimes called the NASA-code. Finally, the shortened Reed-Solomon code *rs32* is one of the types of Reed-Solomon codes used in the cross-interleaved code for the Compact Disk [10].

It can be seen that except for the PAL-like architecture, Reed-Solomon encoders can be realized efficiently with field-programmable logic. Bit-serial encoders exhibit an encoding rate of approximately 30 Mbit/sec. In most cases, parallel encoders are as area-efficient as their serial counterparts or are even smaller. Thus, for bit-serial encoding it would be best to use a parallel encoder with serial-to-parallel and parallel-to-serial conversion at the interface. This would reduce the apparent logic depth to a single functional block and should allow encoding rates near 100 Mbit/sec.

5.3 Dynamic Code Adaption

In case an in-circuit reconfigurable device is used for encoder implementation, FPLDs give rise to a new interesting feature for data transmission over noisy channels. With FPLDs it becomes possible to build encoders with dynamic code adaption based on measurements of the channel quality. Fig. 6 shows a system for data transmission that uses this feature. At the side of the sender, an encoder adds parity symbols to the stream of information symbols, sending codewords to the modulator. The receiver first demodulates the incoming signals from the *data channel* and than decodes a received word $v(x)$ to restore the original information $i(x)$. Decoder and demodulator at the receiving end collect information on the number of detected/corrected errors and on the signal noise. This data is handed over to an unit that evaluates the channel quality. If the ECC currently used on the data channel does no longer match the channel's actual noise level, coding may be switched to another code by reconfiguring the encoder and the decoder. Reconfiguration of the encoder is induced over a second low bandwidth channel

Code Acronym	Code Description	n	k	g	d	t
rs31	Reed-Solomon code over $GF(2^5)$	31	15	8	17	8
rs255	Reed-Solomon code over $GF(2^8)$	255	223	16	33	16
rs32	shortened RS code over $GF(2^8)$	32	28	3	5	2

Code	Altera MAX7000	Altera FLEX8000	Xilinx XC4000	Actel ACT-3
rs31 parallel	90 MCs 45.5 MHz	90 LEs 29 MHz	45 CLBs 28 MHz	116 LMs ca. 45 MHz
rs31 bit-serial	167 MCs 40 MHz	172 LEs 31 MHz	86 CLBs 35 MHz	211 LMs ca. 40 MHz
rs255 parallel	no fit	314 LEs 22 MHz	145 CLBs 27 MHz	478 LMs ca. 40 MHz
rs255 bit-serial	no fit	547 LEs 27 MHz	264 CLBs 28 MHz	645 LMs ca. 42 MHz
rs32 parallel	78 MCs 40 MHz	55 LEs 28 MHz	26 CLBs 34 MHz	79 LMs ca. 42 MHz
rs32 bit-serial	74 MCs 45 MHz	77 LEs 33 MHz	38 CLBs 40 MHz	91 LMs ca. 50 MHz

Table 5. Code parameters and mapping results for some Reed-Solomon encoders.

which may carry the configuration data as well. Of course, the configuration channel has to be protected against errors, too. If enough energy is available at the sender, this may be easily accomplished by a high signal-to-noise ratio (SNR). Such a situation is assumed in fig. 6 where the sender may be earthbound while a satellite receiving the configuration data has very limited power. The configuration channel needs to sustain a much smaller bandwidth than the data channel. This eases error protection and allows the use of an inefficient but simple ECC that is trivial to decode (e.g. a repetition code).

6 Conclusion

The implementation results presented show that FPLDs are suitable for the implementation of finite field arithmetic. In this application area, PAL-like devices are significantly inferior to MUX-based FPLD architectures. FPLDs which use LUTs in their functional blocks give the highest capacity at an acceptable speed. The mapping results in section 5 show that even complex and powerful error-correcting codes may be encoded with FPLDs. It remains to be seen whether FPLD-based *decoding* is feasible as well. Besides facilitating rapid prototyping and low-volume production in the field of error control coding, in-circuit reconfigurable FPLDs give rise to data transmission systems with dynamic code adaption based on channel measurement information.

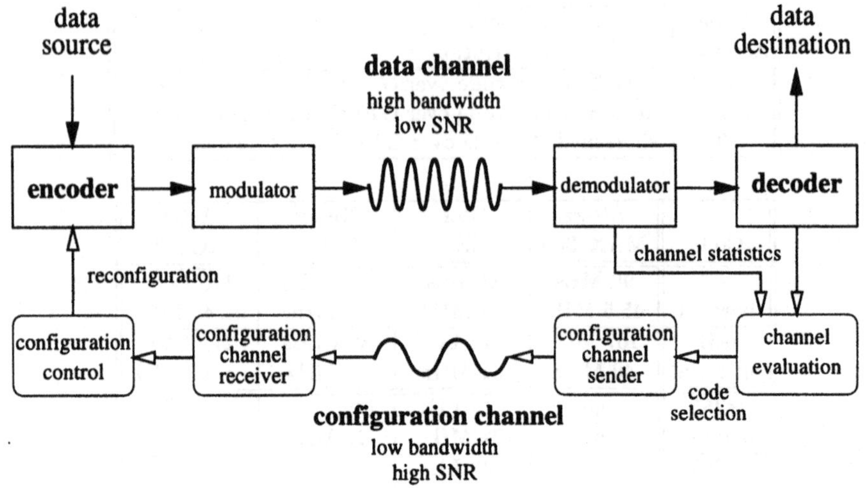

Fig. 6. A data transmission system with dynamic code adaption.

References

1. T. C. Bartee and D. I. Schneider, *Computation with finite fields*, Inform. Contr., vol. 6, pp.79-98, Mar. 1963.
2. E. R. Berlekamp, *Algebraic Coding Theory*, New York: McGraw-Hill, 1968.
3. E. R. Berlekamp, *The Technology of Error-Correcting Codes*, Proc. IEEE, vol. 68, pp.564-593, May 1980.
4. R. E. Blahut, *Theory and practice of error control codes*, Addison-Wesley, 1983.
5. R. K. Brayton et al., *MIS: A Multiple-Level Logic Optimization System*, IEEE Trans. on Computer Aided-Design, CAD-6(6), pp. 1062 - 1081, Nov. 1987
6. R. Lidl and H. Niederreiter, *Introduction to finite fields and their applications*, London, New York: Cambridge University Press, 1986
7. E. D. Mastrovito, *VLSI designs for computations over finite fields* $GF(2^m)$, Ph.D. Thesis No.159, Dept. of Electr. Engineering, Linköping Univ., Sweden, Dec. 1988.
8. R. J. McEliece, *Finite Fields for Computer Scientists and Engineers*, Boston,MA: Kluwer Academic Publishers, 1987.
9. P. J. McWilliams and N. J. A. Sloane, *The Theory of Error-Correcting Codes*, Amsterdam: North-Holland, 1978.
10. H. Hoeve, J. Timmermans and L. B. Vries, *Error correction and concealment in the Compact Disk system*, Philips tech. Rev. 40, No. 6, 1982.
11. J.-L. Politano and D. Deprey, *A 30 Mbits/s (255,223) Reed-Solomon Decoder*, EUROCODE '90, Int't Symp. on Coding Theory and Applications, Udine, Italy, Nov. 1990.
12. H. C. A. van Tilborg, *An Introduction to Cryptology*; Boston,MA; Kluwer Academic Publishers, 1988.
13. C. C. Wang et. al., *VLSI Architectures for Computing Multiplications and Inverses in* $GF(2^m)$, IEEE Trans. Comput., vol. C-34, pp.709-717, Aug. 1985.

Implementation of Fast Fourier Transforms and Discrete Cosine Transforms in FPGAs*

G. Panneerselvam, P. J.W. Graumann and L. E. Turner

Department of Electrical and Computer Engineering
University of Calgary, 2500 University Drive N. W.
Calgary, Alberta, Canada T2N 1N4
Phone: (403) 220-5810 Fax: (402) 282-6855
email: turner@enel.uclagary.ca

Abstract. Fast Fourier Transform (FFT) and Discrete Cosine Transform (DCT) processors are designed and implemented using a Xilinx Field Programmable Gate Array (FPGA) device XC 4010. This device allows a 16-point FFT/DCT processor implementation. To design the CLB-efficient FFT/DCT processor in FPGAs, a pipelined bit-serial architecture with bit-parallel input data format is employed. These processors operate with a 20 MHz bit-clock and 16-bit system word length, and compute an entire 16-point DFT/DCT transform for every 16-bit clock cycle.

1 Introduction

The Discrete Fourier Transform (DFT) and Discrete Cosine Transform (DCT) have a wide range of applications in Digital Signal Processing (DSP), image processing, power spectrum analysis and filter simulation [1,2]. The DFT is a well-known classical algorithm. A fast and efficient algorithm, known as Fast Fourier Transform, was proposed by Cooley and Tukey in 1965 [3] to compute the DFT of a data set. Algorithm, architectures, discrete hardware and, recently, Application Specific Integrated Circuits (ASICs) have been developed to implement the FFT. Although DCT was originally discovered by Ahmed, Natarajan and Rao [4] in 1974, it has provided a significant impact on DSP and related fields. Various fast algorithms and architectures for DCT have been presented based on FFT, real arithmetic and recursive algorithms. Hardware and ASICs have been developed. Field Programmable Gate Arrays (FPGA), the fast emerging microelectronic technology, permit prototyping and cost-effective, high-performance ASICs. This paper presents the design and implementation of the FFT and DCT processors using a single Xilinx FPGA device XC4010 which provides approximately 400 Configurable Logic Blocks (CLBs)[5], and is utilized in these FFT and DCT processor implementations. The available CLBs permit a 16-point FFT/DCT processor, in a bit-serial, fully pipelined architecture using a bit-parallel input/output

*. This work was supported in part by funding from the National Sciences and Engineering Research Council of Canada and Micronet, A Canadian Network of Centres of Excellence.

format. To achieve a CLB-efficient implementation, various optimization techniques are used.

This paper presents the FFT and DCT algorithms and their respective data flow graphs. Design criteria for a CLB-efficient implementation of the FFT and DCT processors are explained. Architecture design details and the implementation results are presented in the subsequent sections. Further possible research directions are given in the conclusion.

2 Algorithms

2.1 FFT Algorithm

The DFT of the data set $x(N)$ is defined as follows:

$$X(k) = \sum_{n=0}^{N-1} x(n)\, W_N^{kn} \qquad k = 0,1,2, \ldots, N-1$$

where $W_N = e^{-j\left(\frac{2\pi}{N}\right)}$ is the complex N th root of unity.

In the DFT computation, each term involves complex arithmetic operations which requires $4N$ real multiplications and $4N$-2 real additions. The entire DFT computation requires $4N^2$ real multiplications and $4N^2$-$2N$ real additions, resulting in $O(N^2)$ complexity. Decomposing the sequence $x(n)$ into successively smaller sequences (decimation-in-time) reduces the computation from $O(N^2)$ to $O(N \log N)$ for the special case of N, i.e., $N = 2^r$, as described by Cooley and Tukey [1], and known as Fast Fourier Transform (FFT). The computation of the N-point DFT can be obtained by the evaluation of two $N/2$-point sequences in parallel which consists of the even-numbered and odd-numbered points in $x(n)$. These two transforms are combined together to form the N-point DFT. These $N/2$-point DFTs can be further divided into $N/4$,..., until $N = 2$. For example, a *16*-point DFT computation can be divided into *8*-, *4*- and 2-point DFTs. These DFTs can be computed using butterfly structures. These structures result in a parallel architecture with a reduced number of multiplications, providing an efficient realization of the DFT in FPGAs.

2.2 DCT Algorithm

Varieties of fast algorithms have been proposed for the DCT computation of the data set x(N) [2]. Lee [6] proposed a fast algorithm for DCT computation which is similar to the FFT, known as Fast Cosine Transform (FCT). This DCT is described as follows:

$$X(k) = \frac{2}{N} e(n) \sum_{n=0}^{N-1} x(n) \cos\left[\frac{(2k+1)n\pi}{2N}\right] \qquad k = 0,1,2,\ldots,N-1$$

where $e(n) = \begin{cases} (1/\sqrt{2}), & \text{if } n = 0, \\ 1, & \text{otherwise.} \end{cases}$

The DCT algorithm requires only the real arithmetic operations rather than the complex arithmetic operations used in the FFT. This algorithm can be implemented efficiently using the decomposition procedure as explained in the previous section. An *N*-point DCT is evaluated using two *N/2*-point DCTs; these *N/2*-point DCTs can be further divided into *N/4*,..., until $N = 2$. This FCT algorithm can be effectively implemented using the butterfly network with some modifications as explained in subsequent section. This algorithm uses a lower number of real multiplications, providing an efficient architecture which is suitable for FPGA implementation.

3 Architecture

3.1 FFT Architecture

The functional flow-diagram of an N-point FFT is shown in Fig. 1, for $N = 16$. The input and output bins (value of the transform X at the k-th frequency point) are represented by *x(0), x(1), x(2),..., x(15)* and *X(0), X(1), X(2),..., X(15)* respectively. As explained in section 2.1, a higher-order FFT computation can be achieved using the lower-order FFTs. Thus a *16*-point FFT can be computed using two *8*-point FFTs, each of which consists of two *4*-point FFTs. Furthermore, the *4*-point implementation requires two 2-point FFTs.

The *radix-2* butterfly permits parallel data flow and computation for a prescribed data set as shown in Figure 2. It facilitates the complex arithmetic computations in parallel, where *x(0)* and *x(8)* may be real or complex, and *W* represents the weight of the unity magnitude, which are complex data. These data are used in the subsequent sections of the network.

In the first stage of the flow diagram in Fig. 1, the computation is simplified into real additions and subtractions since, for $N = 2$, the complex root of the unity (constant) is W_2^0 which is 1. In the second stage, complex arithmetic operations are involved due to the constant values W_4^0 and W_4^1. The third and fourth stages employ complex multiplications and complex additions. In the third stage, W_8^0, W_8^1, W_8^2 and W_8^3 constants are used. The fourth stage requires the following constants:

$W_{16}^0, W_{16}^1, W_{16}^2, W_{16}^3, W_{16}^4, W_{16}^5, W_{16}^6$ and W_{16}^7. Some of these constants can be simplified using the periodicity relations. The following constants

$W^0, W^0_2, W^0_8, W^0_{16} = 1$ and $W^1_4, W^2_8, W^4_{16} = j$ need no multiplier. Constants $W^1_8, W^2_{16} = \left(\frac{1}{\sqrt{2}} + \frac{1}{j\sqrt{2}}\right)$ and $W^3_8, W^6_{16} = -\frac{1}{\sqrt{2}} + \frac{1}{j\sqrt{2}}$ need two multipliers each. Whereas constants $W^1_{16}, W^3_{16}, W^5_{16}, W^7_{16}$ require four multipliers for every evaluation. The butterfly network results in complex data, which consists of both real and imaginary terms and requires a magnitude output.

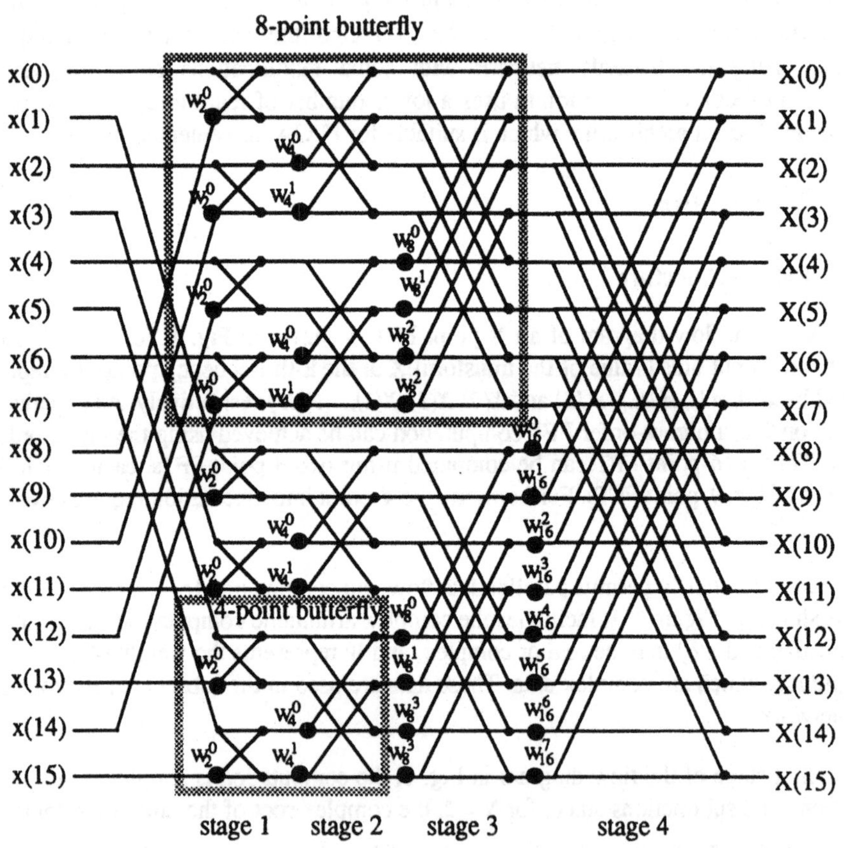

Figure 1. Functional flow diagram of 16-point FFT.

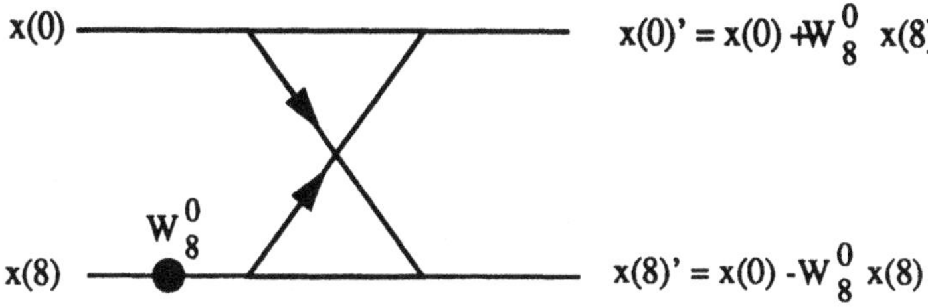

Fig 2. Butterfly representation for FFT.

3.2 DCT Architecture

The signal flow-diagram of an N-point DCT is shown in Fig. 3, for $N = 16$. Unlike FFT computation, the DCT uses only the real inputs and outputs. The inputs and outputs are represented by $x(0), x(1), x(2),..., x(15)$ and $X(0), X(1), X(2),..., X(15)$ respectively. As explained in section 2.2, higher-order computations can be achieved using lower-order DCTs. Thus, a *16*-point DCT can be computed using two *8*-point DCTs. These *8*-point DCTs use *4*-point DCTs, and the *4*-point DCT uses two *2*-point DCTs. The entire computations can be divided into two major parts: butterfly data computation and data manipulation. The butterfly stage consists of four butterfly stages. Each butterfly performs a real addition/subtraction operation followed by data multiplication by a constant, where the constant is determined by the algorithm. In the butterfly part, the first stage performs a higher-order butterfly operation. This butterfly operation is shown in Figure 4. This butterfly facilitates the parallel real data computations, where $x(0)$ and $x(15)$ are real/complex, and $C_{32}^{\ 1} = \cos\pi/32$ represents the constituent constant value determined from the algorithm. The $x(0)'$ and $x(15)'$ data are used in subsequent parts of the network for the DCT computation.

The first stage of butterflies use the following constants: $1/\left(2C^1{}_{32}\right)$, $1/\left(2C^3{}_{32}\right)$ $1/\left(2C^7{}_{32}\right)$, $1/\left(2C^5{}_{32}\right)$, $1/\left(2C^{15}{}_{32}\right)$, $1/\left(2C^{13}{}_{32}\right)$, $1/\left(2C^9{}_{32}\right)$ *and* $1/\left(2C^{11}{}_{32}\right)$. In the second stage of butterflies perform the *8*-point DCT computations which use $1/\left(2C^1{}_{16}\right)$, $1/\left(2C^3{}_{16}\right)$, $1/\left(2C^5{}_{16}\right)$ *and* $1/\left(2C^7{}_{16}\right)$ constants. The third stage performs the *4*-point DCT computations and uses the following constants $1/\left(2C^1{}_8\right)$, $1/\left(2C^3{}_8\right)$. The final stage of butterflies computes the *2*-point DCT and uses $1/\left(2C^1{}_4\right)$ as the constant value. Each constant requires one multiplication. After these computations, the real data are manipulated with data addition as shown in the data flow graph. Once these additions are carried out, the data sets are scaled down by the prescribed values.

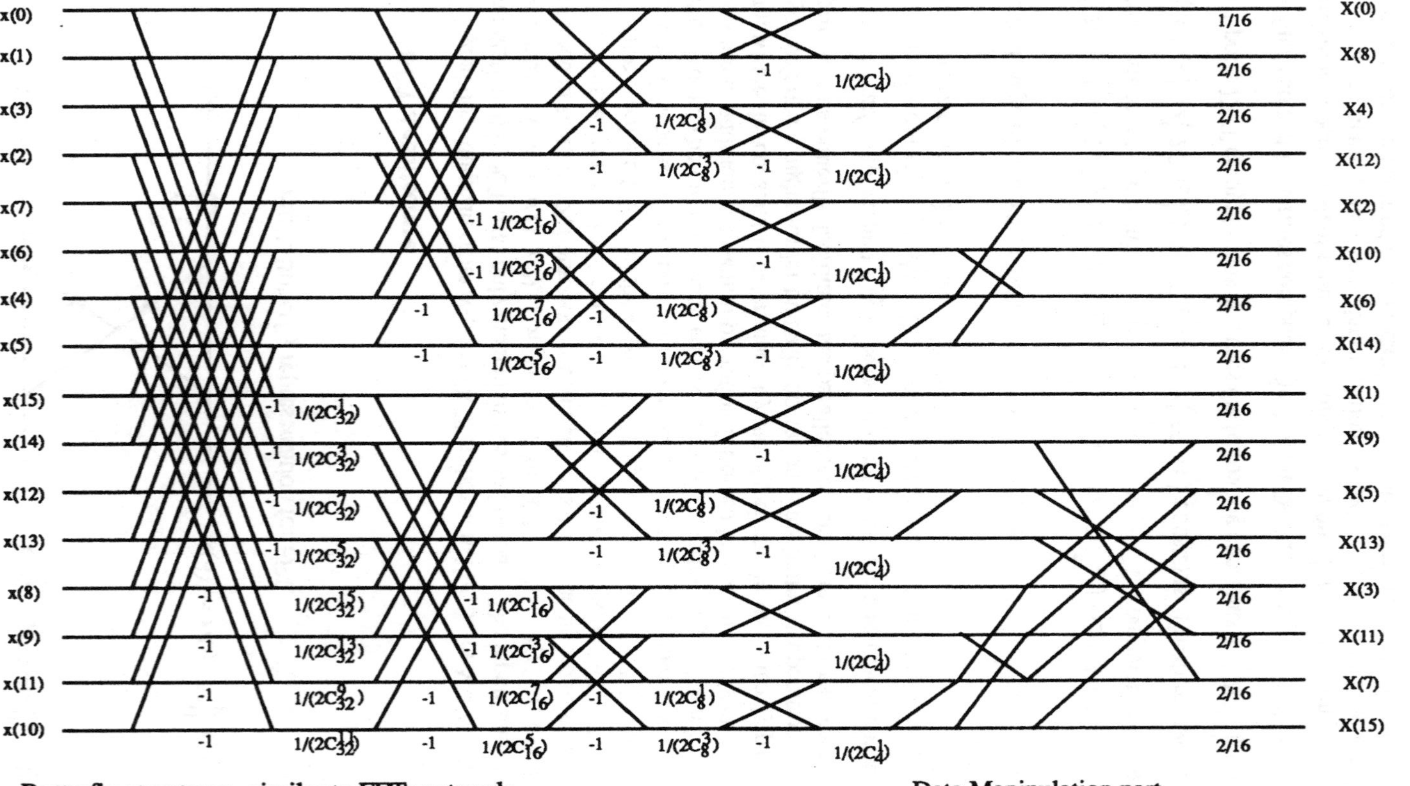

Figure 3. Signal flow diagram for 16-point DCT computation

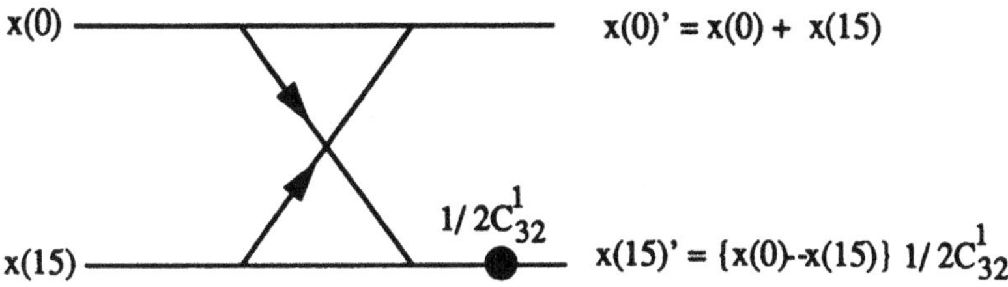

Fig 4. Butterfly representation for DCT.

3.3 Design Architecture Considerations

The functional flow graphs as shown in Figures 1 and 3 are directly implemented as architecture for the FFT and DCT, respectively. Since the available number of CLBs in an FPGA device restricts the size of the processor, each stage requires a different number of multipliers, and pipelined latches. In order to realize the largest possible FFT/DCT processor, the following design criteria were used.

Bit-parallel and bit-serial [7] architectures were examined and bit-serial architecture were used. The butterfly architecture allows parallel data flow and parallel computation. A fully pipelined bit-parallel input/output interface was used to achieve a high throughput rate.

The finite precision effect of data has to be considered in order to determine the system's word length and input/output data format. These effects include roundoff noise in truncating the result of a multiplication, scaling the data to prevent overflows, resulting in inaccurate transforms. Therefore, a 16-bit system word length was selected with an 12-bit input/output data format that allows 4-guard bits.

4 Processor Implementation

4.1 System Architecture

The FFT and DCT processor system architecture is shown in Fig. 5. The analog input signal is converted into a *12*-bit digital signal and fed into as a parallel data format. This input is converted into 16-bit bit-serial data, and transmitted to the delay network, so that all *N* inputs are fed together in parallel. The FFT and DFT compute engines represent the entire 16-point butterfly architectures which are shown in Figure 1 and 3, respectively. It is implemented in a pipelined manner. It allows the FFT/DCT engine to compute *N* inputs in parallel as well as in pipelined fashion. Once the pipeline is filled with valid data, all outputs emerge one set after another with a constant system latency. However, the bin-selector allows the selection of a specific output from the output-set. The output of the FFT engine is complex data and the corresponding magnitude is obtained by the magnitude approximator. Whereas, the DCT engine outputs real data, and the magnitude approximator is not needed. The multiplexed parallel output port allows single output bin data in a *12*-bit word. Thus the same system architecture can be utilized for both FFT and DCT computation.

4.2 System Implementation

Tools The FFT/DCT processors have been designed, implemented and simulated using the *DFIRST* RTL language and the *dsim* simulator [8,9]. The *DFIRST* description can be translated into the gate-level netlist using the *trans* compiler. This gate-level netlist format is appropriate for the Xilinx implementation tools. To achieve a CLB-efficient implementation, sub systems and components are designed and implemented using the minimum number of CLBs. In order to save hardware, the redundant elements in the circuit are removed, and the loads of the removed devices were driven by the remaining elements. This requires control of fan-in and fan-out factors. In order to reduce the gate-counts for these pipelined latches, a RAM-based design was used. In order to increase the number of data flip-flops, unused input/output ports are utilized two data flip-flops.

Multiplier The *SHIFTMULT* primitive performs multiplication on a signal of bit-serial data with a fixed coefficient value. This multiplier is much smaller than other designs due to the fixed canonic signed digit coefficient (CSD) representation [10]. These coefficients are represented in two's complement format. In the CSD format the weight of each bit is $\pm 2^i$. The coefficient of *98* can be easily set using these three parameters, namely, number of bits, number to be set and sign (negative number). For example, parameters *[8,162,32]* results in the coefficient value 98. The latency is given as the sum of bits and number of bits to be set minus one. The above coefficient set results in 10 as latency.

Magnitude Approximator The magnitude of a complex number is the root of the sum of the squares of the real and imaginary parts. A four-region approximation algorithm is $mag = Max\{A, B, \frac{7}{8}A + \frac{1}{2}B, \frac{1}{2}A + \frac{7}{8}B\}$, where A and B are positive numbers [7]. The DCT computation does not require this sub system.

4.3 16-Point FFT Processors

The *16*-point FFT processor consists of both the system architecture and the FFT engine. The entire system has been implemented into single Xilinx *XC4010* device. The first and second stage of the FFT engine uses no multiplier and their latency is one (due to adder). The third and fourth stage uses four and ten multipliers respectively, and results in a latency of 13-bit. The butterfly engine's latency is 28 bits. The system's latency is 52 bits which includes the latency of the I/O interface, parallel to serial, magnitude approximation and serial to parallel operations and the butterfly engine. Once the pipeline has been filled a new *16*-point FFT is computed and the output data is available every *16* clock cycles. This system occupies *96%* of the CLBs, I/O blocks *130* of *160* (*81%* utilization), function generators *738* of *800* (*92%*) and data flip-flops *768* of *800* (*96%*), and the total utilized CLBs are *389*. This system operated at *20* MHz bit-clock.

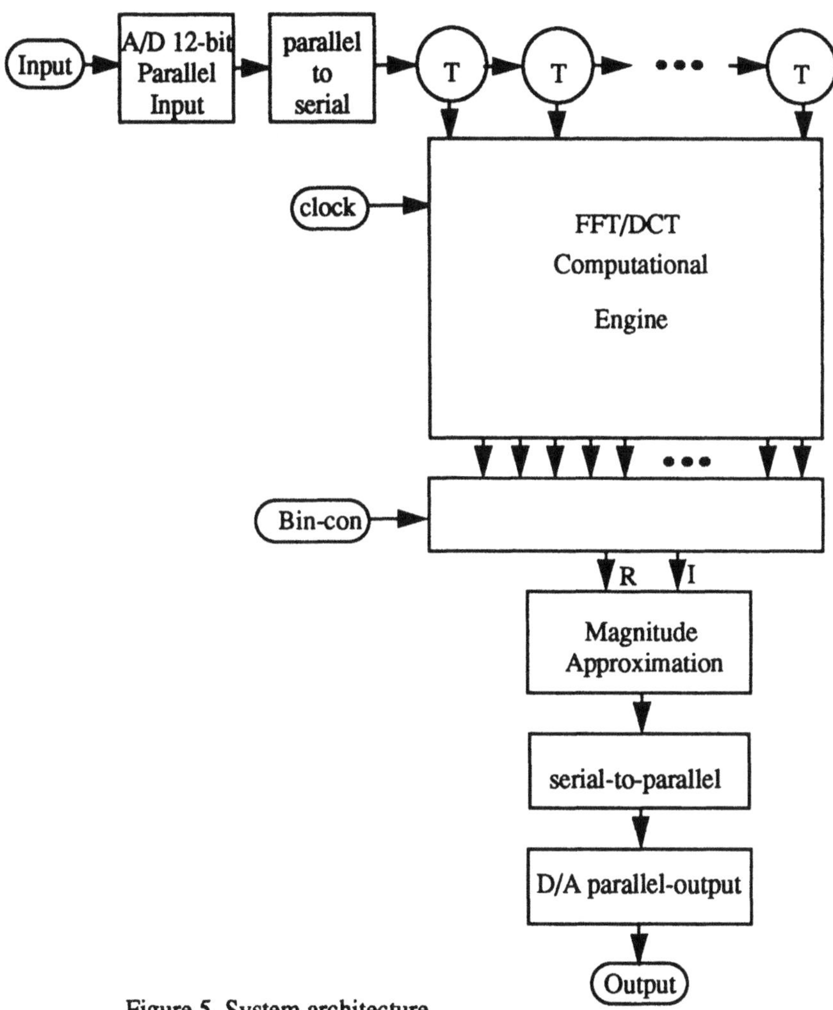

Figure 5. System architecture.

4.4 16-Point DCT Processors

The *16*-point DCT processor consists of both the system architecture and the DCT engine. The whole system has been implemented into single Xilinx *XC4010* device. In the DCT engine, the butterflies stage consists of four stages. Each stage uses eight multipliers, and 32 multipliers were utilized for the entire network. The longest latency of the multiplier is 15 bits. The system's latency is 75 bits. This system occupies *96%* of the CLBs, I/O blocks *136* of *160* (*81%* utilization), function generators *551* of *800* (*69%*) and data flip-flops *770* of *800* (*96%*), and the total utilized CLBs are *385*. This system operated at *20* MHz bit-clock.

5 Conclusions

16-point FFT and DCT processors were designed and implemented in single Xilinx XC4010 device. A bit-serial architecture with a pipelined arithmetic scheme was employed. To fit the entire design in a single device, hardware efficient CSD multipliers were utilized. The 16-point FFT/DCT processor implementations utilize 389 and 385 CLBs out of 400, respectively. Both systems operate at *20* MHz bit-clock. The system implementation allows a new FFT/DCT computation every *16* clock cycles. To reduce the hardware effectively only a single output is available. However, the DCT processor design could be easily modified to observe all the bins together in parallel for various applications.

6 References

[1] Oppenheim, A.V. and R. W. Schafer, "Digital Signal Processing," Prentice-Hall, Inc., Englewood Cliffs, New Jersey, 1975.

[2] Rao, K.R. and P. Yip, "Discrete Cosine Transform Algorithms, Advantages Applications," Academic Press, Inc. 1990.

[3] Cooley, J.W. and J.W. Tukey, "An Algorithm for the Machine Calculation of Complex Fourier Series," Math. of Comput., Vol 19, pp 297-301, April 1965.

[4] Ahmed, N., T. Natarajan, and K.R. Rao, "Discrete Cosine Transform," IEEE Trans. on Computers, Vol C-29, pp 90-94, Jan. 1974.

[5] XILINX, The Programmable Logic Data Book," Xilinx Inc., 1994.

[6] Lee, B. G., "A New Algorithm to Compute the Discrete Cosine Transform," IEEE Trans. on Acoustics, Speech, and Signal Processing, Vol ASSP-32, pp 1243-1245, Dec. 1994.

[7] Denyer, P. and D. Renshaw, "VLSI Signal Processing: A Bit-Serial Approach," Addison-Wesley Publishing Company, 1985.

[8] Graumann and L. Turner, "TRANS User's Guide," Dept. of Electrical and Computer Engg. University of Calgary, 1992.

[9] Graumann, P., "DFIRST User's Guide," Dept. of Electrical and Computer Engg. University of Calgary, 1992.

[10] Oberman, R.M.M., "Digital Circuits for Binary Arithmetic," Macmillan Press Ltd. 1979.

[11] Kassem, A., J. Davidson, and J.L. Houle, "FPGA implementation of Discrete Cosine Transform", 3rd Canadian Workshop on Field-Programmable Devices: Technology, Tools and Applications, Montreal, Canada, pp 222-225,May 29-June 1, 1995.

Implementation of a 2-D Fast Fourier Transform on an FPGA-Based Custom Computing Machine

Nabeel Shirazi, Peter M. Athanas, and A. Lynn Abbott
Virginia Polytechnic Institute and State University
Bradley Department of Electrical Engineering
Blacksburg, Virginia 24061-0111

Abstract. The two dimensional fast Fourier transform (2-D FFT) is an indispensable operation in many digital signal processing applications but yet is deemed computationally expensive when performed on a conventional general purpose processors. This paper presents the implementation and performance figures for the Fourier transform on a FPGA-based custom computer. The computation of a 2-D FFT requires $O(N^2 \log_2 N)$ floating point arithmetic operations for an NxN image. By implementing the FFT algorithm on a custom computing machine (CCM) called Splash-2, a computation speed of 180 Mflops and a speed-up of 23 times over a Sparc-10 workstation is achieved.

1 Introduction

Two dimensional convolution is a fast and simple way of filtering an image in the spatial domain if the template being used is relatively small (i.e., 8x8 pixels). As the template grows in size, the computational burden increases geometrically. Convolution of larger templates can be done much faster by converting an image in the spatial domain to the frequency domain and then applying a filter by doing point-by-point multiplication[6]. The filtered image in the frequency domain is then converted back to the spatial domain by doing an inverse Fourier transform.

Image and digital signal processing (DSP) applications typically require high calculation throughput [4,10]. The 2-D fast Fourier transform application presented here was implemented for near real-time filtering of video images on the Splash-2 FPGA-based custom computing machine (CCM). This application requires the ability to do floating point arithmetic. The use of floating point allows a large dynamic range of real numbers to be represented and helps to alleviate the underflow and overflow problems often seen in fixed point formats. An advantage of using a CCM for floating point is the ability to customize the format and algorithm data flow to suit the application's needs.

An overview of the FFT algorithm and the method used for filtering video images are given in Section 2. A description of the floating point format used in the application is given in Section 3. The implementation of the 2-D FFT on the Splash-2 architecture is shown in Section 4. In Section 5, error analysis is presented to show that the chosen floating-point format used is adequate for this application. The performance of this implementation of an FFT was compared to a wide range of architectures in Section 6.

2 Image Filtering using the Fourier Transform

An example illustrating the application of the Fourier transform to images is shown in Figure 1. An exponential highpass filter is used to attenuate the low frequency components in order to perform edge detection. The black pixels of the filter in Figure 1 correspond to zero values and the white pixels correspond to values of one with the remaining gray pixels ranging between 0.0 and 1.0. The four corners of the images in the frequency domain are the locations of the low frequency components. The high-frequency components are located near the center of the image.

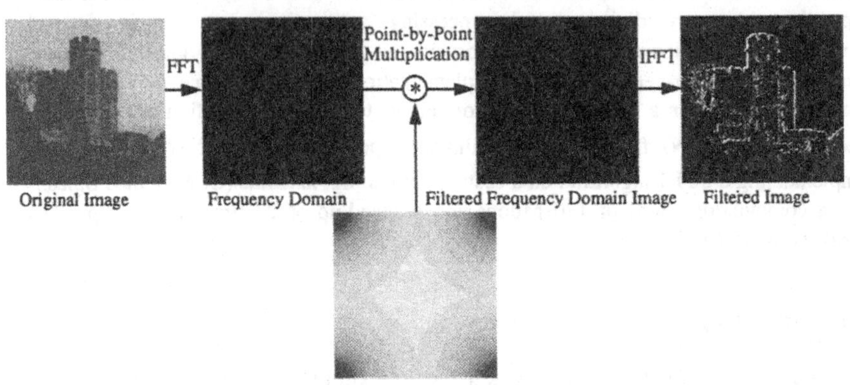

Figure 1: Fourier Transform Filtering Method.

2.1 Discrete Fourier Transform

The 2-D DFT of a NxN image, *f(x,y)*, is defined by the expression,

$$F(u) = \frac{1}{N}\sum_{x=0}^{N-1}\sum_{y=0}^{N-1} f(x,y)\, e^{-j2\Pi(xu+vy)/N}$$

for $u, v \in [0, N-1]$

The 2-D DFT expression can be decomposed into multiple 1-D Fourier transforms. The above equation can be expressed in the form:

$$F(u,v) = \frac{1}{N}\sum_{x=0}^{N-1} F(x,v)\, e^{-j2\Pi ux/N}$$

for $u \in [0, N-1]$

and,

$$F(x,v) = \frac{1}{N}\sum_{y=0}^{N-1} F(x,y)\, e^{-j2\Pi vy/N}$$

for $x, v \in [0, N-1]$

where $W^k_N = e^{-j2\pi x/N}$ or $e^{-j2\pi y/N}$ and is called the twiddle factor.

This shows that an NxN 2-D DFT can be computed by first performing N 1-D DFTs (one for each row), followed by another N 1-D DFTs (one for each column).

2.2 Fast Fourier Transform

The fast Fourier transform algorithm (FFT) consists of a variety of tricks for reducing the computation time required to compute a DFT[10]. The number of complex multiplications and additions required to implement an N-point DFT is proportional to N^2. The 1-D DFT can be decomposed so that the number of multiply and add operations is proportional to N $\log_2 N$. The FFT algorithm achieves its computational efficiency through a divide and conquer strategy. The essential idea is a grouping of the time and frequency samples in such a way that the DFT summation over N values can be expressed as a combination of two point DFTs. The two point DFTs are called butterfly computations and requires one complex multiply, and two complex additions to compute. The notation used for a butterfly structure is shown in Figure 2. By using the FFT partitioning scheme, an 8 point FFT can be computed as shown in Figure 2. Each stage of the N point FFT is composed of N/2 radix-2 butterflies and there are a total of $\log_2 N$ stages. Therefore there are a total of $(N/2)\log_2 N$ butterfly structures per FFT. In addition, the input is in bit-reverse order and the output is in linear order. A 2-D FFT can be decomposed into two arrays of 1-D DFTs, each of which can be computed as a 1-D FFT.

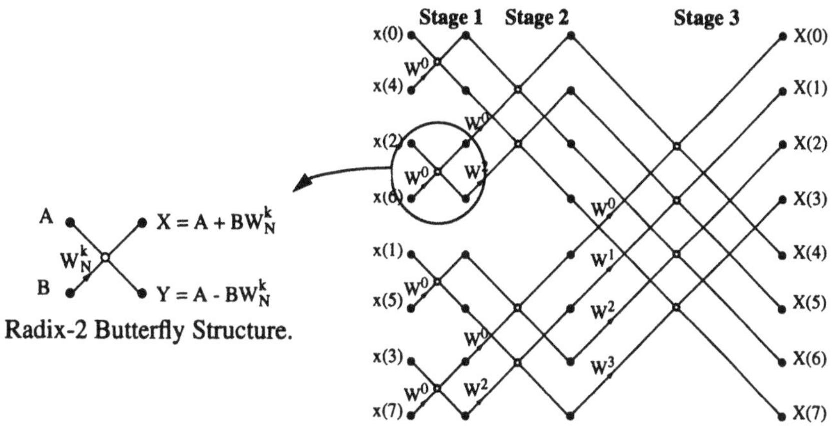

Figure 2: Decimation-in-Time Eight Point FFT.

3 Floating Point Arithmetic

In order to implement an FFT on Splash-2, [1,2] floating point arithmetic adder/subtracter and multiplier units were selected to satisfy the numerical dynamics of this application [12]. Until recently, any meaningful floating point arithmetic has been virtually impossible to implement on FPGA based systems due to the limited density, routing resources and speed of older FPGAs. In addition, mapping difficulties occurred due to the inherent complexity of floating point arithmetic. With the introduction of high level languages such as VHDL [9], rapid prototyping of floating point formats has become possible making such complex structures more feasible to implement. Although low level design was possible, the strategy used in all application development was to specify all aspects of the design in VHDL and rely on automated synthesis to generate the FPGA mapping.

3.1 Floating Point Format Representation

The floating-point format used in this application is similar the IEEE 754 standard for storing floating point numbers [7]. For the FFT implementation presented here, a smaller 18-bit floating-point format was developed. The format was chosen to accommodate two specific requirements: (1) the dynamic range of the format needed to be quite large in order to represent very large and small, positive and negative real numbers accurately, and (2) the data path width into one of the Xilinx 4010 processors [14] of Splash-2 is 36 bits wide and real and imaginary operands of a complex number are needed to be input on every clock cycle. Based on these requirements the format in Figure 3 was used.

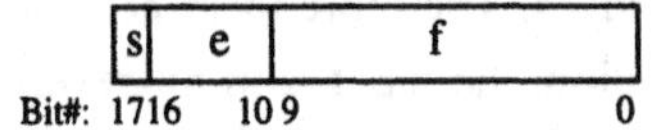

Figure 3: 18 Bit Floating-Point Format.

The 18 bit floating point value (v) is computed by:

$$\mathbf{v = -1^s\, 2^{(e - 63)}(1.f)}$$

The range of real numbers that this format can represent is $\pm 3.6875\text{x } 10^{19}$ to $\pm 1.626\text{x } 10^{-19}$.

3.2 Floating-Point Addition/Subtraction and Multiplication

The aim in developing a floating point adder/subtracter routine was to pipeline the unit in order to produce a result every clock cycle. The floating point addition and subtraction algorithm that was implemented is similar to what is done in most traditional processors; however, the computation is performed in three stages to improve performance. A summary of the resulting size and speed of the 18-bit floating point units is given in Table 1.

Floating point multiplication is much like integer multiplication. Because floating-point numbers are stored in sign-magnitude form, the multiplier needs only to deal with unsigned integer numbers. Like the architecture of the floating point adder, the floating point multiplier unit is a three stage pipeline that produces a result every clock cycle. The bottleneck of this design was the integer multiplier. For more information regarding the algorithms used the reader is referred to [12].

The Synopsys Version 3.0a VHDL compiler was used along with the Xilinx 5.0 tools to compile the VHDL description of the floating point arithmetic units. The Xilinx timing tool, *xdelay*, was used to estimate the speed of the designs.

	Adder/ Subtracter	Multiplier
FG Function Generators	28%	44%
Flip Flops	14%	14%
Stages	3	3
Speed	8.6 MHz	4.9 MHz
Tested Speed	10 MHz	10 MHz

TABLE 1. Summary of Properties of 18-bit Floating Point Units.

The floating point arithmetic units have also been incorporated in another application: an FIR filter [13]. The FFT application operates at 10 Mhz and the results of the transform are stored in memory on the Splash-2 array board. These results were checked by doing the same transform on a Sparc workstation and we noted the results matched. Therefore, the maximum clock speed of the arithmetic units given by the *xdelay* program is conservative and we conclude that the arithmetic units can operate at least at 10 MHz.

4 FFT Implementation

Implementing filtering method discussed in Section 2 involved mapping a 2-D FFT, a filter, and a 2-D IFFT to a two-board, Splash-2 system [1, 2]. The filtering method was constructed in such a way that the FFT and the IFFT are computed in parallel and are continuously provided video images from a frame buffer. This section discusses the recirculation method used to implement the FFT, the butterfly operator used in the FFT, and the filtering process.

4.1 FFT Recirculation Method

To calculate a 2-D FFT, a method requiring the recirculation of data through a butterfly operator was implemented. A block diagram of this method is shown in Figure 4. Two banks of memory are used to store the input and output data of each stage of the FFT. A bank of memory consists of three processing elements that store the real and imaginary components of the two 18-bit floating point numbers into their local memories. Since the local memories are only 16 bits wide, the two 18-bit floating point values are divided between the three memories. To compute an FFT on an input image, a frame of data is accepted from the frame buffer, converted from 8-bit integer values to 18-bit floating point values, and stored into Bank 1. The 2-D FFT is computed by first computing a 1-D FFT of each row of the image and then a 1-D FFT on each column of the row transforms. The 1-D FFT is computed in the same manner as shown in Figure 2. The first stage of the FFT is computed by reading each row of data points in bit-reversed order from Bank 1 and passing it to the butterfly operation. The results of the butterfly operation are stored in the second bank of memory. Once each butterfly is computed in the first stage, the second stage is computed by first reading the data out of the second bank of memory in linear order, and then into the butterfly operator. The results of this stage are stored in Bank 1. The recirculation method continues by reading data out of one bank of memory while the other bank of memory is storing the results. The recirculation terminates when a 1-D FFT is calculated on each row of the image. The second set of 1-D FFTs are computed in the same manner except it is done on each column of the result of the first set of 1-D FFTs. Once the final stage of the last FFT is calculated, the data is passed over the crossbar from X11 to X15 where the data is filtered. The complete 2-D FFT process involves $2N^2 \log_2 N$ passes through the butterfly operator.

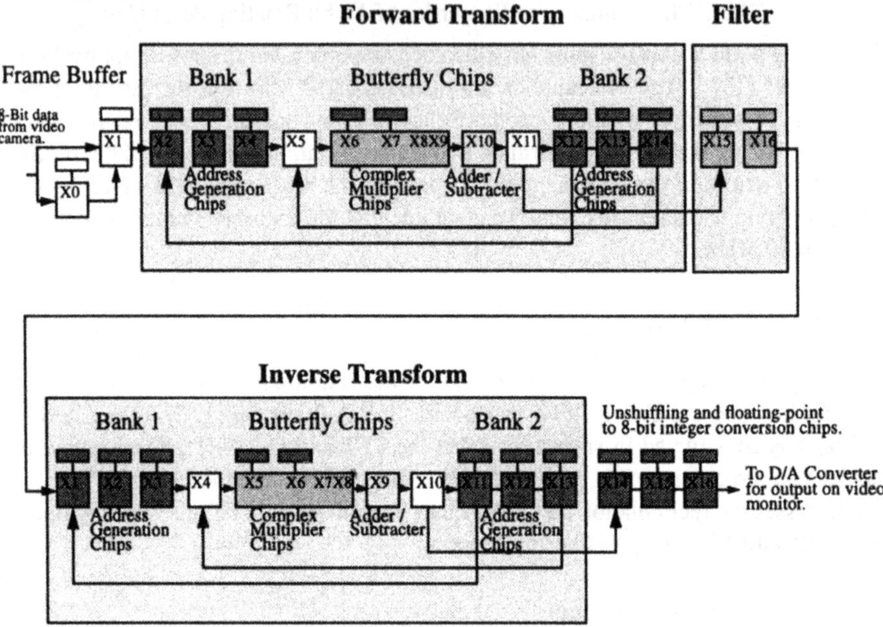

Figure 4: Splash-2 FFT Image Filtering Method.

4.2 Butterfly Implementation

The butterfly operation is the heart of the FFT algorithm. It is pipelined in order to compute a real and complex result every clock cycle. The butterfly diagram shown in Figure 2 involves calculating a complex floating point multiplication and two floating point additions/subtractions. The complex multiply involves four multiplications and two additions/subtractions. In total, eight floating point operations are calculated every clock cycle at 10 Mhz. The throughput of the butterfly operation is therefore 80 Mflops.

Figure 5, shows a block diagram of how the butterfly operation was partitioned between five processing elements on Splash-2. The real and imaginary parts of the complex multiplication of $BW^k{}_N$ is given respectively by the equations:

$$BW_N^k.re = B.re\, W_N^k.re + B.im\, W_N^k.im \qquad (4.1)$$

$$BW_N^k.im = B.re\, W_N^k.im - B.im\, W_N^k.re \qquad (4.2)$$

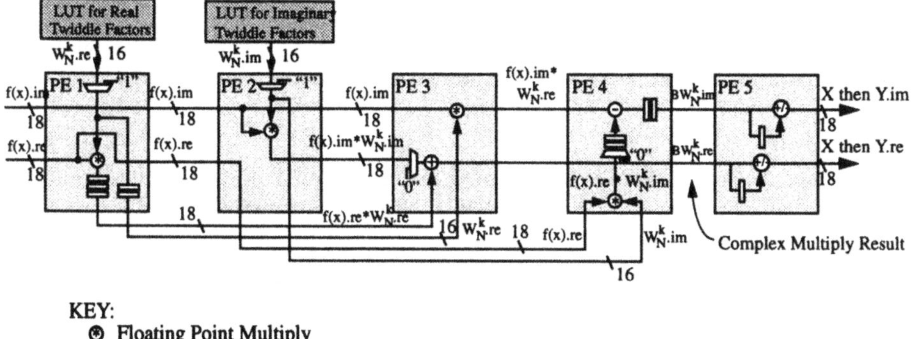

KEY:
- Floating Point Multiply
- Floating Point Add
- Floating Point Subtract
- 16 or 18 Bit Mux
- 18-Bit Delay Register

Note: f(x) = Input Data, A in the first clock cycle and then B in the second clock cycle.

Figure 5: Block diagram of a five PE Splash-2 design for a butterfly operation.

Both the A and B inputs of the butterfly operation shown in Figure 2 are denoted in Figure 5 as *f(x)*. The *A* value is inserted into the pipeline followed by the *B* value on the next clock cycle. The A input is not multiplied by the twiddle factor. In order to pass the *A* value through the pipeline without changing its value, multiplexers are used to multiply it by one and add zero to it. When the real and imaginary values of B are inserted into the pipeline, these pass through four processing elements in order to calculate the complex multiply of BW^k_N. The first processing element (PE 1) reads the real component of the appropriate twiddle factor and multiplies it by real component of B. The result and the twiddle factor is passed via the crossbar to PE 3, and the real and imaginary components of B are passed to PE 2. The second PE multiplies the imaginary component of B and the appropriate twiddle factor, and the result is passed to the third PE. The third PE reads the result from PE 1 ($B.reW^k_N.re$) off the crossbar and adds it to the result from PE 2 ($B.imW^k_N.im$) to produce the final result of the real component of the complex multiply ($BW^k_N.re$). The imaginary component, $BW^k_N.im$, of the complex multiply is computed in the same manner in PEs 3 and 4. The butterfly operation is completed by adding A to BW^k_N in the first clock cycle to produce X, and subtracting BW^k_N from A in the second clock cycle to produce Y.

The 18-bit format was not used to store the twiddle factors in the local memories of PE 1 and 2 since the memory data bus width is only 16 bits wide. A smaller 16-bit format was created by decreasing the exponent field of the 18 -bit floating point number by 2 bits. Since twiddle factors can be expressed in terms of sine and cosine functions by using Euler's rule, the value of the floating point number will never have an exponent greater than 0 (because the value will always be less than or equal to one). Because of this, the exponent field was changed to range from 0 to -31 instead of 63 to -63 in order to decrease the size of the exponent field from 7 bits to 5 bits. When the twiddle factor is read into the processing element, a conversion is done from the 16-bit format to the 18-bit format used in the arithmetic units.

4.3 Filtering

Once the input image is transformed to the frequency domain, point-by-point multiplication of the matrix filter coefficients, H(u,v) and the transformed image can be computed to filter the image. The values of the elements of the matrix H(u,v), range between 0 and 1.0 and are stored in the local memories in the same manner as the twiddle factors of the butterfly operation. The filter coefficients are calculated before run-time and stored in the local memories of chips X15 and X16 as indicated in Figure 4. Filter chips consist of a floating point multiplier unit and filter coefficient addressing logic. X15 and X16 are used to filter, respectively, the real and imaginary components of the transformed image.

Many different types of filters have been calculated, such as, ideal, Butterworth, exponential and trapezoidal filters[6]. These filters can be down-loaded on the fly from the Sparc-2 host to the local memories of the Splash board in approximately 400ms.

5 Error Analysis

To test round-off error associated with 18-bit floating point format, a forward FFT followed by an inverse FFT was calculated without doing any filtering This process should ideally result in an image which is exactly the same as the original image. However, due to round off error the output image differed slightly.

Statistics such as RMS and absolute error were calculated to quantify the error. Equations used for calculating the RMS and absolute error are:

$$RMS\ Error = \sqrt{\frac{1}{N^2}\sum_{x=0}^{N-1}\sum_{y=0}^{N-1}\left(I(x,y)-I'(x,y)\right)^2} \quad (5.1)$$

$$Absolute\ Error = \frac{1}{N^2}\sum_{x=0}^{N-1}\sum_{y=0}^{N-1}\left|I(x,y)-I'(x,y)\right| \quad (5.2)$$

where I(x,y) is the original image, and I'(x,y) is the output image.

Multiple images were tested and the average RMS error was 0.4% and the average absolute error was 0.2%. Each pixel value can have a gray-scale value from 0 to 256. The output image had a maximum deviation of 2 gray-scale values from the corresponding pixel in the original image. Subjectively, no difference could visually be seen between the original and output images.

The calculated statistical values indicate that the smaller 18-bit floating-point format is adequate for this application. By down-sizing the floating point format we were able to do more floating-point operations per Splash board resulting in increased performance.

6 Performance

In order to compare the performance of this application thoroughly, a wide range of architectures were selected. The architectures ranged from a general purpose workstation to special purpose DSP processors. Since the FFT is a common DSP algorithm, it is used as a benchmark by many DSP chip manufactures. Two DSP chips were selected; one which has approximately the typical performance of a DSP chip, and one which is representative of high end performance.

The test case used to evaluate the different architectures is a 2-D spatial filter of dimensions 512x512 pixels. This process involves performing a 512x512 2-D FFT, filtering the image by doing 512x512 point-by-point multiplications for each of the real and imaginary values of the image in the frequency domain, followed by a 512x512 2-D IFFT to convert the image back to the spatial domain. The execution time, Mflops rating, and the speed-up factor of the Splash-2 implementation over the given architecture is shown in Table 2.

	Execution Time (sec)	Mflops	Speed-up Factor
Splash-2	**.47**	**180**	**1**
Sparc-2	18	32	38.3
Sparc-10	11	60	23.4
Intel i860	.35	200	.74
TI DSP TMS320C40	1.7	80	3.6
Sharp LH9124 DSP	.08	240	.17

TABLE 2. Comparison of Splash-2 Implementation with Other Architectures.

The performance of the Splash-2 implementation of the FFT was calculated in the following way: The number of clock cycles required to compute the NxN 2-D FFT is $2\ N^2 \log_2 N$. The application was run at 10 Mhz and therefore the execution time for doing a 512x512, 2-D FFT is $2\ (512)^2 \log_2(512)\ /\ 10\text{x}10^6 = .47186$ seconds or 2 frames per second. Since the FFT and the IFFT are pipelined and are being computed concurrently, the time for the complete filtering process is the time for calculating one 2-D FFT. The speed of this application was verified by using a logic analyzer to check the time between output frames. In addition, there are 18 floating point units distributed between the FFT, filter and IFFT designs. These units output a result every clock cycle at 10 Mhz therefore, this application operates at 180 Mflops (integer-to-floating point and floating-point-to-integer operations are not included in this figure).

The Cooley-Tukey FFT algorithim[6] was implemented in C and was compiled using the highest optimization level of the *gcc* compiler on a Sparc workstation.

The execution time for the Intel i860 based processing board, Texas Instruments and Sharp DSP chips was calculated by doubling the time required to do a single 512x512 2-D complex FFT and adding a very small amount of time for the filtering process. The time for doing an FFT was doubled in order to account for the time to do and IFFT. The i860 processing board consisted of two, 50 Mhz, i860 chips and 200 Mbytes of RAM. The Sharp DSP chip was chosen since it was the fastest DSP chip surveyed out of almost 60 DSP chips[3]. The Sharp DSP can calculate a complex multiply in one clock cycle at 40 Mhz[11]. The Texas Instruments DSP chip was selected because its performance was about average for the DSP chips in the survey. The algorithm used in the survey to benchmark the DSP chips was a 1024, one dimensional FFT.

It is essential to note that the other implementations used 32-bit single precision floating point arithmetic, and the Splash-2 design takes advantage of the ability to use a smaller 18-bit floating point format. However, this smaller format requires less computation per floating point arithmetic unit than the 32-bit implementation. To implement single precision floating point arithmetic units on the Splash-2 architecture, the size of the floating point arithmetic units would increase between 2 to 4 times over the 18 bit format. A three stage floating point multiply unit would require two Xilinx 4010 chips and a three stage adder/subtracter unit could fit into a single Xilinx chip. The 24-bit multiplier needed in single precession floating point multiply can be broken up into four 12-bit multipliers, allocating two per chip[5]. We found that a 16x16 parallel bit multiplier was the largest parallel integer multiplier that could fit into a Xilinx 4010 chip. When synthesized, this multiplier used 75% of the chip area. However, there was no need to emulate the 32-bit floating-point arithmetic since the desired accuracy was achieved with an 18-bit format. This illustrates an important advantage of FPGA-based computers over traditional approaches.

The Splash-2 performance is more than an order of magnitude better than a general purpose workstation and is similar to an i860 processing board which is faster than many DSP processors. In addition, the Splash-2 implementation is less than six times slower than one of the fastest DSP processors on the market.

7 Conclusions

Due to the flexibility of a CCM, customization of the floating point format was performed to achieve maximum accuracy with the smallest number of bits. By taking advantage of the parallelism of the Splash-2 architecture, address calculation, butterfly operations and filtering could be done concurrently. By pipelining the butterfly operation, a real and complex result was obtained every clock cycle at 10 Mhz. The performance of this application is much faster than a Sparc-10 workstation and is similar to that of a typical DSP processor.

The Splash-2 architecture has been used to improve the performance of a wide range of applications and can be considered as a general purpose custom computing platform. Applications include pattern matching, text searching and genome data base searching, and many different image processing algorithms[2, 8]. The genome base search implementation has shown a speed-up of three orders of magnitude over the MasPar-1. The Splash-2 implementation of a 2-D FFT has shown that the performance is similar to a DSP chip and has shown that floating point arithmetic can be done on CCMs effectively.

References

[1] J. Arnold, D. Buell and E. Davis, "Splash 2," *Proceedings of the 4th Annual ACM Symposium on Parallel Algorithms and Architectures*, June, 1992, pp. 316-322.

[2] P. Athanas and L. Abbott, "Real-Time Image Processing on a Custom Computing Platform," *IEEE Computer*, Vol. 28, No. 2, February 1995, pp. 16-24.

[3] *Computer Design*, "1995 Product Trends and Resource Guide," February 1995, pp 44-47.

[4] J. Eldon and C. Robertson, *A Floating Point Format for Signal Processing*, *Proceedings IEEE International Conference on Acoustics, Speech, and Signal Processing*, 1982, pp. 717-720.

[5] B. Fagin and C. Renard, "Field Programmable Gate Arrays and Floating Point Arithmetic," *IEEE Transactions on VLSI*, Vol. 2, No. 3, September 1994, pp. 365-367.

[6] R. Gonzalez and P. Wintz, *Digital Image Processing*, Addison-Wesley Publishing Company, 1977.

[7] IEEE Task P754, "A Proposed Standard for Binary Floating-Point Arithmetic," IEEE Computer, Vol. 14, No. 12, March 1981, pp. 51-62.

[8] D. Hoang, "Searching Genetic Databases on Splash-2", *IEEE Workshop on FPGAs for Custom Computing Machines*, 1993, pp. 185-191.

[9] R. Lipsett, C. Schaefer, and C. Ussery, *VHDL: Hardware Description and Design*, Kluwer Academic Publishers, Boston, MA., 1989.

[10] L. Rabiner B. Gold, *Theory and Application of Digital Signal Processing*, Prentice-Hall, 1975.

[11] Sharp Electronics Corp., "Fast Fourier Transform," Sharp Application Note for the LH9124 DSP Chip, November 1992.

[12] N. Shirazi, A. Walters, and P. Athanas, "Quantitative Analysis of Floating Point Arithmetic on FPGA Based Custom Computing Machines," To appear at *IEEE Symposium on FPGAs for Custom Computing Machines*, April 1995.

[13] A. Walters, *An Indoor Wireless Communications Channel Model Implementation on a Custom Computing Platform*, VPI&SU Master Thesis in progress.

[14] Xilinx, Inc., *The Programmable Logic Data Book*, San Jose, California, 1995.

An Assessment of the Suitability of FPGA-Based Systems for Use in Digital Signal Processing*

Russell J. Petersen and Brad L. Hutchings

Brigham Young University, Dept. of Electrical and Computer Engineering, 459 CB, Provo UT 84602, USA

Abstract. FPGAs have been proposed as high-performance alternatives to DSP processors. This paper *quantitatively* compares FPGA performance against DSP processors and ASICs using actual applications and existing CAD tools and devices. Performance measures were based on actual multiplier performance with FPGAs, DSP processors and ASICs. This study demonstrates that FPGAs can provide an order of magnitude better performance than DSP processors and can in many cases approach or exceed ASIC levels of performance.

1 Introduction

To meet the intensive computation and I/O demands imposed by DSP systems many custom digital hardware systems utilizing ASICs have been designed and built. Custom hardware solutions have been necessary due to the low performance of other approaches such as microprocessor-based systems, but have the disadvantage of inflexibility and a high cost of development. The DSP *processor* attempts to overcome the inflexibility and development costs of custom hardware. The DSP processor provides flexibility through software instruction decoding and execution while providing high performance arithmetic components such as fast array multipliers and multiple memory banks to increase data throughput. The FPGA has also recently generated interest for use in implementing digital signal processing systems due to its ability to implement custom hardware solutions while still maintaining flexibility through device reprogramming [1]. Using the FPGA it is hoped that a significant performance improvement can be obtained over the DSP processor without sacrificing system flexibility. This paper is an attempt to quantify the ability of the FPGA to provide an acceptable performance improvement over the DSP processor in the area of digital signal processing.

* This work was supported by ARPA/CSTO under contract number DABT63-94-C-0085 under a subcontract to National Semiconductor.

2 Multiplication and digital signal processing

A core operation in digital signal processing algorithms is multiplication. Often, the computational performance of a DSP system is limited by its multiplication performance, hence the multiplication rate of the system must be maximized. Custom hardware systems based on ASICs and DSP processors maximize multiplication performance by using fast parallel-array multipliers either singly or in parallel. FPGAs also have the ability to implement multipliers singly or in parallel according to the needs of the application. Thus, in order to understand the performance of the FPGA relative to the ASIC and the DSP processor a comparison of FPGA multiplication alternatives and their performance relative to custom multiplier solutions is needed. This section presents the basic alternatives for multiplier implementations and their performance when implemented on FPGAs.

2.1 Multiplier architecture alternatives

When implementing multipliers in hardware two basic alternatives are available. The multiplier can be implemented as a fully parallel-array multiplier or as a fully bit-serial multiplier as shown in Figure 1. The advantage of the fully parallel approach is that all of the product bits are produced at once which generally results in a faster multiplication rate. The multiplication rate for a parallel multiplier is just the delay through the combinational logic. However, parallel multipliers also require a large amount of area to implement. Bit-serial multipliers on the other hand generally require only $\frac{1}{N}$th the area of an equivalent parallel multiplier but take 2N bit times to compute the entire product (N is the number of bits of multiplier precision). This often leads one to believe that the bit-serial approach is thus 2N times slower than an equivalent parallel multiplier but this is not true. The bit-times (clock cycles for synchronous bit-serial multipliers) are very short in duration due to the reduced size and hence propagation paths of the multiplier. This results in a bit-serial multiplier achieving about $\frac{1}{2}$ the multiplication rate of an equivalent parallel multiplier on average, even exceeding the performance of the parallel multiplier in some cases.

2.2 FPGA multiplication results

Table 1 lists the performance of several multipliers implemented on three different FPGAs. The FPGAs used were a Xilinx 4010, an Altera Flex8000 81188, and a National Semiconductor CLAy31. The first two FPGAs can be characterized as medium-grained architectures and are approximately equivalent in logic-density while the last FPGA is a fine-grained architecture utilizing smaller but more numerous cells. The multiplication rate of each multiplier is listed in MHz as well as the percentage of the FPGA required to implement the multi-

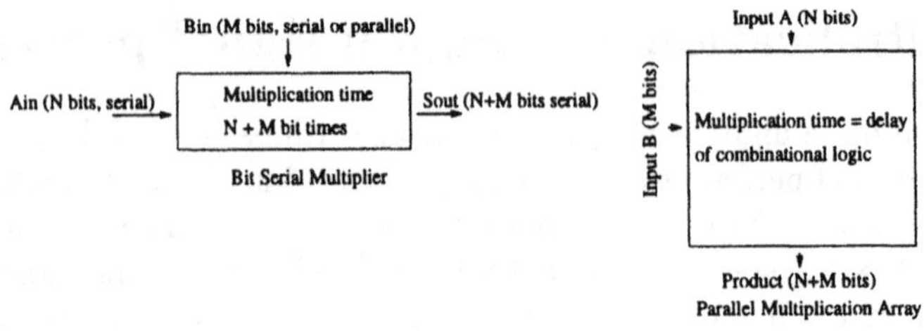

Fig. 1. Block diagrams of basic multiplier alternatives

plier. The bit-serial multipliers have listed both their clock rate (bit-rate) and their effective multiplication rate (clock rate/2N).

2.3 Multiplier table contents

The majority of the multipliers in this study used common architectures such as the Baugh-Wooley two's complement parallel-array multiplier [4] and pipelined versions of the bit-serial multiplier [5] shown in Figure 1. In addition, several custom parallel multipliers were built that take advantage of the special features available on the Altera and Xilinx FPGAs. These are intended to represent near the absolute maximum possible multiplier performance that can be achieved with these current FPGAs. These specific customizations will be discussed below.

Several of the multipliers listed in the tables have the label synthesis attached. This label indicates that the multipliers were created by synthesizing simple high-level hardware language (VHDL) design statements (`z <= a * b`). These multipliers were included so as to allow a comparison between hand-placed multipliers using schematics and high-level language designed multipliers. The table results show that the synthesized multipliers performed very favorably as shown in the Xilinx 4010 parallel multiplier table section. The 8 and 16-bit unsigned and signed array multipliers listed first were designed with schematics and were hand placed onto the FPGA. However, their performance was nearly identical in terms of both speed and area required to the multipliers synthesized from VHDL.

2.3.1 Fast carry-logic based parallel multipliers

The Altera 81188 based multipliers labeled *fast adder* refer to the use of the fast carry-logic available on the Altera FPGAs to make fast ripple-carry adders. These adders are then used to build fast multipliers by using the adders to add the successive partial product rows. This technique results in multipliers that are approximately twice as fast on the FPGAs as those not implemented with

Table 1. FPGA Multiplier Performance Results

Type of Multiplier	# CLB/LC's	% of FPGA	Mult. Speed
Altera 81188 Parallel Multipliers			
8 bit unsigned fast adder	133	13	14.8 MHz
8 bit signed fast adder	150	14	12.8 MHz
8 bit unsigned synthesis	129	12	7 MHz
8 bit signed synthesis	135	13	6.84 MHz
8 bit signed complex synthesis	584	57	5.86 MHz
16 bit unsigned synthesis	519	51	3.66 MHz
16 bit signed synthesis	535	53	3.4 MHz
Altera 81188 Bit-Serial Multipliers			
8 bit unsigned	29	3	84.03/5.25 MHz
8 bit signed	91	9	69/4.6 MHz
16 bit unsigned	61	7	68.49/2.14 MHz
16 bit signed	186	18	64/2 MHz
National Semiconductor CLAy Parallel Multipliers			
8 bit unsigned	329	11	6.7 MHz
8 bit signed	338	11	6.1 MHz
16 bit unsigned	1425	45	3.1 MHz
16 bit signed	1446	46	3.0 MHz
National Semiconductor CLAy Bit-Serial Multlipliers			
8 bit unsigned	48	1.5	29/1.8 MHz
8 bit signed	48	1.5	29/1.8 MHz
16 bit unsigned	96	3	26.3/.82 MHz
16 bit signed	96	3	26.3/.82 MHz
Xilinx 4010 Parallel Multipliers			
8 bit unsigned	64	16	8.54 MHz
16 bit signed	259	65	4.35 MHz
8 bit unsigned synthesis	61	15	9 MHz
8 bit signed synthesis	61	15	8 MHz
8 bit signed complex synthesis	266	66	7.3 MHz
16 bit unsigned synthesis	242	60	3.8 MHz
16 bit signed synthesis	250	63	3.7 MHz
Xilinx 4010 Bit-Serial Multipliers			
8 bit unsigned	17	4	73.1/4.6 MHz
8 bit signed	32	8	52/3.3 MHz
16 bit unsigned	33	8	62/1.9 MHz
16 bit signed	64	16	50/1.6 MHz
Xilinx 4010 Parallel Constant Multipliers			
8 bit unsigned ROM	22	5.5	21.7 MHz
16 bit unsigned ROM	84	21	11.36 MHz
8 bit unsigned RAM	39	9.75	17.86 MHz
16 bit unsigned RAM	117	29.3	10.4 MHz

special logic. The disadvantage of this approach is the resulting difficulty that arises with the placement of the multiplier onto the FPGA. The FPGA router is only able to place three of the unsigned 8-bit multipliers on a 81188 FPGA even though they only utilize 13% of the total FPGA resources each.

2.3.2 Constant multipliers and distributed arithmetic

The use of constants (constant multiplicand) in multiplication can significantly reduce the size of a parallel multiplier array. This is because the presence of zeros in the constant can result in the elimination of many partial product terms in the multiplication array. This technique is especially useful in DSP systems since many of the multiplications to be performed can be specified in terms of constant multipliers. For example, with an FIR filter each tap of the filter can be implemented using a multiplier with a constant tap coefficient.

The use of constants in multiplication also makes available another technique that can result in a significant multiplier performance increase. This technique is called the *distributed arithmetic* approach to multiplication and can be implemented by the Xilinx FPGAs due to their ability to provide small blocks of distributed RAM to be used as partial-product lookup tables.

The distributed arithmetic approach to multiplication relies upon the ability to easily precompute all of the possible products of a multiplication when one of the values is held constant. For example, consider an 8x8 bit multiplier implemented with this technique. One, possibility is to break up the 8-bit input word into two nibbles (4 bits) and then use each nibble as the address applied to two separate 12-bit wide, 16-location lookup tables. Two separate 16x12 bit tables are required since each of the nibbles produces 16 possible 12-bit partial products. The partial product outputs of each table are then weighted appropriately and added to produce the product. The method is illustrated in Figure 2. The partial product produced by the high-order nibble of the input word is shifted by 4 bits to the left (a weighting factor of 16) and added to the partial product produced by the lower-order nibble of the input word to produce the 16 bit output [2].

Implementing the 8x8 multiplier on a Xilinx FPGA requires a total of 384 bits of storage along with a 12 bit adder. This results in a minimum of 12 CLBs for the data storage and approximately 12 CLBs for the adder for a total of 24 CLBs. The actual number of CLBs (area) required is dependent upon optimizations that the place-and-route software is able to make and can be seen to be slightly less (22 CLBs) for the ROM-based 8x8 multiplier in Table 1. The difference in size and speed between the RAM and ROM-based versions listed in the table is due to the elimination of the additional inputs on the ROM version and the associated optimizations that the place-and-route software can make. Only unsigned constant Xilinx multipliers are listed in the table but signed versions of the multipliers can also easily be built by sign-extending the partial products and the input to the multiplier.

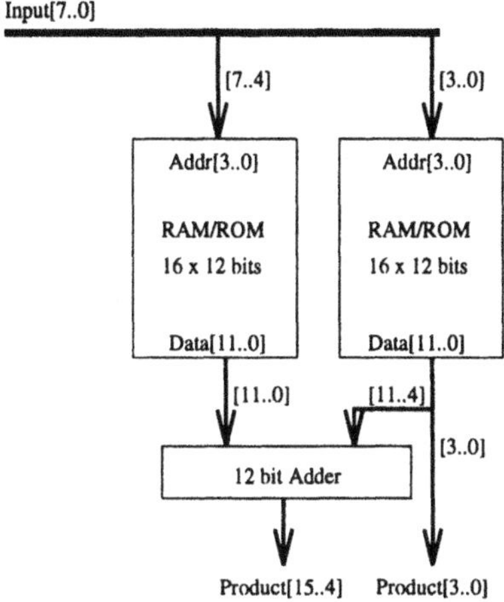

Fig. 2. 8-bit constant unsigned multiplier using distributed arithmetic

2.4 Comparisons to custom multiplier chips

One possible alternative to implementing multipliers on an FPGA is to use external multiplication chips with the FPGA providing the necessary control. This allows the use of multipliers designed in VLSI that are faster, smaller, and less expensive than equivalent implementations on FPGAs. The table below lists several fixed-point multiplication chips available from various manufacturers along with their performance.

Table 2. Custom Multiplier Chip Performance

Part #	Precision	Mult. Speed
Logic Devices LMU08/LMU8U	8x8-16 bit signed/unsigned	28.6 MHz
Logic Devices LMU18	16x16-32 bit signed/unsigned	28.6 MHz
Cypress CY7516/517	16x16-32 bit signed/unsigned	26.3 MHz
GEC Plessey PDSP16116/A	16-64 bit signed/unsigned complex	20 MHz

Disadvantages of using external multipliers include the on/off chip time required for signals between the FPGA and the multiplier and the high I/O pin requirement when interfacing to a multiplication chip. For example, the 16-bit complex multiplier requires 128 pins just for data transfer. Some of the I/O constraints are eased with the 16-bit multipliers by multiplexing the inputs with the output data word but this also requires extra control and adds latency to the multiplier.

As can be seen from Tables 1 and 2 the FPGA-based parallel multipliers obtain approximately $\frac{1}{4}$ to $\frac{1}{3}$ of the performance of the custom multipliers for the 8-bit versions while the 16-bit multipliers obtain only about $\frac{1}{10}$ the performance of their custom counterparts. The only FPGA-based multipliers that come close to matching the custom multiplier performance are the constant multipliers based on the distributed arithmetic approach.

3 Performance comparison of two popular DSP algorithms

Using the previous results for multiplication, rough comparisons can be made between the performance of FPGA-based, DSP processor, and ASIC-based DSP systems. Two popular DSP algorithms that have been chosen for this comparison are a single-dimensional FIR filter and a FFT. Comparisons will be made based on implementations using: FPGAs only, FPGAs combined with external multiplier chips, a single DSP processor, and full custom ASICs. In the comparisons it will be assumed that the multipliers form the limiting path of the system and that an additional 10 ns is required for on/off chip delays between the multiplier and the FPGA when using the external multiplication chips.

Table 3. 20-Tap FIR Filter Performance

System	Precision	# of Chips	Computation Time	Data rate
TI TMS320C5X	16 bit	1	1.0 μs	738.9 KHz
Altera 81188 U-Bit-Serial	8 bit	1	.190 μs	5.25 MHz
Altera 81188 U-Bit-Serial	16 bit	2	.477 μs	2.1 MHz
Altera 81188 S-Bit-Serial	8 bit	3	.227 μs	4.4 MHz
Altera 81188 S-Bit-Serial	16 bit	5	.51 μs	1.96 MHz
Altera 81188 Parallel	8 bit	5	.156 μs	6.4 MHz
Altera 81188 Parallel	16 bit	14	.304 μs	3.28 MHz
LD LMU08	8 bit	2	.9 μs	1.11 MHz
LD LMU18	16 bit	2	.9 μs	1.11 MHz
Altera 81188 Fast Parallel	8 bit	1	1.76 μs	567 KHz
Xilinx 4010 Fast Parallel	16 bit	2	4.8 μs	208 KHz
Xilinx 4010 Constant ROM	8 bit	2	.049 μs	20.5 MHz
Xilinx 4010 Constant ROM	16 bit	5	.1 μs	10.0 MHz
LD LF43881	8 bit	3	.033 μs	30 MHz
PDSP16256/A	16 bit	2	.08 μs	12.5 MHz

3.1 20-tap FIR filter

Performance numbers for a 20-tap FIR filter appear in Table 3. The table entry labeled *TMS320C5X* refers to the popular 16-bit fixed point C5X DSP processors manufactured by Texas Instruments. The benchmark listed is for a C5X with a 35 ns cycle time and a 57 MHz external clock rate [3]. The data throughput rate is less than the inverse of the computation time (1.0 μs) due to the overhead of executing instructions to set up the filter operation and moving data on and off chip.

The entries labeled *Altera U-Bit-Serial* refer to the use of unsigned bit-serial multipliers to build the 20-tap filters while those labeled *Altera S-Bit-Serial* refer to the use of signed bit-serial multipliers. Mapping inefficiencies for signed bit-serial arithmetic resulted in an increase in system chip count for the signed filters by factors of 3 and 2.5 respectively for the 8- and 16-bit 20-tap FIR filters.

The entries labeled *Altera Parallel* refer to the use of signed multipliers synthesized from VHDL, chosen over the fast adder versions (see Table 1) since the fast adder versions create routing difficulties when multiple multipliers are placed on a chip due to their extensive use of the special logic.

The *LD LMU08* and *LD LMU18* entries refer to the use of custom multiplier chips from Logic Devices in conjunction with an FPGA to implement the filter. The FPGA is used to implement the necessary data delays, data path, multiplier chip control, and the product accumulation required for the multiply-accumulate loop of the FIR filter. Again, a 10 ns on-off chip delay time was assumed. For comparison to equivalent implementations using 1-2 FPGAs with one FPGA being possibly dedicated to implementing the multiplier (16-bit version only) the entries labeled *Altera Fast Parallel* and *Xilinx Fast Parallel* were included.

The next entries in the table present the results for the Xilinx constant coefficient distributed arithmetic multipliers discussed previously. The final entries list results for two custom FIR filter ASICs, the Logic Devices *LF43881* 8x8 bit Digital Filter and the Gec Plessey *PDSP16256/A* Programmable FIR filter.

3.1.1 Comparisons and conclusions

Comparing all of the listed filter implementations it can be seen that the ASIC-based implementations obtain the highest performance. Their performance, however, is nearly matched by the Xilinx-based constant multiplier implementations. This clearly indicates the advantage of the use of the distributed arithmetic approach to multiplication. Using this approach the 8-bit and 16-bit versions of the filter obtain speedup factors of 28 and 13 respectively over the DSP processor. The disadvantage of this approach is the need to implement all of the multiplications in parallel since each multiplier is a constant multiplier and is hence dedicated to a particular filter tap. This results in a larger chip count for the 16-bit filter (5 compared to 2 for the ASIC).

The only systems that performed worse than the DSP processor were those using only a single FPGA-based multiplier to perform the entire filter loop (entries labeled *fast parallel*). In these systems a single multiplier is used to compute the entire 20 iteration multiply-accumulate loop of the filter. This method most closely approximates the method used in the DSP processor for performing the filter but results in inferior performance due to the difference in speed of the FPGA-based multipliers and the VLSI-based multiplier on the DSP processor. Hence, when a custom VLSI multiplier chip is used in conjunction with an FPGA (table entries labeled *LD LMU08* and *LD LMU18*), this architecture again surpasses the performance of the DSP processor.

Table 4. Complex Radix-4 FFT Performance

System	Precision	# of Chips	Computation Time		
			64pt	256pt	1024pt
TI TMS320C5x	16 bit	1	108μs	617μs	3.84ms
PDSP16116/A	16 bit	2	11.52μs	61.4μs	.307ms
PDSP16510A	16 bit	1	-	-	98μs
Sharp LH9124	24 bit	2	-	-	80.7μs

3.2 Radix-4 FFT

Comparisons for FFTs using a radix-4 FFT algorithm have also been performed and appear in Table 4. The precision listed for the FFTs in Table 4 gives the number of bits each used for the real and imaginary parts of the input data word to the FFT. The FPGA-based implementation made use of one Altera FPGA and the PDSP16116/A 16-bit complex multiplier chip from GEC Plessey Semiconductors. The FPGA is used for the algorithm control and the implementation of the radix-4 butterfly element. The FFT is computed by calculating each column of the radix-4 FFT successively using the same hardware. Faster implementations can be created by using one FPGA and complex multiplier per FFT column.

As can be seen from the table the FPGA implementation achieves speedup factors of 9.4, 10, and 12.5 over the TMS320C5x DSP processor for FFTs of length 64, 256, and 1024 points respectively. The algorithm as implemented is primarily compute-bound and hence further speedups can be achieved through greater parallelism as mentioned above or through the use of a faster complex multiplier.

For comparison with the performance of ASIC-based implementations the table entries labeled *PDSP16510A* and *Sharp LH9124* are included. These are custom FFT processor ASICs with clock rates of 40 MHz each. The PDSP16510A and LH9124 obtain performance speedup factors of 3.1 and 3.8 respectively over the FPGA FFT processor for a 1024-point FFT.

Using the external multiplier chip along with an FPGA provided an order of magnitude increase in performance over the TI DSP chip. Additional performance increases are possible with ASIC-based systems, however, the performance of ASIC-based systems can be approached or surpassed by the FPGA implementation if one chip set is used per FFT column. For example, using one PDSP16116/A chip and FPGA per FFT column a 1024 point radix-4 FFT could be performed in $307/5 = 61.4\mu s$.

4 Conclusion

The results of implementing multipliers on FPGAs presented in Table 1 indicate that for most types of multipliers FPGAs are significantly slower than custom chips. Thus, in order for FPGAs to obtain a performance increase over DSP processors and ASICs, extensive specialization and concurrency must be employed. The distributed arithmetic approach to multiplication was shown to be a method of specialization giving large performance increases in applications where constant multiplications can be applied. The results for the 20-tap FIR filter and Radix-4 FFT indicated that an order of magnitude performance increase over that possible with a DSP processor is not unreasonable for a FPGA-based DSP system. This can be considered a sufficiently significant increase to warrant further application of FPGAs to DSP. In addition, FPGAs provide a reconfiguration advantage over ASICs. Limited flexibility is possible with proper ASIC design but FPGAs have the ability to completely change in function and I/O through reconfiguration. This makes it possible to customize a design through specialization and increased concurrency to achieve best case performance and to reduce costs through the amortization of the silicon over many different applications.

References

1. N. W. Bergmann and J. C. Mudge. Comparing the performance of FPGA-based custom computers with general-purpose computers for DSP applications. In D. A. Buell and K. L. Pocek, editors, *Proceedings of IEEE Workshop on FPGAs for Custom Computing Machines*, pages 164–171, Napa, CA, April 1994.
2. Kenneth David Chapman. Fast integer multipliers using xilinx fpga's. *Applications note posted to Xilinx BBS*, 1994.
3. Semiconductor Group. *Digital Signal Processing Products and Applications Primer.* Texas Instruments, 1991.
4. Kai Hwang. *Computer Arithmetic: Principles, Architecture, and Design.* John Wiley & Sons, 1979.
5. R.F. Lyon. Two's complement pipeline multipliers. *IEEE Transactions On Communications*, pages 418–424, April 1976.

An FPGA Prototype for a Multiplierless FIR Filter Built Using the Logarithmic Number System

Peter Lee,

University of Kent at Canterbury,
Canterbury, Kent,
CT2 7NT

Abstract

This paper describes the development of a prototype 64-tap multiplierless FIR filter based on the Logarithmic Number System (LNS). The circuit has been implemented and tested using a single Xilinx X64C64 device with external coefficient memory, data memory, ADC and DAC. The filter samples at 14KHz and runs at a rate of 895KHz (64 x 14KHz). This architecture is suitable for implementation using custom VLSI and can provide a compact, low-power solution to a number of simple filtering problems. It can also be expanded or cascaded to produce higher order filters.

1. Introduction

There are a number of applications, such as digital hearing aids, where the requirement for low power digital signal processing precludes the use of standard DSP chips and techniques. For such applications a number of alternative solutions have been considered. One such approach is to use the Logarithmic number system (LNS) which makes it possible to develop multiplierless filter structures [1,2,3]. This paper presents a novel architecture for a generic multiplierless FIR filter which uses both linear and logarithmic arithmetic and describes the implementation of a prototype Log-filter using a Xilinx FPGA. It begins with a brief overview of the LNS and includes an analysis of the performance of the Log-filter compared with equivalent linear and floating point solutions.

2. The Logarithmic Number System

In the logarithmic number system a binary number x is used to represent the exponential function

$$y = 2^x.$$

Or

$$x = \mathrm{Log}_2 y.$$

For signal processing applications the major advantages of using the LNS are:

(i) Multiplication of two linear numbers is achieved by addition of their exponents; division is achieved by subtraction. For example the product P of two binary numbers A and B can simply be calculated by summing the exponents of the two numbers.

if $A = 2^{x_1}$ $B = 2^{x_2}$

then $P = A \bullet B = (2^{x_1}) \bullet (2^{x_2}) = 2^{(x_1+x_2)}$

This property can be used to develop multiplierless filter architectures.

(ii) If signals have a large dynamic range but the overall system can tolerate a lower relative accuracy (or signal to noise ratio) they can be represented in the LNS using fewer bits. Bit reduction is usually achieved by using a piecewise linear approximation to the exponential function [4,5].

For some applications (i) and (ii) can result in a significant reduction in overall power consumption. Unfortunately a major disadvantage in using the LNS is that the process of linear addition (and subtraction) is both difficult and approximate. Addition in the LNS is generally achieved by using the following approximation [5,6]:

$$S = A + B$$

is replaced by

$$S = A(1 + B/A)$$

where S = Sum and A and B are two numbers in the LNS. The factors (1 + B/A) are usually stored in some form of look-up table or PLA. Although a considerable amount of work has been done to minimise the hardware overhead required to implement this function [4,5] this drawback has limited the use of the LNS in many practical systems. However for the special case of a generic FIR filter it is possible to develop an efficient solution based on the LNS..

3. A Generic FIR Filter Architecture

The basic architecture of a linear phase FIR filter is shown in Figure 1. It evaluates the following expression:

$$\mathrm{Dout}(n) = \sum_{i=0}^{i=m} a_i d[n-i]$$

By simply placing Log-to-Linear and Linear-to-Log converters before the input and output of the filter it is possible build such filter where all the arithmetic functions are performed using the LNS.

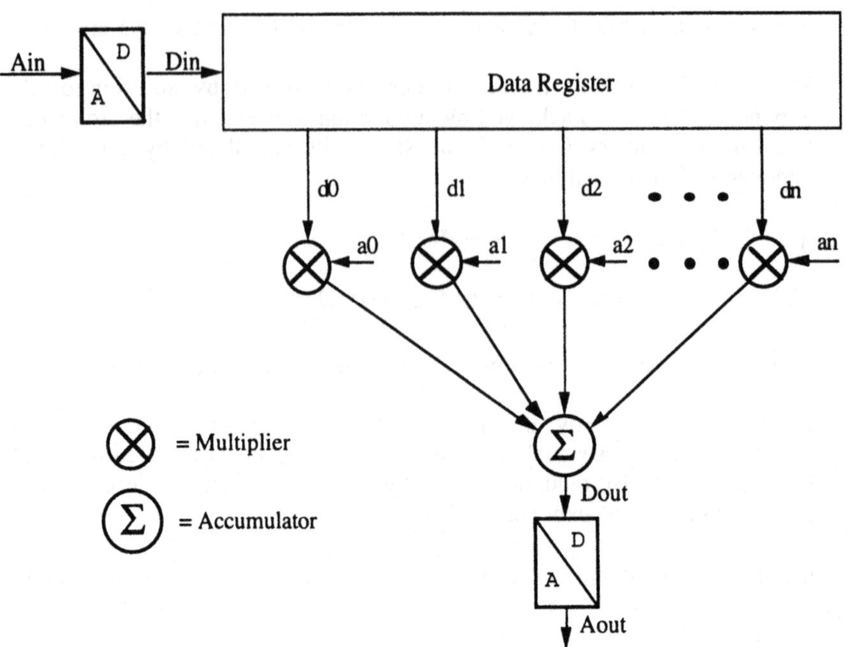

Fig 1. Generic FIR Filter

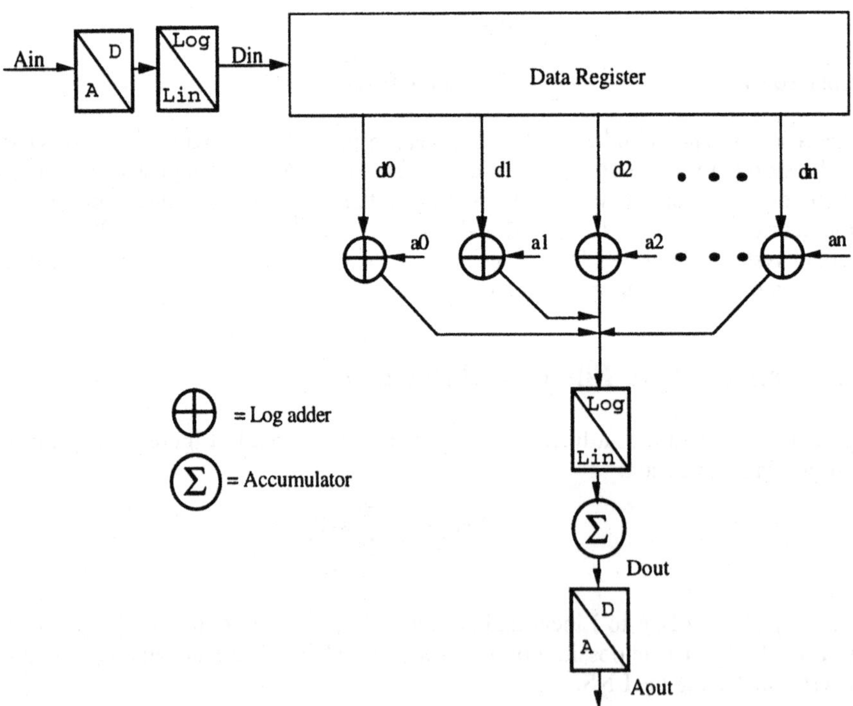

Fig 2. Enhanced Log-FIR structure

However, for such Log-filters another "mixed" approach which uses both the LNS and linear-binary arithmetic is more efficient. Consider the filter shown in Figure 2. The products $a_i x_i$ are still calculated using the LNS. But now the sum of products is calculated by first translating each product $a_i x_i$ back into the linear-binary domain and then using a simple accumulator to calculate the sum of products. This removes the requirement for a look-up table which would be needed to perform addition in the LNS. Furthermore the final output of the FIR is already in a linear-binary format which can be sent directly to a standard DAC.

By using this new mixed architecture, it is possible to build a multiplierless FIR filter where multiplication is performed in the logarithmic domain and addition is performed in the linear domain. This mixed approach may also be used in applications other than in the design of FIR filters.

4. Filter Design

A prototype 64-tap FIR filter was used to investigate the properties and performance of the Log-filter. This architecture was chosen because of its simplicty and linear phase characteristic. Before building the prototype a number of software simulations were performed using MATLAB to make a comparison between the Log-filter and equivalent binary fixed-point and floating point filters. A number of different compression ratios were tried to observe the effect on the Log-filter characteristics. A compression from 12 to 10 bits was finally chosen because this most closely matched

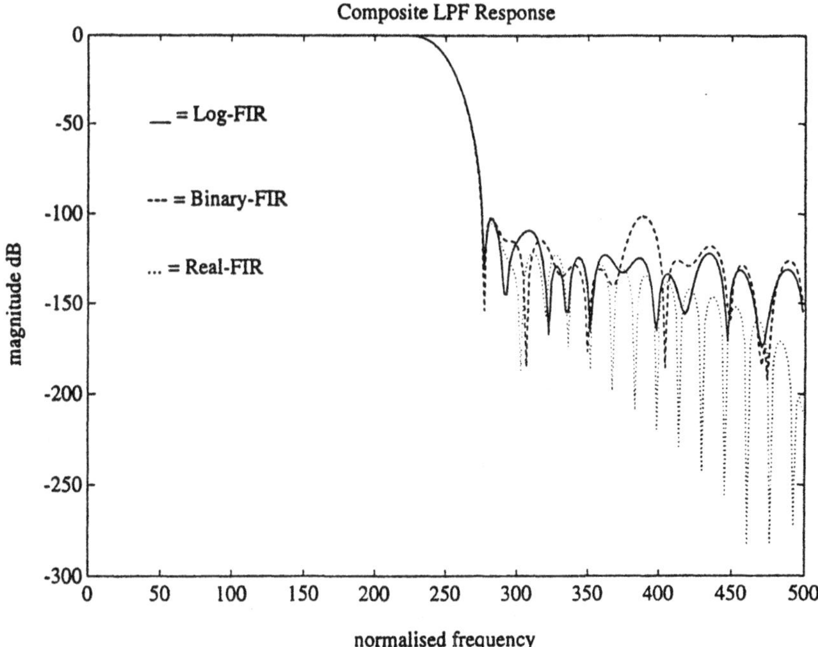

Fig 3. Log-Filter response Impulse Function = 1024

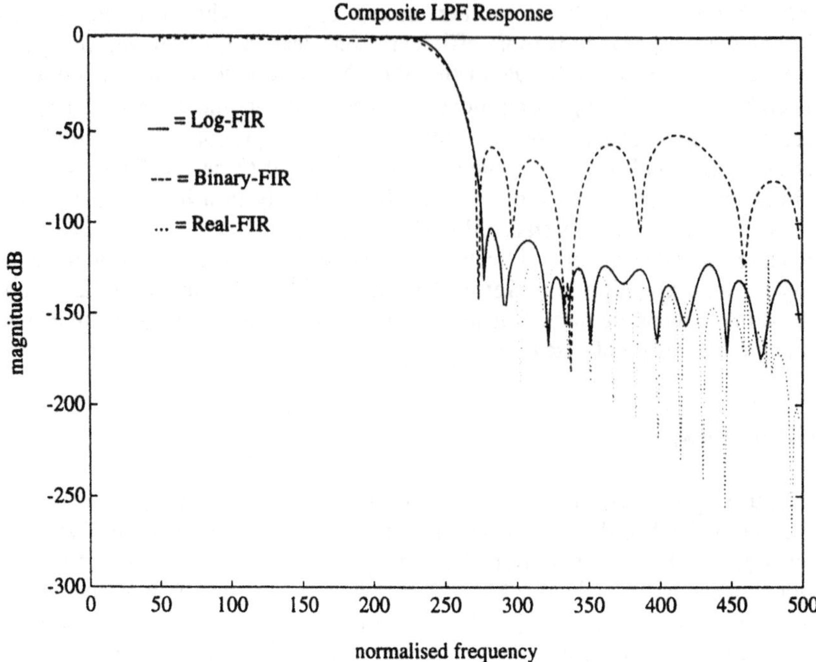

Fig 4. Log-Filter response Impulse Function = 32

the performance of the equivalent fixed-point binary filter. The 10 bits comprised 1 sign bit, 1 zero bit 3 segment bits and 5 fractional bits. A similar performance can also be achieved with a greater compression ratio by increasing the order of the filter. As expected, the log-FIR filter had a far superior dynamic range than the equivalent fixed-point FIR filter. This is mainly due to the effects of truncation of the multiplier output in the fixed-point implementation of the filter for low input values. The performance of the filter for impulse functions of different magnitudes is shown in Figures 3 and 4.

5. Circuit Design

A prototype circuit has been built using a Xilinx FPGA to implement the FIR filter. External EPROM and RAM were used to store filter coefficients and data. The Xilinx device also interfaced to serial 12-bit ADC (LTC1291) and DAC (MAX539) devices. For this solution a standard linear ADC was used, although it would also be possible to use an equivalent logarithmic ADC. This would remove the necessity for a Linear to Log converter. The whole system was built to run off a single 5V supply. A block diagram and photograph of the test board are shown in Figures 5 and 6.

A Xilinx 30C64 was used a total of 175 out of a possible 224 configurable logic blocks (CLBs) was required. For the Linear to Log (Lin2Log) and Log to Linear (Log2Lin) functions simple and efficient circuits which form piecewise linear approximations to this function have been used [6]. Because of the mismatch in data throughput rates of the Lin2Log converter (14KHz) and the Log2Lin converter (895KHz) a serial implementation of the Lin2Log converter and a parallel implementation of the Log2Lin converter was used. The Lin2Log converter is shown in Figure 7. It is similar to the circuit described in [6] except that this circuit can convert 2's complement binary values into logarithmic values. A zero detect circuit has also been added to remove the problem of calculating the logarithm of zero. The Log2Lin circuit is shown in Figure 8. In [6] it has been shown that parallel implementation of such circuits are both regular and efficient and require far fewer transistors than a multiplier This makes them suitable for implementation using VLSI. Similar circuit can also be implemented on FPGAs using multiplexors instead of transistors as switches.

Acknowledgements

This work was supported through the Nuffield Award to Newly Appointed Lecturers number SCI/180/91/430/G.

References

[1]. Integrated Circuit Logarithmic Units
Jeffrey et al IEEE Trans on Computers Vol C-34 No 5 May 1985

[2]. Digital Filtering using Logarithmic Arithmetic
N.G Kingsbury P.J.W Rayner
IEE Electronics Letters 28th Jan 1971 Vol.7 NO.2 pp 56-58

[3]. Error Analysis of Recursive Digital Filters Implemented with Logarithmic Number Systems. T. Kurokawa, J. A. Payne, S.C. Lee
IEEE Transactions ASSP Vol. ASSP-28 No. 6 December 1980.

[4]. Computation of the Base Two Logarithm of Binary Numbers.- M. Combet, H.Van Zonneveld, L Verbeek
IEEE Transactions on Electronic Computers Vol EC-14 No. 6 Dec 1963 pp 863-867

[5]. New Algorithms for the Approximate Evaluation in Hardware of Binary Logarithms and Elementary Functions. - D. Marino
IEEE Transactions on Computers December 1972 pp1416-1421

[6]. Efficient VLSI Digital Logarithmic Codecs - B. Hoefflinger
IEE Electronics Letters 20th June 1991 Vol 27 N013 pp1132-1134

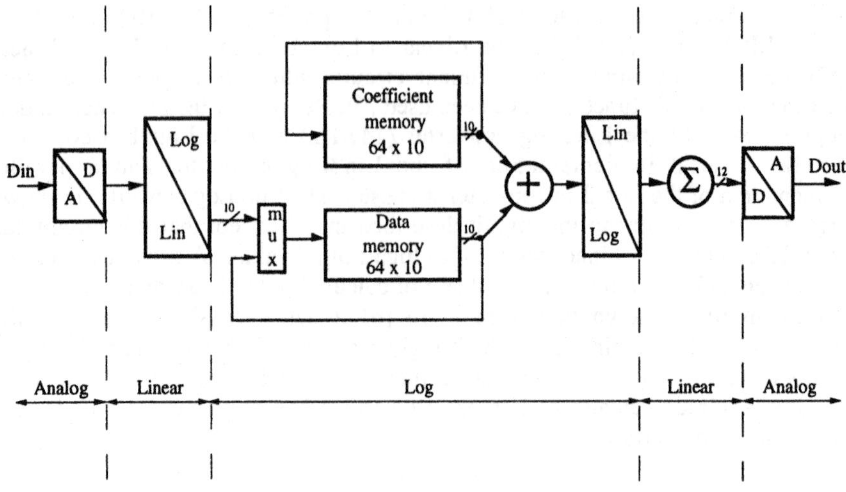

Fig 5. Log-Filter Block Diagram

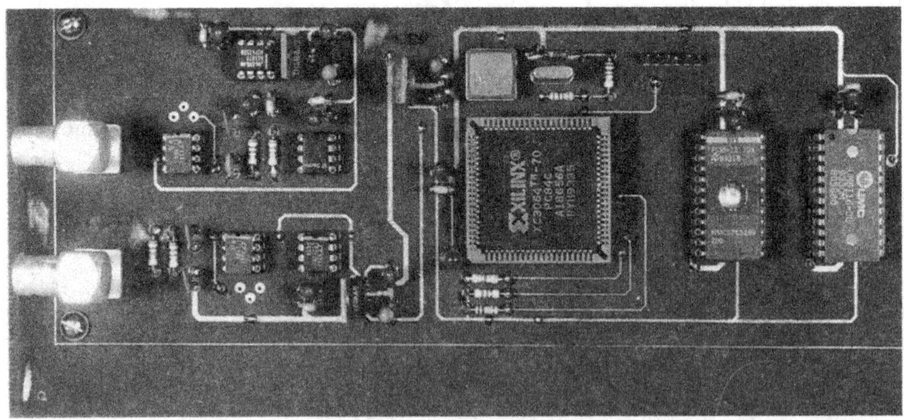

Fig 6. Log-Filter Prototype Board

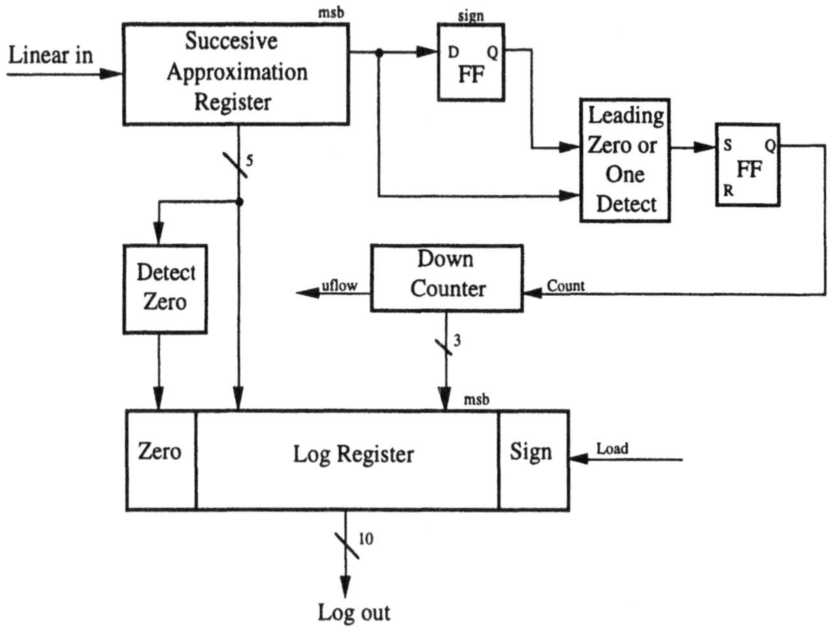

Fig 6. Serial Linear to Log Converter

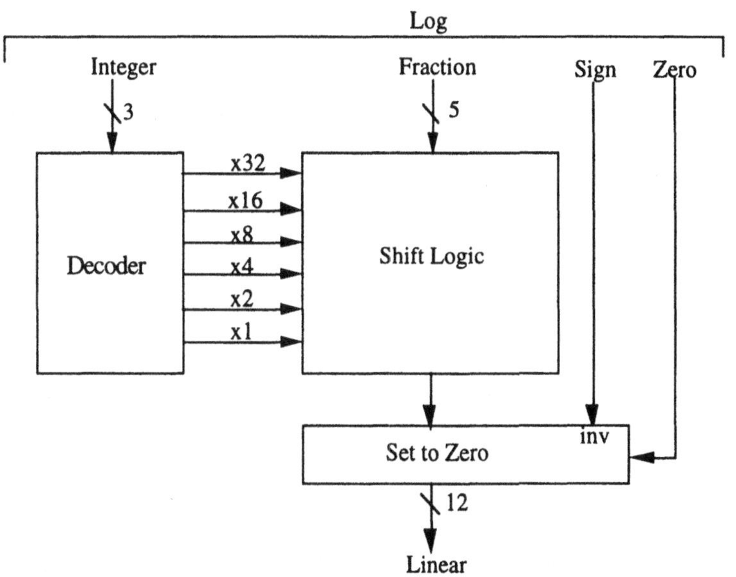

Fig 7. Log to Linear Converter

Bit-serial FIR Filters with CSD Coefficients for FPGAs*

L.E. Turner, P.J.W. Graumann and S.G. Gibb

Department of Electrical and Computer Engineering
University of Calgary
Calgary, Alberta, Canada T2N 1N4
email: turner@enel.ucalgary.ca

Abstract. The implementation of finite impulse response (FIR) digital filters using pipelined bit-serial arithmetic and canonic signed digit (CSD) coefficient coding can be an effective use of hardware resources. However, the necessary time alignment of all data and control signals can be a tedious process. A methodology for implementing FIR filters using pipelined bit-serial arithmetic and field programmable gate arrays (FPGAs) is described.

1 Introduction

Finite Impulse Response (FIR) digital filters are a common component in Digital Signal Processing (DSP) systems. The advantages of the FIR filter include guaranteed stability for all realizable filter coefficient values, freedom from limit cycle or overflow oscillations due to finite precision effects and the ability to implement filters with a linear phase frequency response.

The digital arithmetic methodology chosen to implement an FIR filter has a direct impact on the filter data throughput rate and the hardware size. Full bit-parallel arithmetic designs[1, 2] have the highest data throughput rate with a corresponding large hardware size. Bit-serial arithmetic based filter designs[3, 4, 5] process data one bit at a time resulting in a factor N decrease in the data throughput rate (where N is the data word size) and a corresponding decrease in hardware size. The performance of digit-serial arithmetic[6] systems is between that of bit-serial and bit-parallel systems.

One particular advantage of bit-serial (and digit-serial) arithmetic implementations is the low demand placed on routing resources. The single wire datapath connection for bit-serial arithmetic is particularly attractive when a technology with limited routing resources (such as an FPGA) is used.

In this paper we describe the implementation of FIR filters using Xilinx[7] Field Programmable Gate Arrays (FPGAs), pipelined bit-serial arithmetic and Canonic Signed Digit (CSD)[8] coefficient coding.

* This work was supported in part by funding from the National Sciences and Engineering Research Council of Canada and Micronet, A Canadian Network of Centres of Excellence.

The implementations of two FIR filters with 16 bit parallel input and output data and 16 bit pipelined bit-serial internal arithmetic are described. These filters are a 61 tap bandpass filter (with 12 bit effective coefficient wordlength) and a 53 tap lowpass filter (with 13 bit effective coefficient wordlength). The measured and ideal frequency response characteristics of each of these filters implemented using one Xilinx XC4005 FPGA (1.625 MHz. maximum sample rate) are presented.

2 Pipelined Bit-serial Arithmetic

In a pipelined bit-serial arithmetic implementation, N bits of a data word are transmitted over a single wire in N clock cycles. This leads to arithmetic operators which are typically a factor of N smaller in size than equivalent parallel arithmetic versions. For a common technology and clock rate, bit-serial and bit-parallel arithmetic systems will have roughly the same $size(area) \times time$ (ST) product. However, for a bit-serial system pipelined at the bit level, it is likely that a shorter clock period can be used. Consequently a pipelined bit-serial implementation can have a lower ST product than a comparable (non-pipelined) parallel implementation.

The pipelined bit-serial arithmetic conventions defined for the silicon compiler FIRST[9] are used in the FIR filters described in this report. The arithmetic elements required to implement an FIR filter are addition(subtraction), multiplication and delay (storage).

2.1 Addition

The addition element in a bit-serial arithmetic implementation is the carry-save adder. A carry-save adder consists of a full adder, a flip/flop to save any carry generated until the next bit add time and circuitry which resets the carry at the least significant bit add time. A block diagram of a carry-save adder is given in figure 1. In a Xilinx XC4000 series FPGA implementation, a carry-save adder (or subtracter) will fit within one Configurable Logic Block (CLB). The latency (delay between arrival of the least significant bit (LSB) of the input and generation of the LSB of the output) for each pipelined bit-serial adder is one bit clock time.

2.2 Multiplication

A size-efficient fixed-coefficient bit-serial multiplier can be implemented by coding the multiplier coefficient in a canonic signed-digit (CSD) form. Each digit in the CSD word is allowed to take on a value of 1, 0, or -1. A CSD coded coefficient will have at most one-half as many non-zero digits as the equivalent binary coded coefficient. This translates directly into a reduction in the size of the multiplier implementation. A CSD coded bit-serial multiplier made from bit-serial adders, subtracters and shifters is implemented for each different multiplier

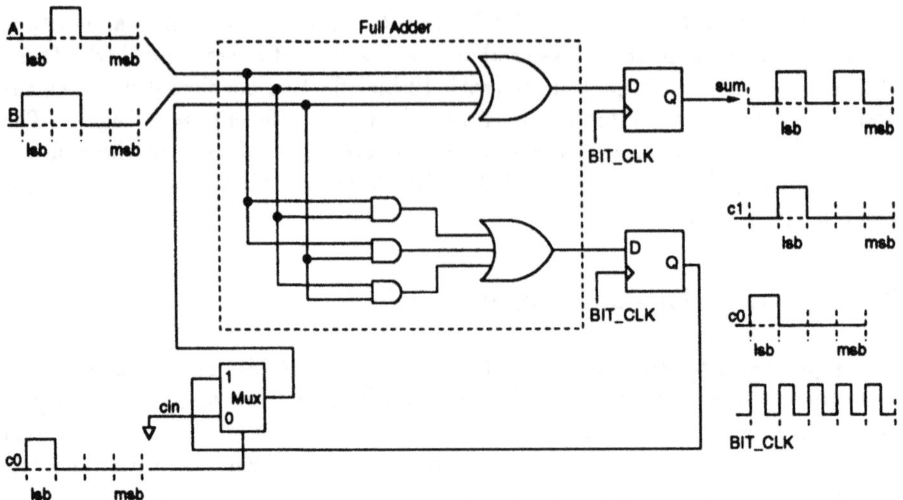

Fig. 1. Block diagram of a carry-save adder.

coefficient in the FIR filter. The logic diagram of a CSD bit-serial multiplier is given in figure 2. The latency for each bit-serial CSD multiplier depends on the coefficient value and the CSD coding used.

2.3 Delay

The bit-serial delay chain in an FIR filter with a uniform N bit data wordlength will require N flip/flops for each delay element in the chain. If random access memory (RAM) is available, it is possible to use the RAM as shift register memory by adding an address counter and read/write control logic. This strategy can result in a smaller implementation in a technology where RAM cells are smaller than individual flip/flops. This is the case in the Xilinx XC4000 FPGA where each CLB contains two four-input lookup tables (LUTs) and two flip/flops. Each LUT can be used as a 16×1 bit RAM. Neglecting the size of the address generator (which can be shared), two 16 bit shift registers can be implemented in one CLB by using the RAM cells in the LUTs. Two 16 bit shift registers would require 16 CLBs if the LUTs were not used.

3 Automated Bit-serial FIR Filter Design: *sfirgen*

The detailed design of a pipelined bit-serial digital signal processing system can be a tedious procedure, as proper operation of the system requires careful time alignment of all data and control signals. In addition, the latency of a CSD coded bit-serial multiplier depends on the coefficient value and the particular CSD coding used. Thus, the number and position of bit delays required in each

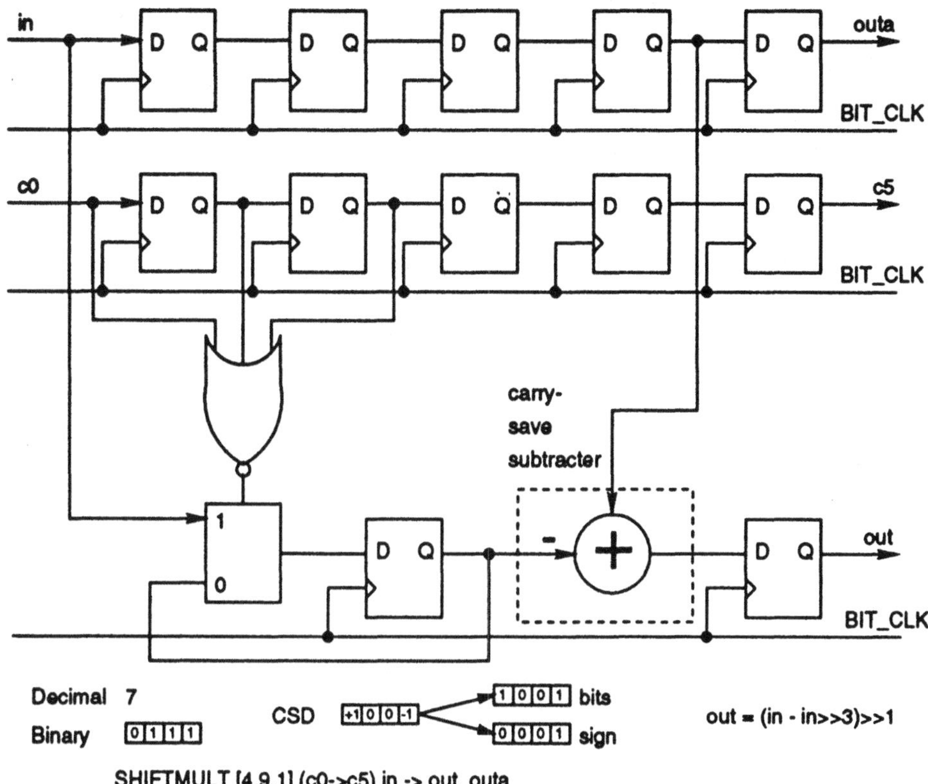

Fig. 2. Block diagram of a CSD multiplier.

data path for time alignment can only be determined after the coefficient values are known. Whenever the FIR filter multiplier coefficient values are changed, the CSD multipliers and the internal bit timing of the filter must be changed as well.

A computer program (*sfirgen*) has been implemented to automate the process of time alignment of data and control signals. In addition, this program:

1. determines the CSD coefficient coding for each multiplier coefficient in the coefficient input file (and thus the latency for each multiplication),
2. optionally combines, using an adder tree, all data signals which would be multiplied by the same coefficient value (The use of this option can in some cases result in a larger filter as it may prevent the sharing of flip-flops between the input delay chain and the CSD coded multipliers. However for linear phase FIR filters with symmetrical coefficients this option normally reduces the filter size.),
3. optionally includes a divide by 2^N (`DSHIFTA`) operation in the input delay chain whenever every following coefficient is less than $1/2^N$,

4. generates CSD multipliers (SHIFTMULT) for each different CSD coded coefficient,
5. generates an adder tree to form the final sum of products output taking into account the latency of each CSD multiplier so as to minimize the need for additional delay elements required for time alignment, and
6. optionally reduce the level of pipelining in the adder tree, to allow further hardware savings if the highest data throughput rate is not required.

The output of the *sfirgen* program is a register transfer level (RTL) description of the FIR filter implementation in the DFIRST[4, 5] language. DFIRST is an extension of the language FIRST[9]. New operators in DFIRST include digit-serial operators, CSD multipliers (SHIFTMULT) and serial↔parallel conversion operators. The DFIRST netlist generated by *sfirgen* for an 8 coefficient FIR filter design with coefficients: 0.25, 2.44140625e-6, -0.0625, -0.0625, -0.0625, -0.0625, -0.1454375, and 0.1267089844 is given in Figure 3. This filter has a input/output wordlength of 16 bits and an internal system wordlength of 20 bits.

A block diagram of this FIR filter is given in figure 4. Note that DSHIFTA (divide by 2^n) operators have been included in the input delay stage. This reduces the number of bits used in each SHIFTMULT operator which results in a reduced latency and a reduction in hardware requirements. The coefficient multipliers have been implemented as SHIFTMULT operators (CSD multipliers) or as DSHIFTA operators when only one bit is set in the CSD coefficient. This also leads to reduced hardware requirements as the latency of the DSHIFTA operator is less than that of the SHIFTMULT operator. In this example, data signals which would be multiplied by the same coefficient value have not been combined. The node naming convention includes latency information. A node named nXX_YY is signal number XX where the least significant bit of the signal occurs at bit time YY at this node.

The event driven simulator DSIM[4, 5] can be used to verify the function of the filter. This allows the FIR filter performance (unit sample response and finite precision effects) and time alignment of all data and control signals to be verified before an FPGA implementation is attempted.

4 Automated Gate Level Implementation: *trans*

The DFIRST RTL filter netlist generated by *sfirgen* is translated into a gate level netlist using *trans* [4, 5, 10]. *trans* is a software tool used to translate digital circuit descriptions into gate level digital logic netlists. An overview of the function of *trans* is shown in figure 5. *trans* assembles the set of logic elements used in each input netlist from a library of parameterized macro elements. The macro elements are described in technology files. *trans* converts an input netlist format into an equivalent representation in the output format using the technology file information. *trans* optionally performs a variety of circuit level optimizations on the gate level design.

```
OPERATOR FILTER[](c0->c17)n0_0->n22_17

    SIGNAL n0_2, n0_20, n1_2, n2_2, n3_2, n4_2
    SIGNAL n5_2, n6_2, n7_2, n8_4, n9_13, n10_5
    SIGNAL n11_5, n12_5, n13_5, n14_13, n15_14, n8_5
    SIGNAL n16_6, n17_6, n13_13, n18_14, n14_14, n19_15
    SIGNAL n20_7, n18_15, n21_16, n20_16
    CONTROL c2, c5, c6, c13, c14, c15, c16
    DSHIFTA[1](c0)n0_0->n0_2
    BITDELAY[18]n0_2->n0_20
    DSHIFTA[1](c0)n0_20->n1_2
    BITDELAY[20]n1_2->n2_2
    BITDELAY[20]n2_2->n3_2
    BITDELAY[20]n3_2->n4_2
    BITDELAY[20]n4_2->n5_2
    BITDELAY[20]n5_2->n6_2
    BITDELAY[20]n6_2->n7_2
    DSHIFTA[1](c2)n0_2->n8_4
    DSHIFTA[10](c2)n1_2->n9_13
    DSHIFTA[2](c2)n2_2->n10_5
    DSHIFTA[2](c2)n3_2->n11_5
    DSHIFTA[2](c2)n4_2->n12_5
    DSHIFTA[2](c2)n5_2->n13_5
    SHIFTMULT[8,149,0](c2->NC)n6_2->n14_13,NC
    SHIFTMULT[10,521,1](c2->NC)n7_2->n15_14,NC
    BITDELAY[1]n8_4->n8_5
    SUBTRACT[1,0,0,0](c5)n8_5,n10_5,GND->n16_6,NC
    ADD[1,0,0,0](c5)n11_5,n12_5,GND->n17_6,NC
    BITDELAY[8]n13_5->n13_13
    SUBTRACT[1,0,0,0](c13)n9_13,n13_13,GND->n18_14,NC
    BITDELAY[1]n14_13->n14_14
    SUBTRACT[1,0,0,0](c14)n15_14,n14_14,GND->n19_15,NC
    SUBTRACT[1,0,0,0](c6)n16_6,n17_6,GND->n20_7,NC
    BITDELAY[1]n18_14->n18_15
    ADD[1,0,0,0](c15)n18_15,n19_15,GND->n21_16,NC
    BITDELAY[9]n20_7->n20_16
    ADD[1,0,0,0](c16)n20_16,n21_16,GND->n22_17,NC
    CBITDELAY[2](c0->c2)
    CBITDELAY[3](c2->c5)
    CBITDELAY[1](c5->c6)
    CBITDELAY[7](c6->c13)
    CBITDELAY[1](c13->c14)
    CBITDELAY[1](c14->c15)
    CBITDELAY[1](c15->c16)
    CBITDELAY[1](c16->c17)

END
```

Fig. 3. DFIRST Netlist Specification for an 8 Tap FIR Filter

The optimizations performed by *trans* include:

- *Flatten the netlist.* Selectively remove hierarchy in the design.
- *Hardware reduction.* Remove unused logic and combine multiple instances of parts with identical input nets (subject to a maximum fanout constraint).
- *RAM conversion.* Convert state storage elements (flip/flops) used as shift registers into RAM with address generators and control logic.
- *Hardware mapping.* Convert low-level parts from the flattened netlist into

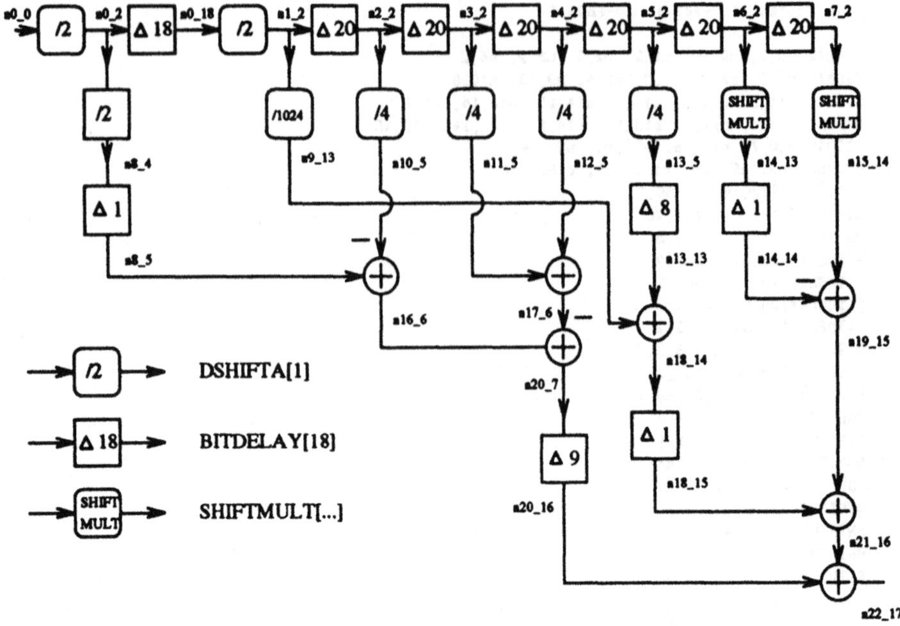

Fig. 4. Block diagram of DFIRST 8 Tap FIR Filter

higher-level parts more suited to a particular technology

- *Delay estimation.* Estimate signal propagation times for each logic path in the design (based on delay versus loading data supplied by the user).
- *Board file generation.* Produce a printable board connection diagram of the implementation showing the I/O signal names.
- *Gate count.* Provide a worst case estimate of the number of gates used.

5 FIR Filter Examples

The implementation of a 61 tap FIR bandpass digital filter and of a 53 tap FIR lowpass digital filter are described in this section. The transfer functions of these linear phase filters are given in figures 6 and 7. The CSD coefficients for these filters were determined using the digital filter design program NOMAD[11].

Using *trans* , the DFIRST RTL description of the filter generated by *sfirgen* is translated into a Xilinx eXternal Netlist Format (XNF) listing and then processed by the Xilinx Partition, Place and Route (PPR) tool. This initial raw translation of the bandpass filter design to the XC4000 architecture required 431% of the D flip/flops (DFF) in an XC4005 (192 CLBS, 2 DFFs each) and 83% of the function generators in the device. This is a requirement for more CLBs than are available in two XC4010s. However, using *trans* the following circuit optimizations can be performed

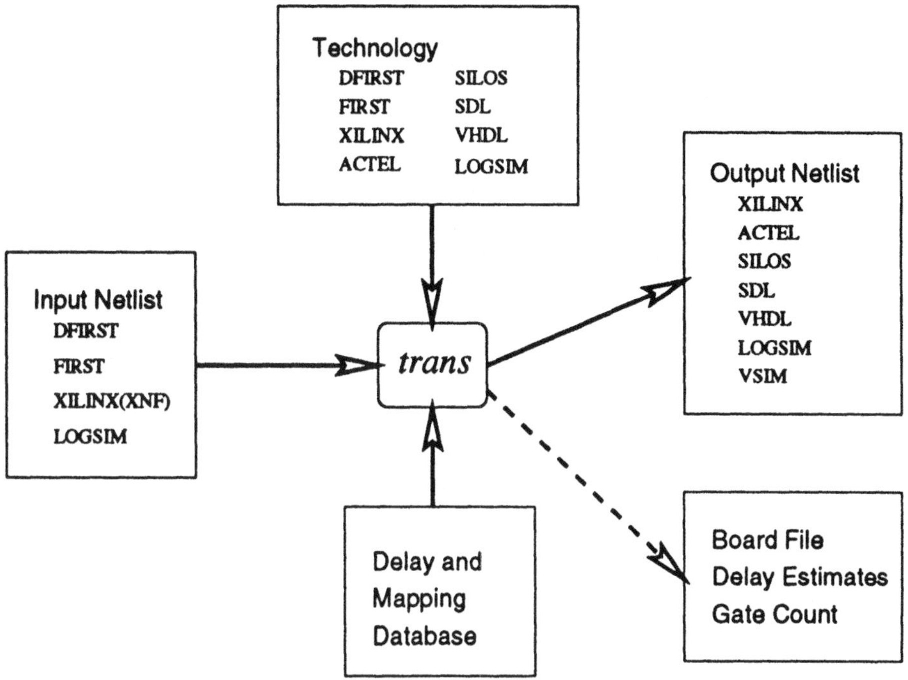

Fig. 5. *trans* input/output.

1. common control and/or internal data paths in the multipliers can be combined (the fanout is increased to 20),
2. convert the FIR state storage shift registers into RAM storage using the Xilinx XC4000 LUTs.
3. unused I/O blocks are used to implement DFFs.

After applying these optimizations, the final implementation of this 61 tap FIR filter for a XC4005 FPGA uses 91% of the DFFs, 92% of the function generators and 84% of the I/O blocks. This filter has been implemented and the measured maximum sample rate is 1.625 MHz. The measured filter frequency response is shown in figure 6.

Using the same process as for the bandpass FIR filter, a linear phase lowpass filter with 53 taps and 13 bit effective coefficient wordlength (CSD coefficients with three non-zero digits) has been implemented using an XC4005 FPGA and uses 86% of the DFFs, 87% of the function generators and 87% of the I/O blocks. This filter has been implemented and the measured maximum sample rate is 1.625 MHz. The measured filter frequency response is shown in figure 7.

Each LUT in the Xilinx XC4000 series FPGA can be used to implement a 16×1 RAM. For this reason, shift register lengths which are multiples of 16 bits can be effectively implemented using the RAM. If the shift register length is less than a multiple of 16 bits, the address generator will not use all of the RAM

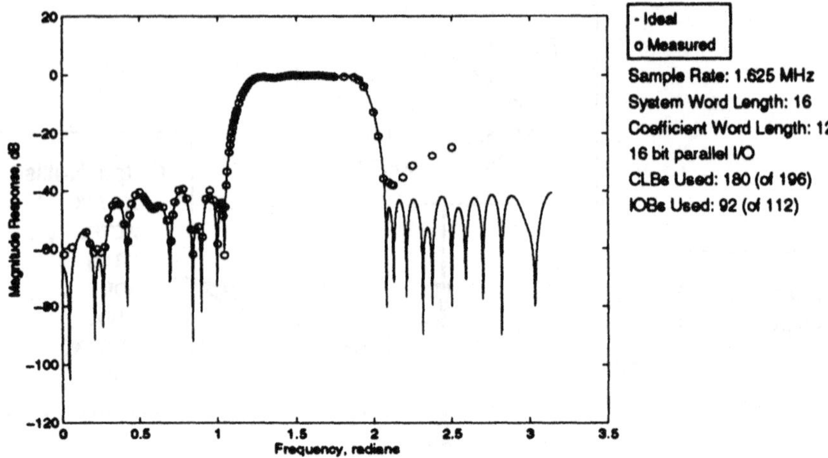

Fig. 6. 61 Tap FIR filter Ideal and Measured Magnitude Frequency Response

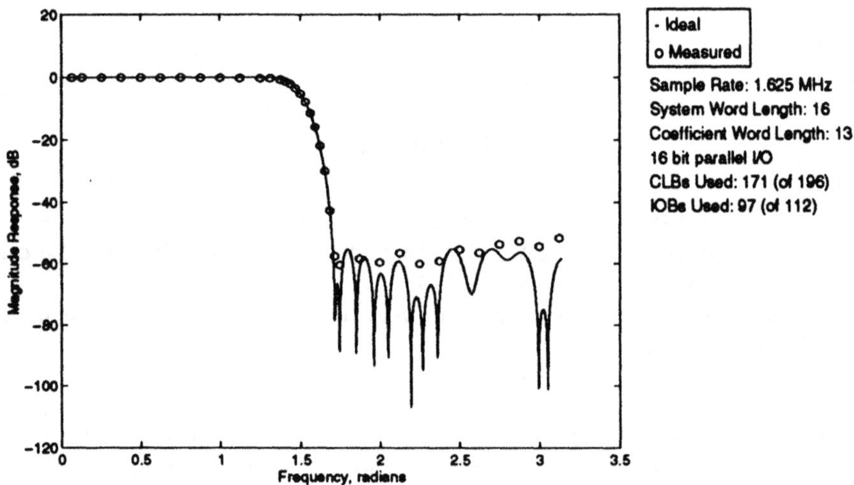

Fig. 7. 53 Tap FIR filter Ideal and Measured Magnitude Frequency Response

memory. The *trans* program compensates for this by selecting a set of RAM shift register lengths which can share a single address generator and extends the shift registers when necessary with flip/flops from the CLBs when the cost of doing so is less than creating an additional address generator. The effect of changing the data wordlength in the FIR filter has been examined for the 53 tap lowpass FIR filter. In this case, only the option to add data signals which would be multiplied by the same coefficient value before multiplication has been enabled, all other options have been turned off.

The XC4000 series DFFs, LUTs and I/O Blocks (with no flip/flops mapped into the I/O Blocks) required for 14, 16, 18 and 32 bit data wordlengths are:

Data Wordlength (bits)	14	16	18	32
XC4000 DFFs	406	406	438	488
XC4000 LUTs	373	342	417	428
XC4000 I/O Blocks	31	27	36	59
XC4000 CLBs	203	203	219	244

6 Conclusion

A method of implementing FIR digital filters in Xilinx XC4000 series FPGAs using pipelined bit-serial arithmetic and canonic signed digit (CSD) coefficient coding has been presented. In these filters the XC4000 series LUT RAM has been used to implement shift register memory thus reducing the number of CLBs required for each implementation.

References

1. J.B.Evans. Efficient fir filter architectures suitable for fpga implementation. *IEEE Transactions on Circuits and Systems-II*, pages 490–493, 1994.
2. P.T. Yang R. Jain and T. Yoshino. Firgen: A computer-aided design system for high performance fir filter integrated circuits. *IEEE Transactions on Signal Processing*, pages 1655–1668, 1991.
3. D.R. Bull and G. Wacey. Bit-serial digital filter architecture using ram-based delay operators. *IEE Proc.-Circuits Devices Syst.*, pages 4–19, 1994.
4. P.J. Graumann and L.E. Turner. Implementing dsp algorithms using pipelined bit-serial arithmetic and FPGAs. *First International ACM/SIGDA Workshop on FPGAs*, pages 123–128, 1992.
5. P.J. Graumann and L.E. Turner. Specifying and hardware prototyping of dsp systems using a register transfer level language, pipelined bit-serial arithmetic and fpgas. *2nd Canadian Workshop on Field Programmable Devices*, 1994.
6. P. Jacob R. Hartley, P. Corbett and S. Karr. A high speed fir filter designed by compiler. *Proc. Custom Integrated Circuits Conf.*, pages 20.2.1–20.2.4, 1989.
7. XILINX. *The Programmable Logic Data Book*. Xilinx Inc., 1994.
8. R.M.M. Oberman. *Digital Circuits for Binary Arithmetic*. Macmillan Press Ltd., 1979.
9. P. Denyer and D. Renshaw. *VLSI Signal Processing: A Bit Serial Approach*. Addison-Wesley Publishing Company, 1985.
10. L.E. Turner P. Graumann and S. Barker. TRANS User's Guide. *Department of Electrical and Computer Engineering, University of Calgary, Internal Report*, 1992, 1993.
11. L.E. Turner P. Graumann and M. Svihura. NOMAD User's Guide. *Department of Electrical and Computer Engineering, University of Calgary, Internal Report*, December 1992.

A Self-Validating Temperature Sensor Implemented in FPGAs

M. Atia, J. Bowles, D.W. Clarke, M. Henry, I. Page*, G. Randall* and J. Yang

Oxford University Department of Engineering Science, Parks Road, Oxford, OX1 3PJ.
*Oxford University Computing Laboratory, Parks Road, Oxford, 0X1 3QD.

Abstract. A SEVA sensor is an intelligent device which monitors its own performance and generates quality indices for each measurement value, including its level of uncertainty. This paper describes a SEVA temperature sensor in which all the measurement uncertainty and validation calculations are carried out in FPGAs. Hardware compilation is used to configure the FPGAs to provide the required functionality.

1 Introduction

Industrial instrumentation is currently undergoing a revolution in response to the availability of cheap and reliable microprocessors and the increasing use of digital communications. Traditionally, measurement and control actuation have been carried out in devices using at best analogue electronics. A widely adopted communication standard has been to use 4-20mA over a twisted pair to convey a single analogue variable.

Today, a series of standards for digital communications in industrial plants, known collectively as Fieldbus, are nearing completion. These will for the first time allow open two-way communication between sensors, actuators and control systems.

These developments have led to considerable interest in extending the functionality of sensors and actuators, including in the area of validation. In the past, sensor data has either simply been assumed to be correct, or expensive plant-based modeling has been employed to detect faults. With internal processing power, intelligent sensors are now able to perform self-diagnostics, and attention has moved to focus on standard generic descriptions of sensor fault states.

2 SEVA

Henry and Clarke [1] describe the SEVA (SElf-VAlidating) Sensor concept, in which a sensor monitors its own performance and generates quality indices for each measurement value. Figure 1 shows the principal parameters generated by the SEVA sensor.

The Validated Measurement Value (VMV) corresponds to the conventional measurement. It represents the best estimate of the true process value taking into account all the relevant factors, including the presence of faults. The Validated

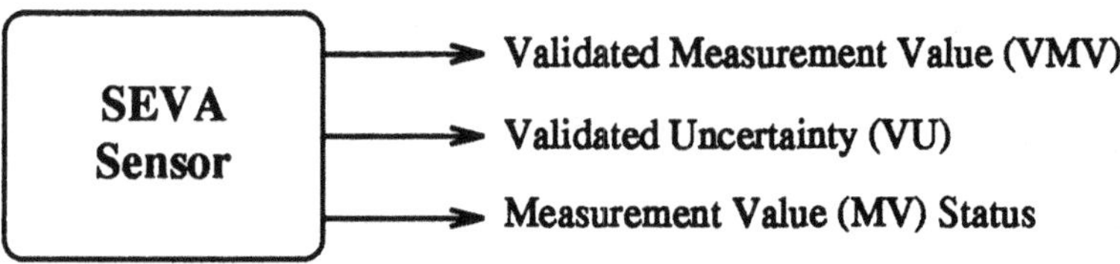

Fig. 1. The principal parameters generated by the SEVA sensor.

Uncertainty (VU) gives the probable error of the VMV. This is always greater than zero, and is calculated for each sample to take into account the time varying influence of each error source. The MV status is a discrete parameter describing how the current VMV was calculated. It has four principal values taken from a visual analogy:

- CLEAR: VMV calculated normally; sensor is fault free.
- BLURRED: VMV corrected for a fault, but is still based on live data.
- DAZZLED: VMV projected from past history as raw measurement is not credible; the effect is likely to be temporary.
- BLIND: VMV projected from past history as raw measurement is not credible; the effect is permanent.

Prototype systems have been developed using a range of commercial sensors, including pressure, mass flow, dissolved oxygen and temperature. At present all prototypes are PC-based, which require an expensive and labour intensive development programme.

The next generation of the SEVA prototypes are being developed using an FPGA-based flexible architecture implementation [2]. It can take two forms. The first uses a transputer as a processor, with standard FPGAs controlling sensor-specific i/o requirements and low level processing. The second uses FPGAs to perform all sensor measurement and validation activity. Both forms are currently implemented as PC cards so that the PC host can provide a graphical interface.

This paper describes a thermocouple-based temperature sensor implemented using FPGAs. The Handel programme carries out all the SEVA validation calculations and presents the measurement and validity data to the PC bus for display and storage.

3 Hardware Compilation

It is a common experience that producing hardware solutions for application problems is fraught with difficulties. Traditional hardware design techniques lack the flexibility of introducing design changes and improvements after implementation. Moreover, debugging the design is a painful process of repeated re-design and re-build.

The emergence of dynamically reconfigurable FPGA (Field Programmable Gate Array) chips has changed the nature of hardware design implementation.

Reconfiguring an FPGA takes only a few milliseconds by downloading parameterization data to the chip. This new technology has led to considerable interest in searching for high-level hardware design tools which speed up the time to produce new systems. Hardware compilation is a technique for reducing the entire hardware design and implementation to a purely software process. Coupled with the use of FPGAs, Hardware compilation gives the designer the flexibility to adopt new design improvements and changes even after implementation.

In this paper, Handel [3] is the hardware compilation language used to implement the sensor. It is a variant of Occam2. Its advantages are as follows:

- *Efficiency.* As programming languages are normally intended to be executed on a processor, they assume the availability of such facilities as an instruction set, RAM for code and data storage and a fetch-execute cycle. Handel does not make use of a processor, but implements its own functions only as required. This approach is particularly silicon efficient.
- *Expressive power.* Typically programming languages prescribe sequential steps, while FPGA operations are largely carried out in parallel. Handel's ability to express parallel processing makes it especially suitable for FPGA design.
- *Correctness.* As a variant of Occam, Handel inherits Occam's algebraic semantics. Therefore the transformation of a Handel program to a net list is provably correct [4].

4 SEVA Temperature Sensor

The SEVA temperature sensor was originally designed and built by Yang [5]. The sensor described in this paper is a simplified version based on FPGA implementation. This section describes the basic principles of operation which the FPGA implementation inherits from the original SEVA temperature sensor.

4.1 Temperature measurement

Temperature is one of the most widely measured engineering variables, providing the basis for a variety of control and safety systems. The thermocouple is the most commonly used temperature sensor, accounting for 37% of the market.

The thermocouple operation exploits a phenomenon called thermoelectricity. It consists of two electrical conductors made of different types of metal connected together at one end -*The sensing junction*- while the other end is left unconnected -*The reference junction*- as shown in figure 2. If the two junctions are held at different temperatures T_{sens} and T_{ref}, a finite open-circuit electric potential will be produced between the two metals at the reference junction [6]. The magnitude of this EMF corresponds to the difference in temperature ($T_{sens} - T_{ref}$). Although the relation is non-linear, it is highly repeatable and the EMFs for different values of temperature (assuming the reference junction is at ice point) are well documented in thermocouple lookup tables.

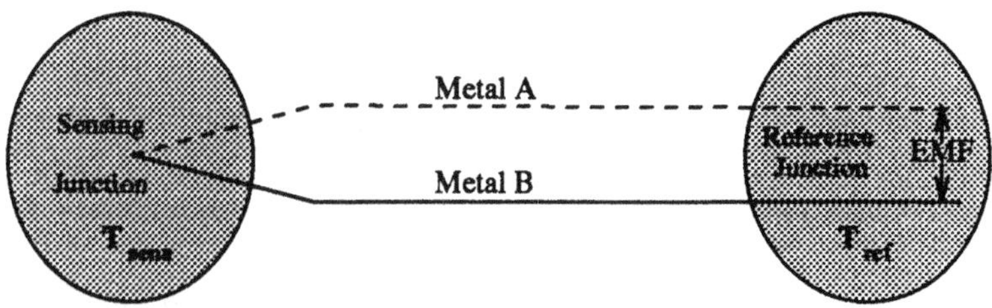

Fig. 2. A thermocouple outputs a voltage corresponding to the temperature difference between its two junctions.

To determine the absolute sensing junction temperature (T_{sens}), one possibility is to measure the reference junction temperature (T_{ref}) and add it to the thermocouple temperature reading (T_{emf})

$$T_{sens} = T_{emf} + T_{ref} \tag{1}$$

This arrangement is also advantageous in industrial practices: the reference junction can be placed in a benign environment so that its temperature may be determined by a number of means (e.g. using a temperature sensing integrated circuit), while the sensing junction can be placed at the point whose temperature needs to be determined, which could be in extreme, dangerous or remote conditions.

4.2 Uncertainty Analysis

The aim of any sensor is to obtain a value representative of a certain physical, chemical or electrical property. Each measurement is influenced by a number of factors such as equipment limitations or environmental effects, so that the true value can never be determined. The measurement is thus only an estimate of the true value, and has an associated error. Uncertainty is a parameter which characterizes the spread of values that could reasonably be attributed to the true value. An uncertainty δM associated with a measurement M implies that the true value M_{true} could lie anywhere within the range of

$$(M - \delta M) \leq M_{true} \leq (M + \delta M)$$

at a certain probability (typically 95%).

For a measurement M that is dependent on several other factors m_1, m_2, ... and m_n, its uncertainty δM is estimated using the uncertainty propagation method [7]

$$(\delta M)^2 = \left(\frac{\partial M}{\partial m_1}\delta m_1\right)^2 + \left(\frac{\partial M}{\partial m_2}\delta m_2\right)^2 + \ldots + \left(\frac{\partial M}{\partial m_n}\delta m_n\right)^2 \tag{2}$$

where δm_1, δm_2, ... and δm_n are the uncertainties of the variables.

The SEVA sensor calculates the on-line uncertainty of each measurement. According to the current state of the sensor, it outputs the Validated Uncertainty (VU) of the measurement as described in section 2.

4.3 Fault Detection

All sensors are susceptible to device failures which usually affect the output measurement. In an industrial plant, failing to identify a faulty sensor could result in wrongly judged control action which might compromise safety and efficiency.

The SEVA temperature sensor tests itself for the most common faults. The faults considered by the FPGA implementation are:

- Loss of thermal contact with the process fluid. This fault results in the sensor giving a temperature different from the actual process temperature. Yang [5] argues that the Loop Current Step response test (LCSR) can give an indication of whether or not the thermocouple is in adequate thermal contact with the process. A simplification of the test is implemented on the FPGA version as described in section 5.3.
- A loose connection on the thermocouple head or an open circuit. As a fault detection circuit, Yang [5] suggests a pull-up resistor connected between one of the thermocouple wires and the power supply.

5 FPGA Implementation

A simplified version of the SEVA temperature sensor is implemented as a PC card. The card circuit includes several FPGAs which are configured using Handel code to perform the SEVA calculations. The output of the card is written to the PC via the PC Bus. A program running on the PC simply reads the values and displays them in a real time graph.

The card hosts two FPGAs (Master and Slave 1) and a daughter board as shown in figure 3. Slave 1 is configured to calculate the temperature and uncertainty (unvalidated). Section 5.1 gives details of the temperature and uncertainty calculations. The Master FPGA is configured to trigger the fault detection test and validate the temperature and uncertainty. Details of the fault detection test are given in section 5.3.

The daughter board hosts the thermocouple, the reference temperature sensor, the sigma-delta A/D converter ($\Sigma\Delta$) and an FPGA (Slave 2). The $\Sigma\Delta$ is programmed through Slave 2 to collect data from both the thermocouple and the reference temperature sensor. Slave 2 has several other functions:

- Downloading the configuration data to the $\Sigma\Delta$ to specify its operation parameters.
- Instructing the $\Sigma\Delta$ to multiplex between the two input channels.
- Collecting the $\Sigma\Delta$ data and converting it from serial to parallel form in order to facilitate its usage in subsequent Handel calculations.
- Sending data to the temperature calculation routine.

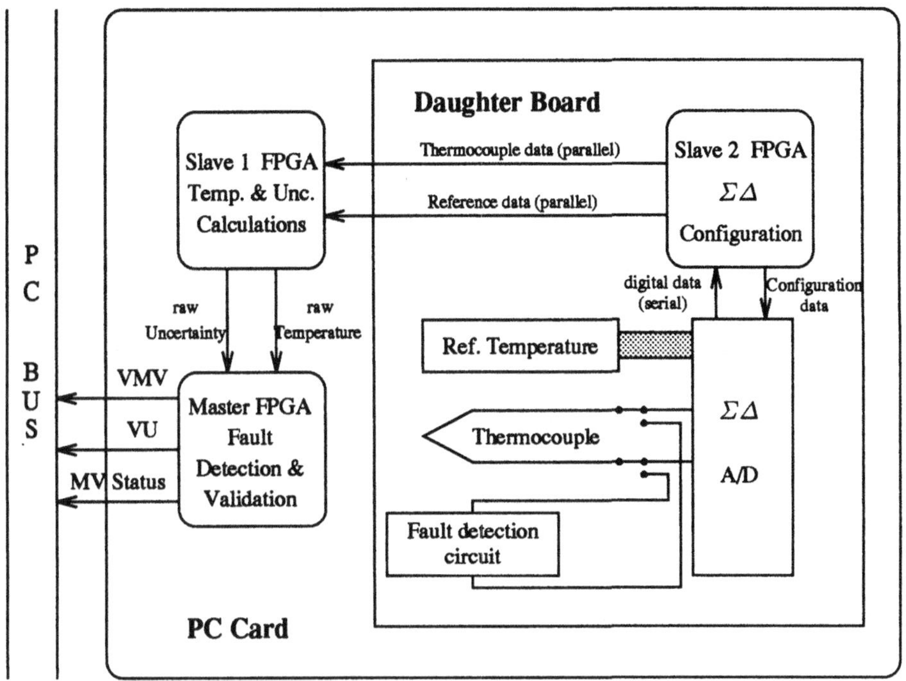

Fig. 3. Block diagram of the SEVA sensor circuit FPGA implementation.

5.1 Temperature calculation

The analogue voltages produced by the thermocouple and the reference temperature sensor are fed into a two channel $\Sigma\Delta$. Before converting the voltages, the $\Sigma\Delta$ multiplies them by a user specified gain in order to make use of its entire range in improving the accuracy of the reading. It then converts the voltages to 16 bits words.

In order to estimate the temperature of the sensing junction, the voltages of the thermocouple and the reference temperature sensor have to be calculated from the $\Sigma\Delta$ outputs. Each voltage (V_{sens}) is calculated by multiplying the ratio between the corresponding $\Sigma\Delta$ output data (D) and the maximum data value (D_{max}) by the full scale voltage (a) of the $\Sigma\Delta$. A half has to be subtracted from the ratio because the $\Sigma\Delta$ is operating in the bipolar mode, as shown in equation (3).

$$V_{sens} = a[\frac{D}{D_{max}} - \frac{1}{2}] \tag{3}$$

V_{sens} is then converted to the corresponding temperature (T_{sens}). A linear relation is assumed, as shown in equation (4).

$$T_{sens} = C_{sens} V_{sens} \tag{4}$$

where C_{sens} is the conversion factor. Equations (3) and (4) are used to calculate the temperatures of the thermocouple and the reference temperature sensor.

Both temperatures are then substituted into equation (1) to give the temperature of the thermocouple sensing junction

$$t = f_{th}D_{th} + f_{ref}D_{ref} - Const \tag{5}$$

$$f_{th} = 0.24, \quad f_{ref} = 0.63 \quad and \quad Const = 718.89$$

where D_{th} and D_{ref} are the data from the thermocouple and the reference temperature sensor respectively. To implement equation (5), 4 fixed point operations have to be used. These are very silicon expensive as they use many Configurable Logic Blocks (CLBs). To solve this problem the factors in equation (5) are scaled to give

$$t = 0.01(24D_{th} + 63D_{ref} - 71889) \tag{6}$$

Equation (6) consists of 4 integer operations and 1 fixed point. It is easier to implement and employs less CLBs.

5.2 Uncertainty calculation

For the SEVA temperature sensor described in this paper, the main sources of uncertainty are:

- The limited accuracy, nonlinearity and gain errors of the thermocouple and the reference temperature measurement chip.
- Amplifier and A/D converter gain drift and output noise.
- Calibration errors.
- Errors in using a simple linear measurement equation rather than the look-up table to account for the thermocouple non-linearity.

Applying the uncertainty propagation equation (2) to the temperature calculation relation given by equation (5), gives the uncertainty associated with the temperature t

$$(\delta t)^2 = (D_{th}\delta f_{th})^2 + (f_{th}\delta D_{th})^2 + (D_{ref}\delta f_{ref})^2 + (f_{ref}\delta D_{ref})^2 + \delta Const^2 \tag{7}$$

where the uncertainties associated with the factors δf_{th}, δD_{th}, δf_{ref} and δD_{ref} are calculated by applying the propagation equation to the relation between each factor and its elementary components. As δD_{th}, δf_{ref}, δD_{ref} and $\delta Const$ are negligible with respect to δf_{th}, equation (7) reduces to

$$\delta t = 0.0034\delta D_{th} \tag{8}$$

Equation (8) is implemented in Handel to calculate the uncertainty of the output temperature.

5.3 Fault detection test

As mentioned in section 4.3, there are two main tests performed by the SEVA thermocouple. They are the loss of thermal contact and the disconnection of the thermocouple from its transmitter circuit. This section describes the details of the loss of thermal contact test, as implemented in the FPGA version of the SEVA temperature sensor.

The test starts with a learning stage during which the sensor takes the average of the temperature readings T_{av} over half a second (50 readings). A current is then passed through the thermocouple. The sensing junction is usually submerged in silicon paste which acts as a thermal path between the junction and the process fluid. If the paste has leaked away, the thermocouple will be in bad thermal contact with the process fluid and the test will result in a high increase in the thermocouple temperature. If the silicon paste is still in place, the injected current will not result in a significant increase in the thermocouple temperature.

After the current injection, the rise in temperature is compared with T_{av}. If the rise exceeds T_{av} by a certain factor (which is learned by experiments), then loss of contact is flagged.

If the sensor is in bad thermal contact with the process, it outputs the averaged temperature T_{av} as the VMV. The VU is set to be the average of the pre-test uncertainties, but it increases by the fastest observed rate of change of the process. The increase in VU is meant to reflect the reduced accuracy of the substituted measurement. This strategy is explained in more detail in [1].

6 Experimental results

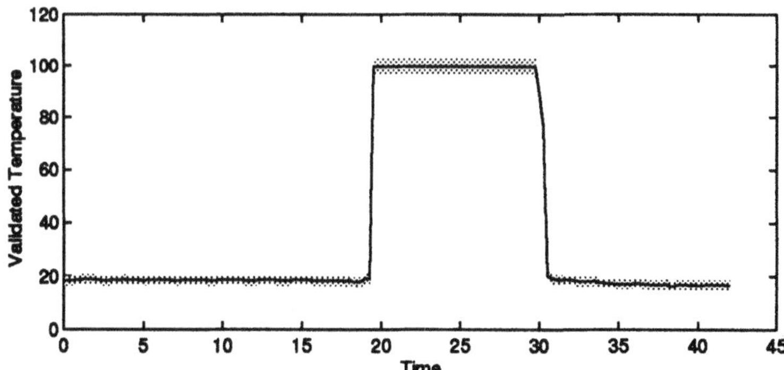

Fig. 4. A real-time graph showing the response of the circuit when the sensing junction was immersed in boiling water.

The circuit design described in section 5 performs the SEVA validation calculations inside a set of FPGAs sitting on a PC card. The calculations and results are then sent to the PC via the PC-Bus. The PC receives the data and displays it, via a graphical host software, on a real time graph as shown in figure 4. The real-time graph is a plot of the temperature VMV in real time. The graph also shows the VU as a range of values lying around the VMV. Figure 4 shows the response of the sensor when the sensing junction was immersed in boiling water. The temperature's VMV has risen to reach $100°C$ and the VU has increased as it is a function of temperature.

The MV status is shown on-line with the VMV and the VU. Figure 5 shows how the MV status behaves during the fault detection test. The graph illustrates the case when the sensing junction is left in the air. After 27 seconds the sensor triggers the fault detection test, the MV status changes to DAZZLED. During the test, the VMV is the pre-test average and the VU is the pre-test average increasing at the fastest observed rate of change. In this example, as the sensor is found faulty, after the test, the MV status changes to BLIND. The sensor continues outputting the substituted VMV and the VU continues to increase until it reaches the maximum value learned from past history.

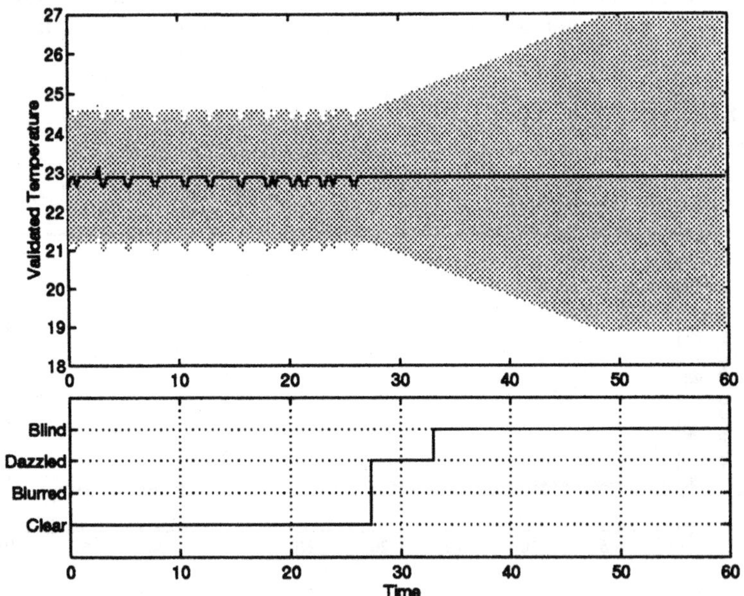

Fig. 5. A real-time graph showing the sensor behavior during the fault detection test. The MV status is shown on-line with the data.

7 Potential for commercial exploitation

This section discusses the likelihood of this technology being used for the development of commercial, as opposed to prototype, sensors. In current practice, low-power sensors (the majority) contain a mixture of analogue and digital circuits including ASICs, while higher-power devices are likely to include a microcontroller or processor. The view of one sensor manufacturer is that FPGA and hardware compilation technologies offer the following advantages:

- Reduced design cost and time to market
- Reduced manufacturing complexity and cost
- More flexible functionality for a given hardware design

while the disadvantages are:

- Power consumption and cost of FPGA devices
- The one-off expense of acquiring new skills, tools, etc

The current cost and power consumption of FPGAs render them unlikely to be used in low-power commercial devices in the near future, but as costs fall and the technology improves, they will become an increasingly attractive option.

8 Acknowledgement

The first author is grateful to Mr. Robin Watts of the Oxford University Computing laboratory for supplying the fixed point macros for Handel. Thanks are also due to Mr. Nigel Archer for the design and built of the PC card.

This project is supported by the Foxboro Company and the EPSRC grants number J44636 and K48884.

References

1. Henry, M.P. and Clarke, D.W., 'The Self-Validating Sensor: Rationale, Definitions and Examples', *Control Eng. Practice*, 1993, **4**:585-610.
2. Henry, M.P., 'Hardware Compilation - A new technique for rapid prototyping of digital systems - Applied to Sensor Validation', to be published in *Control Eng. Practice*, July, 1995.
3. Spivey, M. and Page, I., 'How to Program in Handel', Technical report, Oxford University Computing Laboratory, 1993.
4. Hoare, C. and Page, I., 'Hardware and Software: The Closing Gap', *Transputers Communications*, 1994, **2**:69-90.
5. Yang, J.C., 'Self-Validating Sensors', D.Phil thesis, Oxford University Engineering Department, 1993.
6. Beckwith, T.G., Buck, N.L. and Marangoni, R.D., *Mechanical Measurements.* Addison Wesley, 1982.
7. Kline, S.J. and McClintock, F.A., 'Describing uncertainties in single-sample experiments', *ASME Mechanical Engineering*, 1953, 3-8.

Developing Interface Libraries for Reconfigurable Data Acquisition Boards

Alan Wenban, Geoffrey Brown and John O'Leary

School of Electrical Engineering, Cornell University, Ithaca, New York 14853, USA

Abstract. We describe our suite of design tools tailored for systems with reconfigurable hardware that are intended to be configured by the end user. The reconfigurable system will consist of one or more FPGAs and a number of fixed hardware devices, such as A/D converters and SRAM. We focus on the development of the hardware library functions needed to interface the FPGAs to such fixed hardware devices. We also describe a digital oscilloscope which was implemented with our tools.

1 Introduction

Current technology has enabled the construction of digital systems that are dynamically reconfigurable. We are developing a suite of design tools tailored for systems with reconfigurable hardware that are intended to be configured by the end user. These tools permit a programmer to write software and hardware descriptions in a high-level language, compile, link, and execute these mixed hardware/software descriptions without having to design at the level of current hardware description languages such as VHDL. It is assumed that a reconfigurable system will have one or more field programmable gate arrays (FPGAs) and a number of fixed hardware devices such as A/D converters, RAM, and bus interfaces to host processors. The user will access these fixed components through a library which can be linked to their code. The focus of this paper is on the construction of the hardware library functions needed to interface the compiled hardware to such devices. These functions allow easy access to the external devices from the high-level language while maintaining a high-speed interface. General rules for interfacing to new devices are presented.

A typical application for our system is the support software for a reconfigurable data acquisition card consisting of a fixed analog section and a configurable digital section built from one or more FPGAs and static RAM chips. There are many commercially available data acquisition boards, such as the National Instruments AT-MIO-16X which is an analog, digital, and timing I/O board for PC AT and compatible computers [1]. This board features several analog I/O channels, FIFO A/D and D/A buffers, 16-bit counters, digital I/O lines, PC AT bus interface with DMA and interrupt capabilities, and a special-purpose real-time bus so that multiple data acquisition boards can be synchronized. The software model supported by these boards is one in which transfers of data are set up by the host processor and carried out by the data acquisition hardware according to a rigid schedule. It would not be possible to read (at high speed) analog inputs, modify them, and then write an analog output (for example in a control application). Similarly, it is not possible to scan a digital input stream for a specific pattern and then begin data collection.

A reconfigurable data acquisition card overcomes the limitations of the commercially available cards. However, existing data acquisition cards are generally used by scientists and engineers who are not skilled hardware designers. Currently available tools for field programmable gate arrays require the skills of a hardware designer. For a reconfigurable data acquisition card to be useful, it is necessary that the configuration tools accept as input a high-level programming language which is accessible to a competent programmer. Any custom hardware necessary to interface to the CPU or the analog sections must be provided in the form of libraries which can be conveniently linked to the user's "code." Thus, the primary design requirement for our tools is that the user be removed from making low level hardware decisions. To satisfy this requirement, we are prepared to make some sacrifices in the efficiency of the compiled hardware. Current technology allows the construction of such a data acquisition card with 100,000 digital gate equivalents in the configurable portion. This makes ease of use a more important criterion than efficiency of resource utilization.

In the system we are developing, the design process begins with a concurrent program describing the behavior of the system to be implemented. The concurrent programming language we are using, Hardware Promela (HP), is based upon the language Promela developed by Gerard Holzmann at AT&T Bell Laboratories for use in communication protocol specification and validation [2]. An HP program is compiled into a hardware description which is then linked with pre-designed hardware interfaces. Our hardware compilation technique was inspired by the work of Page and Luk [3] on compiling the Occam language to clocked hardware. Software to access the compiled system can also be written in Promela using the Extended Promela Compiler [4], or it can be written in a language such as C and linked with pre-designed software interfaces.

2 Target Hardware

In order to develop our compilation and simulation tools, we built the data acquisition card illustrated in Figure 1. This card has an Altera EPF8452 FPGA which provides an interface to the PC AT bus and an Altera EPF81188 FPGA which is used to implement custom applications. The Altera EP330 PLD is used to load the initial configuration into the EPF8452. In addition it has a single Hitachi HM628128 128Kx8 SRAM chip. The A/D and D/A functions are provided by an Analog Devices AD7569.

Our data acquisition "card" is physically implemented in two pieces. The reconfigurable hardware, memory, and PC interface reside on a commercial board, the Altera RIPP 10 [5]. The A/D and D/A interfaces are on a separate board. While this configuration has certain limitations (primarily due to the sparse I/O resources of the RIPP 10 board) it was sufficient to perform initial experiments with our tools and to develop a basic hardware interface library.

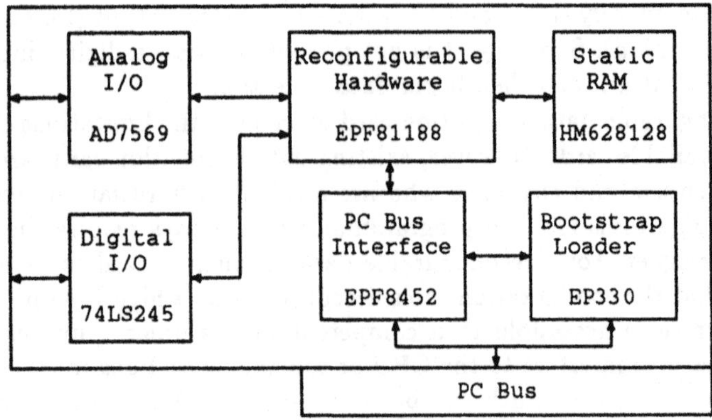

Fig. 1. Reconfigurable Data Acquisition Card

3 Example: Digital Oscilloscope

We have used our system to implement a moderate-sized example, a simple digital oscilloscope. The HP program for the oscilloscope samples the analog input at a programmable rate and stores the values in the SRAM on the RIPP 10. The data in the SRAM is transferred to the PC via DMA. The software running on the PC graphically displays the data in the DMA buffer.

The basic design of the HP program for the digital oscilloscope is shown in Figure 2, where each box represents a different concurrently executing process and the arrows represent message channels. The first process starts A/D conversions at a programmable rate. The next process waits for these conversions to be completed, and then reads the sample data and sends it on internal channel *c1*. The next process receives the sample data from channel *c1* and searches for a triggering condition in the data stream. The process searches for a data point since the last trigger that is below a threshold value and then searches for a point that is above another threshold value. It sends the data and a flag indicating if a trigger has just been found on output channel *c2*.

The next process receives the sample data and the trigger flag from channel *c2*. It waits until the PC is ready for another block of data and until a trigger occurs, and then it stores the next 500 samples in a circular buffer in the SRAM. Another process sends the data from the SRAM to the PC over the DMA channel. After all 500 samples have been sent, a control message is sent to the PC to indicate that a complete buffer has been sent. The interrupt handler on the PC receives the message and informs the main program that it can display another buffer of data. After the data has been displayed, the PC sends a message to the RIPP 10 board indicating that it should start looking for another block of data.

The complete program consists of 166 lines of Hardware Promela and includes 8 concurrent processes. The compiled hardware executes with a clock rate of 5 MHz on an Altera EPF81188GC232-3. The shortest delay allowed between samples is 28 clock cycles, giving a sample rate of 178 Ksamples/second. If we shorten the delay further, the A/D converter cannot keep up with the FPGA.

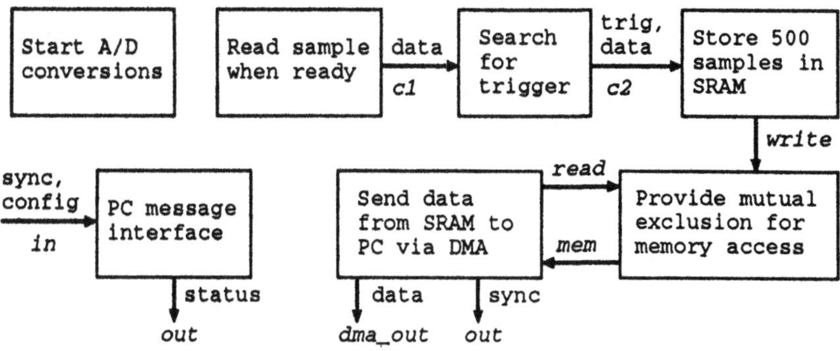

Fig. 2. HP Processes for Digital Oscilloscope

4 Hardware Promela

The source language for our system, Hardware Promela, is based upon the protocol description language Promela. The basic language is similar to Occam [6], i.e. a program consists of a set of concurrent processes communicating through synchronous channels. However, it has a slightly richer set of synchronization primitives including shared variables and output guards. In addition, we have extended the Promela language with arbitrary bit width variables, bit manipulation operators, and parallel assignment statements.

The basic control structure of a compiled HP process is implemented as a one-hot controller, a finite state machine with one flip-flop per state and with exactly one valid state bit per active control thread. HP variables are implemented as arbitrary width registers with a read and write port (Figure 3(a)). The archetypal HP statement has four control signals S, T, F, and B. For example, consider the assignment statement in Figure 3(b). A pulse on the "start" input S triggers execution of the statement. In response, the statement emits a pulse on one of the outputs T, F or B. A response on T, for "true", signifies that the statement has executed and terminated successfully. A response on F, for "false" or "fail", means that the statement was not executable. It is used by the **if** and **do** constructs, which need to choose a single executable statement from a list of options. A response on B, for "break", is a request to exit an enclosing **do** loop. The assignment statement loads a register and terminates successfully after one clock period. Its implementation places a delay element (D flip-flop) between S and T. This implementation is easily generalized to allow multiple assignment.

In HP, several statements may write to each register, so provision is made for shared access to registers. Combinational multiplexers (Figure 3(c)) are inserted automatically by the compiler as needed. In this scheme, contention among multiple concurrent writers is resolved deterministically. To support interfacing

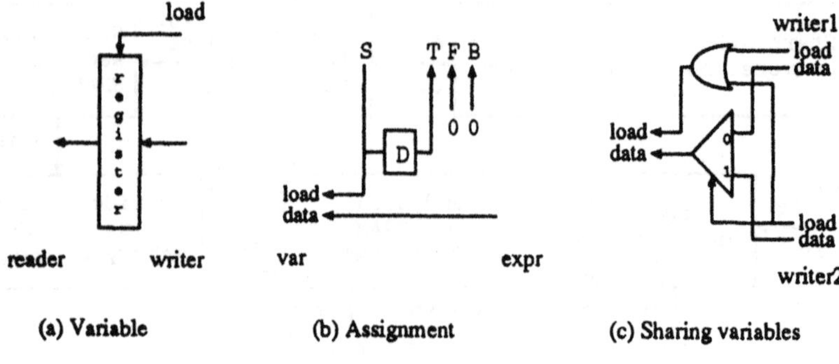

(a) Variable (b) Assignment (c) Sharing variables

Fig. 3. Implementation of Promela Variables

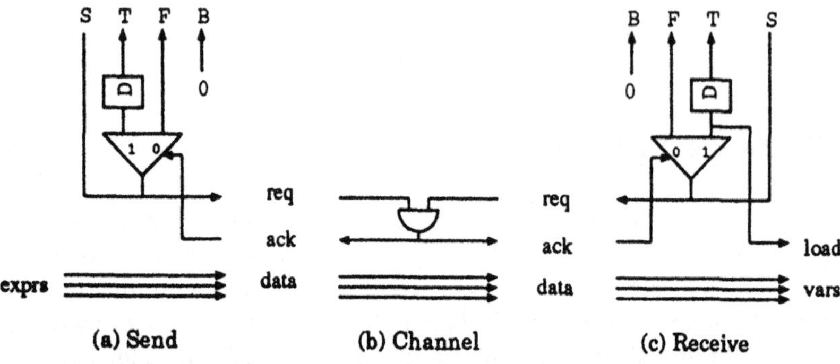

(a) Send (b) Channel (c) Receive

Fig. 4. Implementation of Promela Channels

compiled HP to existing hardware, our compiler permits "external variables" in which the hardware for a variable is assumed to be provided by an interface library and the ports are available to the HP program.

Channels provide synchronization and communication between processes. $c!e^+$ transmits the values of one or more expressions along the channel c. The statement is executable if and only if a receiver is willing to cooperate. $c?v^+$ receives values from channel c and assigns them to the given variables, and is executable if and only if there is a willing sender.

Figure 4 shows the implementation of channels and the send and receive statements. Each end of a channel has request and acknowledge lines, plus one bus for each value to be passed across the channel. To initiate synchronization, a sender or receiver raises its *req* line for one clock period. If its *ack* line goes high, synchronization has occurred. If *ack* remains low, no synchronization has occurred. Send and receive terminate successfully if synchronization occurs and fail otherwise; in the receiver's case, a side effect is that new values are assigned to the target registers. Channels, like variables, require multiplexing to be shared among multiple readers and writers; again, we resolve any contention deterministically. Finally, channels can be declared external to our compiler.

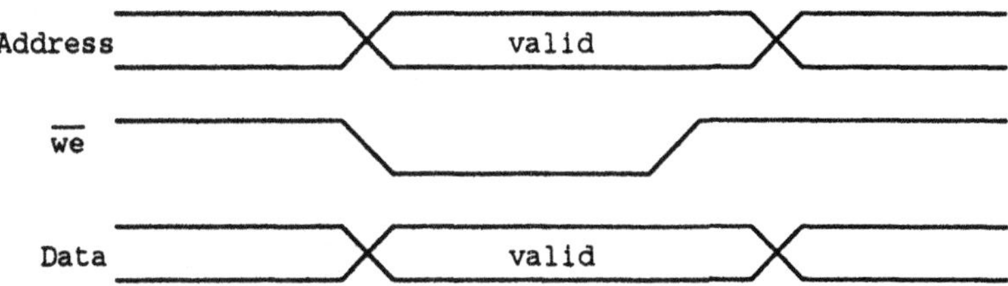

Fig. 5. Simple SRAM Write Timing Diagram

5 Interface Library

An HP program can interact with devices connected to the FPGA through external variables or channels. To the programmer, these appear to be like any other variables or channels. However, in the compiled implementation they are actually signals connected to external components. The hardware interface library provides the link between these external variables and channels and the external devices. The library currently includes interfaces for an SRAM chip, an A/D and D/A chip, and the PC bus.

5.1 Static RAM Interface

A typical static RAM has an interface consisting of an address bus, data bus, chip select ($\overline{cs}$), output enable ($\overline{oe}$), and write enable ($\overline{we}$) signals. Reading from the memory requires placing the address of the desired location on the address bus, activating the chip select and output enable lines, and storing the data returned on the data bus in a register. Similarly, a write is accomplished by driving the address bus, data bus, chip select, and write enable lines (see Figure 5).

We can model the interface for read and write operations with this memory using concurrent assignment statements which reference input and output buses. An output bus consists of a group of data wires with a load signal. An input bus is simply a collection of data wires. This representation exactly matches our implementation of HP assignment statements. We can create a set of external variables and simply connect the ports to the memory with the addition of only a few gates. For a 128Kx8 SRAM, we use the following declarations.

```
extern int:17 address;
extern int:8  mem_in;
extern int:8  mem_out;
```

Figure 6 shows how the data wires and load signals are connected to the SRAM device. With these declarations, we can read memory address i and store the result in variable x with the following concurrent assignment statement.

```
address,x = i,mem_in
```

Similarly, we can write memory address i with the value in variable x with the following concurrent assignment statement.

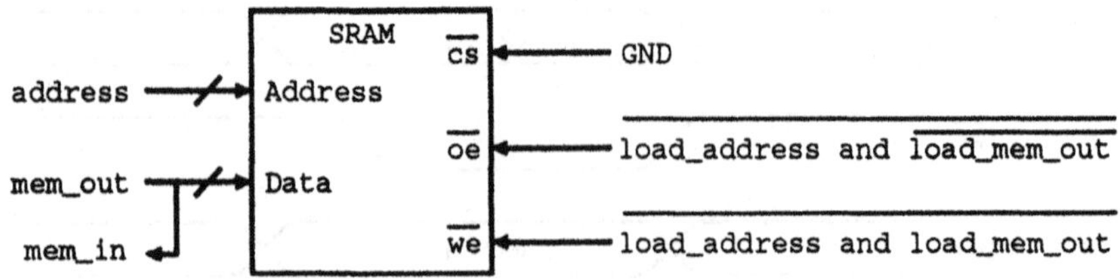

Fig. 6. Hardware Promela to Static RAM Interface

```
address,mem_out = i,x
```

This implementation ignores any hold time requirements after the rising edge of the write enable signal, which has been acceptable on our current testbed. In the future it will not be difficult to add additional clock phases in order to guarantee such timing requirements for more complicated interfaces.

There is one problem with this implementation. We would like accessing the memory to be a safe operation regardless of concurrent use. Currently, if multiple processes access the memory, the memory operations must be placed in a separate process which communicates with the other processes through message channels. The following process guarantees mutual exclusion on memory accesses.

```
do
:: write ? a,d -> address,mem_out = a,d
:: read  ? a   -> address,d = a,mem_in; mem ! d
od
```

5.2 A/D and D/A Interface

The AD7569 analog I/O chip interface consists of a data bus, reset, chip select, start conversion, write data, read data, and interrupt signals. The interface for this chip is similar to that for the SRAM, with load signals from the compiled hardware connected to the chip's control signals and data wires connected to the chip's data bus. With these connections, we can declare the following external variables which will be used to interact with the analog I/O chip.

```
extern int:1 cs_ad;          extern int:1 int_ad;
extern int:8 write_da;       extern int:1 start_ad;
extern int:8 read_ad;        extern int:1 reset_ad;
```

With these declarations, we can create an HP process which will start an A/D conversion once every *delay_count* clock cycles with the following code.

```
interval = delay_count;
do
:: (interval == 2) -> tock; start_ad,interval = 1,delay_count
:: interval = interval - 1
od
```

The **tock** statement above simply delays execution for one clock cycle. It is necessary in this case since conditions (e.g., *interval == 2*) are evaluated combinationally and can cause glitches on the start signal for the next statement. This is acceptable for most HP statements, but in this case the start signal for the assignment to *start_ad* must not have glitches since it is an external variable directly connected to an edge-triggered signal on the AD7569. Any glitches are filtered out by the flip-flop for the **tock** statement. Since processes are implemented as one-hot controllers, the assignment statement as implemented does not create any glitches.

5.3 PC Bus Interface

The PC bus interface is different from the other interfaces in that it is modeled as a pair of external HP message channels rather than as external variables. The PC bus itself consists of data bits, address bits, and a large number of control signals. Each bus transaction can transfer 16 bits of data between the PC and the RIPP 10 board. Since the PC controls the bus, the RIPP 10 must wait for the PC software or DMA hardware to perform the transaction. Since the data transfer does not occur immediately when the RIPP 10 is ready, and since the bus interface logic is more complex than interfacing to a single chip, it was natural to create interface logic which acts as an HP message channel. This implementation allows the interface logic to buffer the message until the message transfer takes place and the HP program can continue execution without blocking. The clock used to latch the data from the PC bus is generated by the PC bus controller, so buffers with a different clock than the compiled HP hardware are necessary. The following external channels are used for communication with the PC.

```
extern chan in  = [0] of {short};       // Channel from PC
extern chan out = [0] of {short};       // Channel to PC
extern chan dma_in  = [0] of {short};   // DMA Channel from PC
extern chan dma_out = [0] of {short};   // DMA Channel to PC
```

The interface logic from the RIPP 10 to the PC receives messages from an HP program on the *out* channel and stores it in a buffer. Then it sets a flag *reqr* that is visible on the PC bus. After the PC has copied the message, it raises the acknowledgement flag *ackr* which informs the interface that it can receive another message from the HP program (see Figure 7). The simplest PC software uses polling to determine when a message is ready to be received. Later we discuss interrupt and DMA driven transfers. The interface logic from the PC to the RIPP 10 receives data from the PC bus and stores it in a buffer. It then attempts to send the data to the HP program on the *in* channel. After the message has been successfully sent, the *acks* flag is raised as an acknowledgement to the PC (see Figure 8). Unless otherwise specified, flip-flop clock signals in the figures are connected to the same clock as the compiled HP hardware.

In order to enhance the PC bus interface to include interrupts, we simply add one flip-flop which is set when the interrupting condition occurs. The output

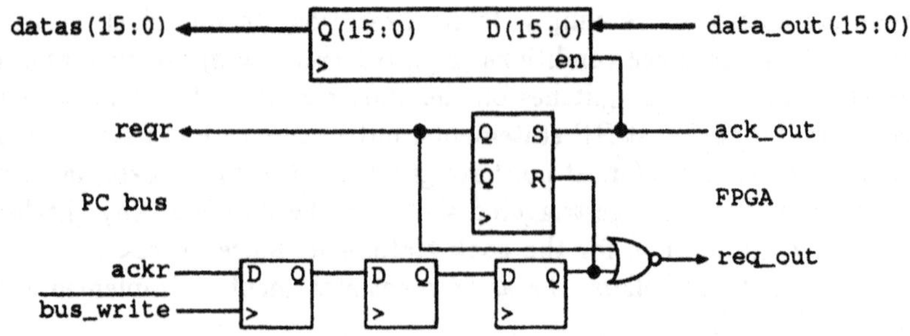

Fig. 7. Hardware to Software Interface

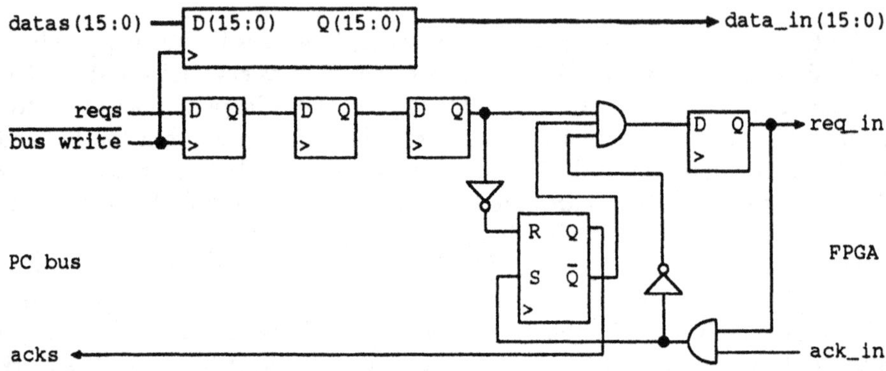

Fig. 8. Software to Hardware Interface

of the flip-flop is connected directly to the appropriate interrupt request line on the PC bus. The interrupt flip-flop is set when either *reqr* or *acks* goes high. The interrupt handler on the PC checks which condition caused the interrupt and performs the appropriate action and then resets the interrupt flip-flop.

In order to use the DMA capabilities of the PC bus, we have added two additional external channels to our HP programs, *dma_in* and *dma_out*. The DMA controller on the PC gains control of the bus when the RIPP 10 raises its DMA request line. It can only perform one transfer before releasing control of the bus again to the microprocessor. The RIPP 10 must then raise the DMA request line again to transfer the next data word. The DMA interface logic from the RIPP 10 to the PC is nearly the same as for the non-DMA case. It receives a message from an HP program on the *dma_out* channel and stores it in a buffer. Then it raises the DMA request signal on the bus and waits for a DMA acknowledge signal, signifying that the data has been copied. Then the interface logic is ready to receive another message from the HP program. The PC software sets up the memory address and the word count for any DMA transfers. The RIPP 10 can initiate a transfer by first sending a standard message on the *out* channel to notify the PC to set up a DMA transfer. The DMA channel from the PC to the RIPP 10 is similar.

Variable interface	Channel interface
FPGA has complete control over device. It can issue commands immediately.	FPGA is slave device and must wait to be granted access in order to communicate.
Clocking determined by FPGA.	Asynchronous clocks may be used to latch data.

Table 1. Choosing Between Variable and Channel Interfaces

6 Conclusion

We have described our design tools for use with reconfigurable hardware, with emphasis on the development of the interface libraries. The implementation of HP variables and channels allows for straightforward control of external devices from an HP program. Table 1 shows interface characteristics that can be used to determine whether a variable interface or a channel interface should be used when connecting to a particular device. In the future, we envision expanding our library with interfaces to devices such as network controllers and the PCI bus.

7 Acknowledgements

We would like to thank Altera Corporation and Intel Corporation for supplying software and equipment. Alan Wenban was supported by a fellowship from Intel Corporation and by NSF grant CCR-9058180. Geoffrey Brown was supported by NSF grant CCR-9058180 and matching funds from AT&T. John O'Leary was supported by NSF grant CCR-9058180 and by a fellowship from Bell-Northern Research Limited.

References

1. National Instruments Corporation. *Instrumentation Reference and Catalogue*, 1995.
2. G. J. Holzmann. *Design and Validation of Computer Protocols.* Prentice Hall, Englewood Cliffs, NJ, 1991.
3. I. Page and W. Luk. Compiling occam into FPGAs. In Will Moore and Wayne Luk, editors, *FPGAs*, pages 271–283. Abingdon EE&CS Books, Abingdon, England, 1991.
4. A. S. Wenban, J. W. O'Leary, and G. M. Brown. Codesign of communication protocols. *Computer*, 26(12):46–52, December 1993.
5. Altera Corporation. *Reconfigurable Interconnect Peripheral Processor (RIPP10) Users Manual*, May 1994.
6. G. Jones. *Programming in OCCAM.* Prentice Hall, Englewood Cliffs, NJ, 1987.

Prototype Generation of Application Specific Embedded Controllers for Microsystems

H.-J. Herpel[2], U. Ober[1], M. Glesner[1]

[1]Darmstadt University of Technology
Institute for Microelectronic Systems
Karlstr. 15, D-64283 Darmstadt

[2]Forschungszentrum Karlsruhe GmbH
Project Microsystems Technology
P.O. Box 3640, D-74021 Karlsruhe

Abstract

Microsystems technology is a rapidly growing field that requires application specific solutions for complex data processing at very low power and area consumption, and cost. In most cases single chip solutions are necessary to fulfill these strong requirements. Several studies showed that the later in the design process an error is detected the higher the costs are to correct it. Rapid prototyping is a proven method to check a design against its requirements during early design phases and thus shorten the overall design cycle. In this paper we present a design environment that supports the designer of application specific embedded controllers during the requirement's validation phase.

1 Introduction

The quality of a product heavily depends on the quality of the process that creates it. In microsystem technology many people from many different engineering disciplines are involved in this creation process. Every engineering discipline has its own language, models, and culture. This makes communication between customers and developers, and also among developers from different disciplines difficult and inefficient. Every product development starts with the definition of requirements.

Pei Hsia [1] pointed out, that without a well-written requirement's specification, developers do not know what to build, customers do not know what to expect, and there is no way to validate that the system as built satisfies the requirements.

Requirement engineering is the disciplined application of proven principles, methods, tools, and notations to describe a proposed system's intended behavior and its associated constraints. The primary output of requirements engineering is a requirement's specification. It must treat the system as a black box. It must delineate inputs, outputs, functional, and nonfunctional requirements. The functional requirements show the external behavior in terms of input, output, and their relationships. The nonfunctional requirements include performance, reliability, and safety.

As proven by many authors (e.g. [2], [3], [4], [5], [6]), *rapid prototyping* is a promising approach in requirements engineering. It means the construction of an executable system model to enhance understanding of the problem, and identify appropriate and feasible external behaviors for possible solutions. Prototyping is an effective method to reduce risk on microsystems projects by early focus on feasibility analysis, identification of real requirements, and elimination of unnecessary requirements.

2 Methodology Basis

A methodology based on rapid prototyping was developed within the framework of the basic research program „Integrated Mechano-Electronic Systems (IMES)" at Darmstadt University of Technology [10]. The main goal was to bridge the previously mentioned communication gap between customers (engineers from the mechanical engineering department) and developers (electronic engineers and computer scientists) through requirement's animation and thus shorten the overall development cycle. The methodology called MCEMS[1] consists of a system description model, a sequence of basic design steps, and a set of tools (**CAP***tools*[2]) for prototype generation. The methodology supports a wide range of target architectures from pure software solutions based on standard processors to application specific hardware solutions.

In this paper we focus on the generation and emulation of application specific embedded controllers (ASEC) as used in microsystems to perform complex data processing and control functions.

The model of the application specific embedded controller is based on the characteristics of the information processing in microsystems. It includes three views of the system: *behavioral* view, *structural* view, and *execution* view.

The behavioral view describes how the application specific embedded controller interacts with its environment. A single model is not sufficient to describe all variants of embedded controllers. The models we use include simple functional relations between inputs and outputs, look-up tables, finite state machines, and fuzzy rule bases. Different models can be combined within the behavioral description.

The structural view describes the architecture of the application specific embedded controller as a network of arithmetic, logic, and memory elements. The transition from behavior to the structural view requires synthesis tools able to generate a datapath and controller under given timing and area constraints.

The execution view describes the emulation board as a network of interconnected FPGAs, interconnection chips, RAM blocks, and I/O elements. The transition from the structural to the execution view requires partitioning and technology mapping tools.

2.1 Basic Steps in the Design Process

Several steps are necessary to validate the user's requirements at each level of description before the actual chip is being produced. The basic design steps in MCEMS (Fig. 1) are a combination of the rapid prototyping approaches used in software engineering [4] and ASIC emulation [6].

The rectangles in Fig. 1 show the activities and the rectangles with rounded corners represent the results of an activity. In the first step the (micro)system is splitted into the embedded controller and the environment in which the ASEC is supposed to operate. The left path in Fig. 1 describes the (simplified) process to create a micro-

---

[1] MCEMS = Methodology for the Conceptual Design of Embedded Microelectronic Systems

[2] **CAP***tools* = Computer Aided Prototyping *tools*

mechanical component. The right path shows the steps from formal specification to an executable model of the ASEC.

Several levels of detail are necessary to completely model the application specific embedded controller and check it against its requirements [7]:

- At the **functional-compliant** level the correctness of the chosen algorithm is tested, and algorithmic parameters are tuned.
- The **bit-compliant** model performs all computing with the same word-length as the final implementation. At this level all finite word-length effects (rounding, truncation, overflow/underflow) are analyzed. Its main goal is to reduce the word-length as much as possible and thus, save silicon area, and cost.
- The **architecture-compliant** model uses the same architectural entities as the final implementation. At this level a trade-off between different architectures can be performed and communication mechanisms can be tested.
- At the **timing-compliant** level all signals between the ASEC and its environment have the same timing as in the final implementation. Its goal is to check and correct timing problems before the actual chip is built.

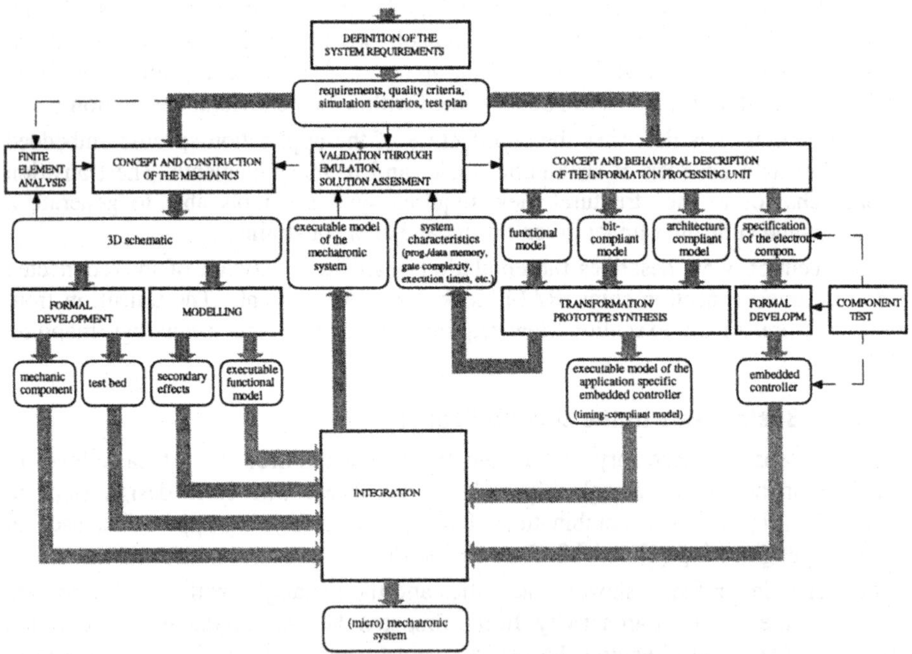

Fig. 1 Life cycle model of MCEMS

The central path in Fig. 1 shows the validation step. The model of the environment and the ASEC model are integrated into a simulation/emulation environment and checked against the user's requirements. Thus, early feedback with the customer is possible.

3 ASEC Design Process

3.1 Functional- and Bit-compliant Model

The first step in the ASEC design process is to describe the behavior of the ASEC. FSMs, high level programming languages, tables, and fuzzy rule bases can be used to describe this behavior (either functional or bit-compliant). Next, the model is converted into an executable model and integrated with the model of the environment where the ASEC is supposed to operate. The simulation environment (**CAP***view*) we developed for that purpose, runs under Microsoft® Windows™ and provides a comfortable way to interact with the simulation models. Fig. 2 shows the user interface of our simulation environment during the simulation of a high precision length meassuring instrument.

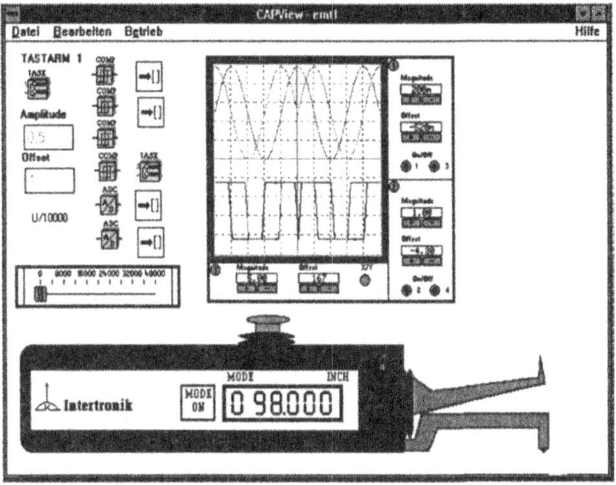

Fig. 2 User interface of the simulation environment

During algorithm development it is common practice to use floating-point arithmetic. This helps to concentrate on the algorithm itself rather than to take care on rounding and truncation effects. But even with 0.8 µm CMOS technology it is costly to implement floating-point arithmetic on a chip. Thus, the conversion into a solution that uses fixed-point arithmetic can help to save Si-area and cost. With this bit-compliant model the designer can study all the effects that appear when a fixed word length is used. After converting the bit-compliant model into an executable model, it can be integrated into the same simulation scenario and checked against the user's requirements.

3.2 Generation of the Architecture-Compliant Model

We defined several steps to convert the bit-compliant model into the architecture-compliant model:

1. The designer has to specify the instruction set for his application based on a generic architecture.

2. The bit-compliant model has to be converted into an intermediate format (`IAL` = IMES Assembly Language) based on the instruction set definition.
3. The architecture has to be generated based on the generic architecture, the instruction set definition, and the IAL-implementation of the data processing algorithm.

The generic processor architecture (Fig. 3) reflects the requirements of our application domain. It has four main building blocks: datapath, controller, memory, and process communication unit. The process communication unit comprises all functions necessary to read data from A/D-converters or digital I/O-ports into the internal memory or to write data to interface elements like D/A- converters or digital ports.

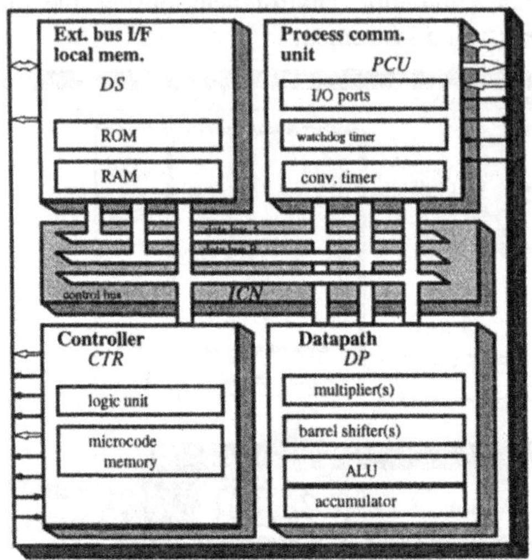

Fig. 3 ASEC building blocks

Fig. 4 shows an example of an instruction set definition. The first two lines define the multiplier type (`LOC=1`: Shift&Add multiplier). The third line specifies the adder type (`LOC=CLA`: Carry-Look-Ahead adder). The mnemonic of the instruction is given in brackets (`[MACCS]`: Multiply ACCumulate and Shift). The lines starting with `S:` address the arithmetic/logic elements necessary to execute the instruction. The lines starting with `P:` define the possible data sources and destinations for the instruction. The lines starting with `V:` specify the control flow during the execution of the instruction.

```
MULOPT1:LOC=I:1
MULOPT2:LOC=I:1
ADDOPT1:LOC=CLA
[MACCS]
S:MUL1,MUX2,MUX3,MUX4,ADD1,REG1,BSH2,BUF3
P: A,BC,C,AB
B:2:3
V: 1: SYSRAM_A_ADR = $VAR1$;
V: 1: SYSRAM_B_ADR = $VAR2$;
V: 1: SYSRAM_A_WE = 0;
```

```
V: 1: SYSRAM_B_WE = 0;
V: 1: goto $2$;
V: 2: SYSRAM_A_ADR = $VAR1$;
V: 2: SYSRAM_B_ADR = $VAR2$;
V: 2: SYSRAM_A_WE = 1;
V: 2: SYSRAM_B_WE = 1;
V: 2: goto $3$;
V: 3: SYSRAM_A_ADR = $VAR1$;
V: 3: SYSRAM_B_ADR = $VAR2$;
V: 3: SYSRAM_A_WE = 0;
V: 3: goto $4$;
#INCLUDE
V: 4: IF SYSMUL_READY == 0 THEN $4$ ELSE $5$
V: 5: SYSRAM_A_ADR = $VAR1$;
V: 5: SYSRAM_B_ADR = $VAR2$;
V: 5: SYSRAM_A_WE = 1;
V: 5: SYSRAM_B_WE = 1;
V: 5: goto $NEXT$;

[BINT0]
S: COMP
P:L
V: 1: if (INT0 == 0) then $1$ else $LABEL$
```

Fig. 4 Example of an instruction set definition

After the designer has specified his specific instruction set, he has to convert the bit-compliant model into the IAL format.

By using the program **IAL2XNF**, the architecture of an application specific processor is generated from the IAL description. The designer controls the transformation process simply by specifying design constraints like max. internal bus width, number, and type of arithmetic blocks, etc. Different design alternatives can be evaluated quickly with regard to complexity and number of clock cycles. **IAL2XNF** analyses the data dependencies and utilizes the parallelism of the data processing algorithm in order to minimize the number of clock cycles.

3.3 Design Partitioning

Once the architectural description is generated, the design has to be partitioned in a way that it can be executed on the emulation board.

Most published partitioning approaches for FPGA based systems perform the partitioning after technology mapping, that is at the level of FPGA building blocks (e.g. [9]). Main goal is to minimize the number of FPGAs and the number of interconnections under given timing constraints. High sophisticated heuristics are necessary to partition a netlist at the level of FPGA building blocks. This usually results in long execution times.

Our application domain has some special characteristics that allows us to follow a totally different strategy:

- Rapid prototyping or ASIC emulation is based on a fixed emulation board architecture with a fixed number of FPGAs and a flexible number of interconnections between the FPGAs on the board.

- The architectures that are to be mapped onto the emulation board mainly consist of quite large arithmetic and/or logic building blocks and internal RAM/ROM. These blocks are connected to one or more internal buses.

This leads to the following partitioning strategy (Fig. 5):

1. Estimate the complexity of all entities in the netlist
2. Find clusters along buses with a complexity less than the complexity of a single FPGA.
3. Cut the nets between the clusters, and generate all necessary I/Os including their local control circuitry.
4. Place all clusters.
5. Generate the output files.

If the complexity of a single element (e.g. 32-by-32 parallel multiplier) exceeds the complexity of a single FPGA than bipartitioning at CLB-level is performed after module expansion.

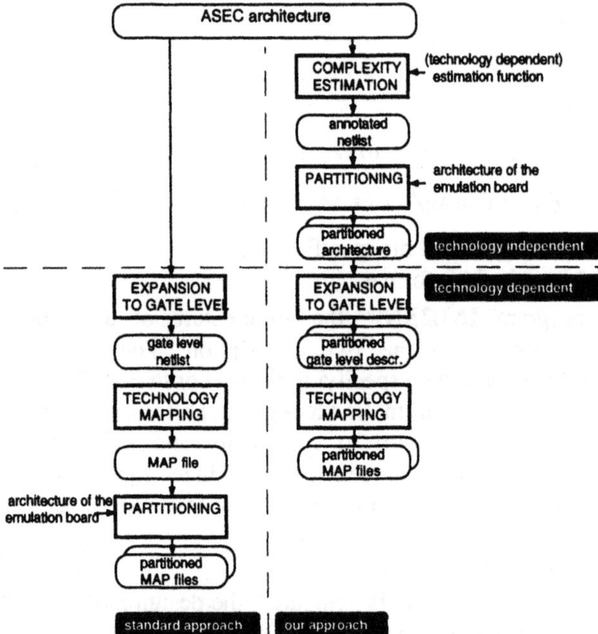

Fig. 5 Partitioning approaches

We developed two programs to support the described partitioning strategy: **CAP***estx*[3] and **CAP***mapx*[4]. **CAP***estx* estimates the number of CLBs for every entity in the netlists based on functions derived from actual implementations of macrocells with

---

[3] **CAP***estx* = Computer Aided Prototyping *est*imation tool for *x*ilinx netlist format

[4] **CAP***mapx* = Computer Aided Prototyping *map*ping of netlists onto multiple *x*ilinx FPGAs

different bit-widths. For example, the complexity functions for parallel multipliers are as follows (see also Fig. 6):

```
fCLB_est = 4.5491·n+0.7917                for m =1
         = 0.983·n²+1.5818·n-4.1265  for m = n
         = a+b·n+c·n²+d·m+e·n·m+f·n²·m+g·m²+h·n·m²
              for m = 1, m = internal bit-width, n = external bit-width
a = 58,86, b = 6,29, c = 0,11, d = -21,33,
e = 0,80, f = -0.02, g = 1,95, h = 0,01

fReg_est = XE·n+YE·n+ar·2·n+3,2419·n+8    for m = 1
fReg_est = XE·n+YE·n+ar·2·n               for n = m
```

Since only simple functions have to be evaluated to calculate the number of CLBs the execution is much faster than the (more accurate) estimation function implemented in the Xilinx PPR tool (see Table 1). In all our test applications the estimated number of CLBs always exceeds the real number of CLBs by less than 10%. This accuracy is good enough for the partitioning process because the user can also adjust the max. number of available CLBs in a single FPGA to guarantee routability.

CAP*mapx* partitions the netlist and introduces additional I/Os where necessary. The user can also preplace certain blocks to speed up the partitioning process.

After finishing partitioning several netlist files have to be expanded to gate level and mapped onto the internal structure of the FPGAs.

Table 1 Comparison of CAPestx and PPR

Example architecture	CAPestx		PPR (Xilinx)	
	Estimation	Runtime	Estimation	Runtime
8 Bit µC-architecture (clutch control)	149 CLBs	2' 33"	141 CLBs	4' 15"
16 Bit DSP architecture (digital controller)	273 CLBs	1' 59"	269 CLBs	11' 22"
32 Bit DSP architecture (IIR-Filter)	676 CLBs	1' 44"	abborted	7' 36"

3.4 Module Expansion

The refinement of an architectural description to gate level is not very well supported in most FPGA design systems. Most of the things have to be carried out by hand. Module generators can help to automate this step. In the frame of our project a module generator (**CAP***blox*[5]) for XILINX® FPGAs was developed. The actual implementation supports 44 parametrisable macro cells:

- Input/output cells (INPUT, OUTPUT, OUTPUT-T, IO, INPORT, OUTPORT),
- select cells (TBUF, MUX, MAXU, MAXS, MINU, MINS),
- comparators (COMP, CLTGT, CLACBE, CLACAE, CLACLE, CLACGE),
- basic logic functions (AND, OR, XOR, NAND, NOR, XNOR, NOT, INV),
- basic arithmetic functions (ADD/SUB (riple-carry), CLAADD/ADC/SUB/SBB (Carry-Look-Ahead), MUL, BRLSHIFT, CONST),
- memories (RAM, ROM, REGISTER, SHIFTREG),
- inference cells for hardware implementations of fuzzy logic (FBIU, FBIS),

[14]**CAP***blox* = Computer Aided Prototyping by *blo*ck synthesis to *x*ilinx netlist format

- special cells (CONTROLLER) to include HDL descriptions (e.g. ABEL-HDL).

It should be noted here that **CAP*blox*** includes generators for different types of multipliers and for fuzzy logic cells. Both are not supported in the actual version of XBLOX. Fig. 6 shows the complexity and cycle times of different multipliers generated by the **CAP*blox*** multiplier generator. It shows that the complexity and cycle time of a shift-and-add multiplier grows linear with the input bit- width while the parallel multiplier shows a quadratic behavior.

CAP*blox* accepts XILINX® netlist format (XNF) as input format, which may be either generated from a graphical description or from synthesis tools (e.g. **IAL2XNF**).

When module expansion is finished Xilinx tools can be used to perform the technology mapping and to generate the FPGA configuration bit stream.

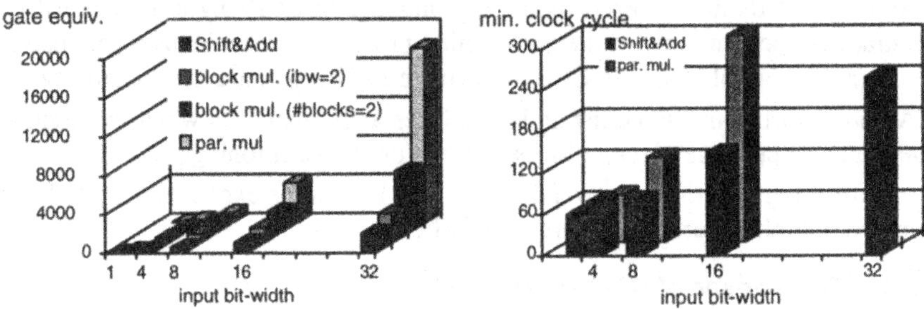

Fig. 6 Cost functions for different types of multipliers

3.5 ASEC Emulation

The emulation board is embedded in a heterogeneous multiprocessor environment [8] which provides the interface to the development computer (PC-AT). This interface is used to download the configuration file to the board, to transfer data from the development computer, and to read back results from the emulation board. Therefore, the architecture-compliant model can be integrated into the simulation scenarios developed for the functional- and bit-compliant models, and checked against the user's requirements. The emulation speed is much faster than any simulation but usually slower than the real chip. For many applications, where the process time is slow (in the order of ms) the architectural model is equivalent to the timing-compliant model (not internally, but at the interface between ASEC and its environment!).

The emulation board itself consists of three main building blocks (Fig. 7): *logic emulation unit*, *process communication unit*, and *memory unit*. The memory unit provides buffers for the communication with other processors and local memory (up to 256 Kbytes) that can be used either as microcode or data memory. Different types of communication buffers (FIFO, dual port RAM, direct path), and synchronization mechanisms (mailbox, semaphore) are integrated on a single chip (**IMES-*ICM***) to support different communication mechanisms.

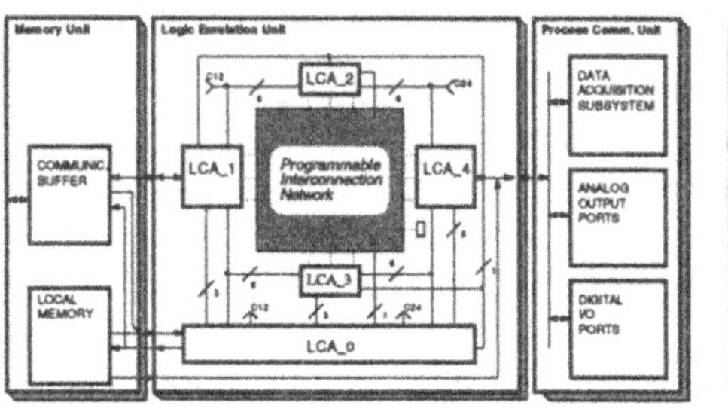

Fig. 7 Emulation Board Architecture

The process communication unit supports the communication with external sensors and actuators. The logic emulation unit mainly consists of five FPGAs (XC4005) and four programmable interconnection chips (**IMES**-*CP*516). The combination of fixed and programmable interconnections between the FPGAs allows for a high utilization and thus, a net gate capacity of approx. 20 k gates.

4 Example

The example we present here demonstrates the usage of **CAP***tools* in a mechatronics project. One goal of the mechatronics project „*Active pulsation absorber for hydraulic systems*" is to develop a highly integrated system for on-line computation and control of the volume flow in hydraulic systems. After simulation at functional- and bit-compliant level an assembly language implementation of the measurement and control algorithm was used to generate an application specific processor architecture. Architecture variants could be generated within minutes, simply by specifying constraints like max. internal bus width, or type of multiplier. From the same high-level description, code for different DSPs was generated and analyzed for comparison. Thus, an early trade-off between speed and area was possible (Table 2).

In this particular case, a single chip solution with various interfaces (analog and digital) was required. The solution with a single add&shift multiplier fulfills the timing requirements (< 100 μs) at the lowest possible complexity. We mapped this architecture onto the emulation board and performed intensive testing before the actual chip was built.

Table 2 Complexity and execution times of different design alternatives

No.	Design alternative	Gate complexity	# cycles	Execution time/ emulation clock
1	1 parallel-multiplier	532 CLBs	50	25 μs / 2 MHz
2	1 block multiplier (2 blocks)	468 CLBs	90	45 μs / 2 MHz
3	1 Add&Shift multiplier	322 CLBs	210	42 μs / 5 MHz
4	TMS320C25		80	8,0 μs / 20 MHz
5	TMS320C30		80	4,8 μs / 33 MHz

The transformation of the emulated architecture into a standard cell design with the ES2 1.5 μm CMOS library had to be carried out by hand. At this point the design cycle is broken, but a shift from netlist formats to VHDL will allow to automate this design step in the future.

5 Conclusions

MCEMS was successfully used in several mechatronic projects like self-supervising clutch, on-line process identification for combustion engines, tire with integrated sensor for friction monitoring, high precision measurement instrument. Now, we adapt the models and tools to the requirements in microsystems technology. We actually use **CAP***tools* in the definition phase of a battery operated high precision length measuring instrument.

The software tools and the emulation hardware need to be improved to handle more complex designs (> 50 k gates). The next step will be to fully automate the transition from the bit-compliant model to the architecture-compliant model by completion of the C to IAL compiler. Further work will concentrate on the development of synthesis tools supporting VHDL rather than netlist formats. This will make the transition from prototype to product much easier.

6 References

[1] P. Hsia, A. Davis, D. Kung: Status Report: Requirements Engineering. In: *IEEE Software, Vol. 10, No. 6*, Nov. 1993

[2] V. S. Gordon, J.M. Bieman: Rapid Prototyping: Lessons Learned. In. *IEEE Software*, Jan. 95, pp. 85-95

[3] B.W. Boehm, T.E. Gray, T. Seewaldt: Prototyping Versus Specifying: A Multiproject Experiment. In: *IEEE Transactions on Software Engineering*, Vol. SE-10, No. 3, May 1984, pp. 290-302

[4] Lugi, M. Ketabchi: A Computer Aided Prototyping System, *IEEE Software*, March 1988, pp. 66-72

[5] H. Gomaa: The Impact of Prototyping on Software System Engineering. In: R.H. Tayer, M. Dorfman (ed.): *System and Software Requirements Engineering*, Washington D.C.: IEEE Computer Society Press, 1990, pp. 543-552

[6] S. Walters: Computer-Aided Prototyping for ASIC-Based Systems. In: *IEEE Design & Test of Computers*, Juni 1991, pp. 4-10

[7] M. Adé, P Wauters, R. Lauwereins, M.Engels, J.A. Peperstraete: Hardware-Software Trade-Offs for Emulation. In: L.D.J. Eggermont, P. Dewilde, E. Deprettere, J. van Meerbergen (eds.): VLSI Signal Processing, IV. New York: IEEE, 1993

[8] H.-J. Herpel, M. Held, M. Glesner: Real-Time System Prototyping Based on a Heterogeneous Multi-Processor Environment. In: *Proc. 5th Euromicro Workshop on Real Time Systems*, Oulu, Juni 1993, pp. 62-67

[9] R. Kuznar, F. Brglez, K. Kozminski: Cost Minimization of Partitions into Multiple Devices. In: Proc. of the 30th Design Automation Conference, Dallas, June 1993, pp.

[10] H.-J. Herpel, M. Held, M. Glesner: MCEMS Toolbox - A Hardware-in-the-Loop Simulation Environment for Mechatronic Systems. In: *Proc. IEEE International Workshop on Modeling, Analysis and Simulation of Computer and Telecommunication Systems*, Durham, Jan. 1994, pp. 356-357

A Hardware Genetic Algorithm for the Traveling Salesman Problem on Splash 2

Paul Graham and Brent Nelson*

Reconfigurable Logic Laboratory
Brigham Young University, Provo UT 84602, USA
801-378-4012
grahamp@fpga.ee.byu.edu
nelson@ee.byu.edu

Abstract. With the introduction of Splash, Splash 2, PAM, and other reconfigurable computers, a wide variety of algorithms can now be feasibly constructed in hardware. In this paper, we describe the Splash 2 Parallel Genetic Algorithm (SPGA), which is a parallel genetic algorithm for optimizing symmetric traveling salesman problems (TSPs) using Splash 2. Each processor in SPGA consists of four Field Programmable Gate Arrays (FPGAs) and associated memories and was found to perform 6.8 to 10.6 times the speed of equivalent software on a state-of-the-art workstation. Multiple processor SPGA systems, which use up to eight processors, find good TSP solutions much more quickly than single processor and software-based implementations of the genetic algorithm. The four-processor island-parallel SPGA implementation out performed all other SPGA configurations tested. We conclude noting that the described parallel genetic algorithm appears to be a good match for reconfigurable computing machines and that Splash 2's various interconnect resources and support for linear systolic and MIMD computing models was important for the implementation of SPGA.

1 Introduction

The development of reconfigurable computers such as Splash [1], Splash 2 [2], and PAM [3] makes it possible to quickly create custom hardware implementations of a wide range of algorithms. In some cases, classes of algorithms not previously suitable for ASIC realization can now benefit from hardware implementation on these machines. Splash 2 and PAM applications have been shown to often run orders of magnitude faster than equivalent software implementations[4][5]. In this work we describe the result of mapping a genetic algorithm for solving the traveling salesman problem onto the Splash 2 system. We demonstrate the

* This work was supported by ARPA/CSTO under contract number DABT63-94-C-0085 under a subcontract to National Semiconductor

effectiveness of Splash 2 for this and the performance advantages of using a reconfigurable computer for this computation.

We begin with a brief overview of genetic algorithms, providing a description of our specific algorithm for the traveling salesman problem. We then describe a basic implementation using four FPGAs and compare its speed to an equivalent software implementation. Two parallel versions of the algorithm, taking full advantage of the parallel capabilities of Splash 2, are then presented along with results and conclusions.

2 Genetic Algorithms

In 1975, Holland [6] developed an optimization technique, based on the process of natural selection and evolution, which he called the *genetic algorithm* (GA). Follow-on work since that time has shown its usefulness for optimization problems requiring the search of large and complex problem spaces from engineering design to combinatorial optimization to control [7][8][9].

During its operation the GA maintains a collection (*population*) of candidate solutions. Associated with each candidate is a *fitness* or measure of its quality. The algorithm proceeds by selecting candidates from the current generation to propagate into the next generation. In the process of this propagation the algorithm may simply copy the selected candidates to the new generation or it may combine pairs of candidates through a *crossover* operation, reminiscent of mating in natural systems. In this case, the newly created solutions have characteristics taken from both parents. The selection of candidates for copy and crossover is randomized but biased toward candidates with higher fitnesses, thus, more fit individuals are more likely to be used to produce future generations of solutions. As a means of preventing premature convergence to local minima, an operation known as *mutation* randomly perturbs solutions to yield new ones not otherwise related to existing solutions.

2.1 A Genetic Algorithm for the Traveling Salesman Problem

SPGA (Splash 2 Parallel Genetic Algorithm) is a hardware GA which searches for optimal solutions to symmetric traveling salesman problems (TSPs). This family of problems involves finding the shortest path through a collection of n cities, visiting each city exactly once and returning to the starting city. Software implementations of this algorithm have previously been studied—this is the first known hardware implementation of a GA for this problem.

Each candidate solution in the population consists of an ordered list describing the sequence in which each city is visited—a list referred to as a *tour*. Since the object of optimizing a TSP is to find the shortest tour, the fitness of a given tour is related to its length.

Crossover is performed in the following way: once a pair of tours (call them tours A and B) have been selected for crossover a cut point is selected at random and the two tours are cut at that point. The head of tour A is used as the head of offspring #1 and the head of tour B is used as the head of offsprint #2. The tail of offspring #1 is formed by taking, in order, the cities from tour B not contained in the head of tour A. The tail of offspring #2 is formed in a similar fashion. Mutation is performed on selected tours by reversing the order of cities visited within a sub-tour contained within the original tour. The endpoints of this sub-tour are chosen randomly for each tour mutated. These crossover and mutation operations closely follow those described in [10]. The reader is directed there for details on their operation.

The above algorithm readily lends itself to parallelization in at least two ways. The brute force approach is to run multiple independent copies of the algorithm, selecting the best result produced by the collection of runs. In contrast, a cooperative parallel model can be employed where *islands* of computation are executed in parallel but periodically exchange solutions with one another through *migration*. This often seeds searches into new and better areas of the search space and has been shown to be robust in dealing with a variety of difficult optimization problems due to its tendency to avoid premature convergence to local minima[11].

3 Splash 2

SPGA was developed on Splash 2, a reconfigurable computer developed at the Center for Computing Sciences (formerly the Supercomputing Research Center). Splash 2 was designed to support linear systolic, SIMD, and MIMD processing styles. Splash 2 is hosted on a Sun Sparc and consists of an interface board and a collection of processor array boards as shown in Figure 1. The interface board provides support for the host to program, control, and exchange data with the Splash 2 array boards.

Splash 2 is "programmed" using VHDL[12]. Commercial synthesis tools are used to map the VHDL design onto the FPGA resources of Splash 2, while auxiliary text files provide configuration information for the memories and crossbars. VHDL models for the entire Splash 2 system including the interface board, buses, crossbars, and local processor memories are available and provide a complete and accurate pre-synthesis simulation environment.

4 Basic Algorithm Implementation

SPGA's basic unit of computation is a *processor*, consisting of four FPGAs (Xilinx 4010s) and their associated memories. The FPGAs are arranged in a bidirectional pipeline, as shown in Figure 2. During execution the memories hold

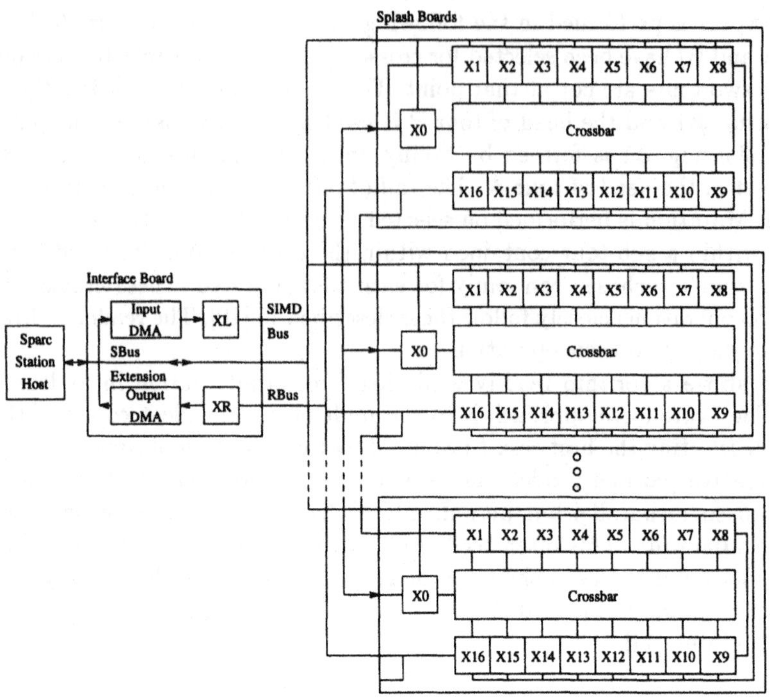

Fig. 1. The Splash 2 System

the current generation, the new generation as it is constructed, tour fitnesses, the inter-city distance matrix, and operational parameters. Initial population data is provided by the host driver program.

The function of the FPGAs is as follows:

- *FPGA 1* performs the biased selection, choosing tour pairs from memory. This is done using a hardware-pipelined *roulette wheel* algorithm. Initially, a random number is generated called the *target*. Tour fitness values are then sequentially accumulated until the target value is exceeded. The tour causing the overflow is the next tour selected. The width of each tour's roulette wheel slot is proportional to its fitness, therefore, highly fit tours are more likely to be chosen. Once a pair of tours is selected, their index numbers are passed to *FPGA 2* via the bottom pipeline path.

- *FPGA 2* has two choices when given a tour pair: it can simply copy the tours to the right unchanged or it can combine them via crossover and send the new offspring to the right. This decision is made based on a random number generated on chip. Crossover probabilities of 10% to 60%

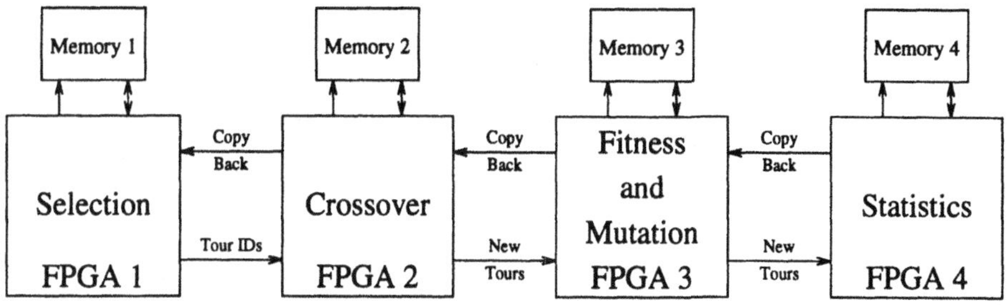

Fig. 2. A Single Genetic Algorithm Processor

have been shown to give good results for TSPs.

- The role of *FPGA 3* is threefold. First, it computes new fitness values for tours formed by crossover. Second, it randomly selects tours for mutation and performs the mutation. Finally, it sends the tour pairs and their fitnesses to *FPGA 4*.
- *FPGA 4* writes the new population to memory, determining and recording the best and worst tours for this generation and the best tour to date.

The above process is repeated until a population equal in size to the original population is collected in the memory of *FPGA 4*. The upper pipeline path in Figure 2 is then used to copy the new population and fitnesses back to the memories of *FPGA 1* and *FPGA 2*. This ends the computation of a complete generation. Execution terminates when the desired number of generations has been reached.

The implementation required 3,500 lines of VHDL code. The maximum clock frequency of 11 MHz was dictated by the arithmetic sections of FPGA's 1 and 4. CLB utilizations for the four chips range from 37% to 60%. Also, a family of C-based driver programs were written for the SPARC host to control the computation and retrieve the results.

The performance of a software implementation executing on a 125 MHz HP PA-RISC workstation is compared to SPGA in Table 1. All numbers are averages across a collection of runs. The 24 city problems ran for 120 million cycles, completing 2,000 generations, while the 120 city problems ran for 3.4 billion cycles, completing 20,000 generations. The software and hardware implementations found similar quality solutions in our tests.

The significant advantage of the FPGA design over software was initially surprising, given the 11x difference in clock rates. We attribute the modest software performance to operating system overhead, TLB and cache misses, complex addressing calculations, and strict sequential thread of control. However, we have little quantitative information on these effects. In contrast, SPGA employs a

Num. of Cities	*Pop. Size*	*Crossover Probability (%)*	*Mutation Probability (%)*	*Ave. Exec. Time (sec) Hardware*	*Software*	*Soft./ Hard*
24	128	10	10	4.38	43.7	9.97
24	256	10	10	11.23	118.7	10.57
120	256	60	10	295	1999.9	6.78

Table 1. Comparison of Hardware and Software Execution Times

custom 4-stage pipeline, achieves nearly 100% memory bandwidth utilization, and incurs no overhead for the operating system, address calculations, or cache misses.

5 Two Parallel Implementations

The implementation described above uses only 4 FPGAs on a single Splash 2 array board—the remaining 30 FPGAs and memories in our two-board Splash 2 are idle. In fact, there was no need for Splash 2 at all for the basic implementation. Four FPGAs and memories would have been sufficient. However, given a two-board Splash 2, it is a simple extension to run seven additional copies of the algorithm—no additional hardware design is required. Since the copies of the algorithm do not interact, the only added overhead is the small amount of time spent initializing the additional memories. Thus, an eight-fold increase in search rate is possible with this approach which we call the *trivially parallel* model.

A more interesting approach to parallelizing the algorithm applies a cooperative or *island* model of parallel computation. In this model several searches are conducted simultaneously with periodic migration of solutions between search islands. Figure 3 gives a diagram of a single array board design for this model.

During migration each island, in turn, broadcasts a subset of its tours to the other islands via the crossbar. The receiving islands replace random solutions in their own populations with these broadcast tours. Unlike the trivially parallel model described above, modifications to the original SPGA design were required for this. First, the islands employ dedicated lines in Splash 2 to signal to X0 when they are ready to perform migration. Second, X0 coordinates these migration transfers by configuring the crossbar and handshaking with each island in turn as required. This added functionality required the creation of an X0 design, significant additions to FPGA 4's design, and the creation of multiple crossbar configurations.

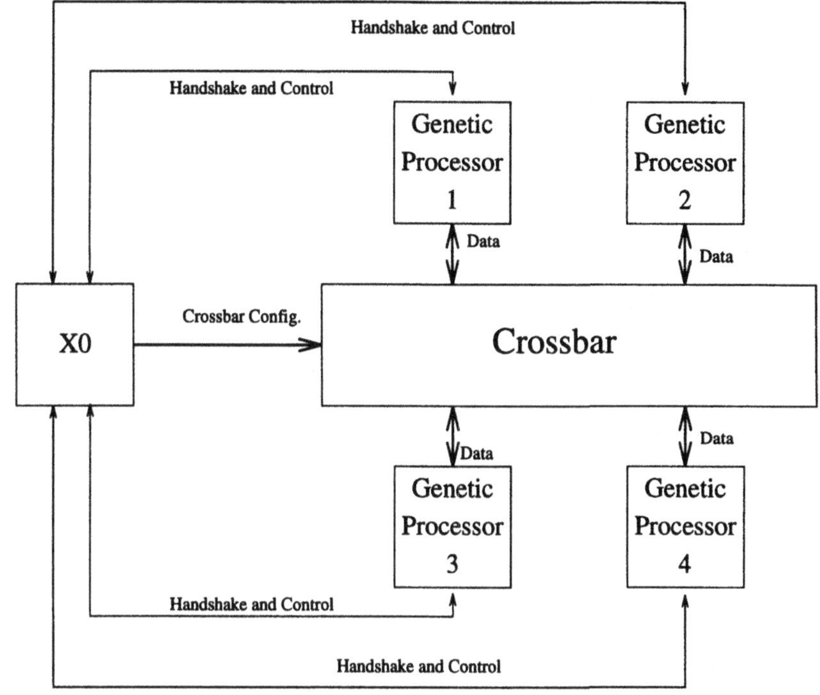

Fig. 3. The Island Parallel Model

5.1 Performance Results

Figure 4 compares the Splash 2 performance of the three computation models described above. These include a 1-processor design, 4-processor and 8-processor trivially parallel designs, and a 4-processor island model. All results are averages across many executions.

To a first order approximation, the rate of search is proportional to the number of search engines. Thus, using the numbers from Table 1 the 8-processor trivially parallel version would be expected to search faster than the software by factors from 54 to 85. However, these speedups do not translate directly to better solutions since the quality of solution obtained isn't linearly proportional to the number of solutions evaluated.

If one's goal is to find the best solution possible we have found that more processors are always better, but marginally so. For long runs (3.5 billion cycles) the 8-processor parallel version gives an answer about 4% better than a single processor and the 4-processor island model is better by about 6%.

However, if the goal is to find a good solution quickly, the data shows that the island model is far superior to the others. At 500 million cycles the 4-

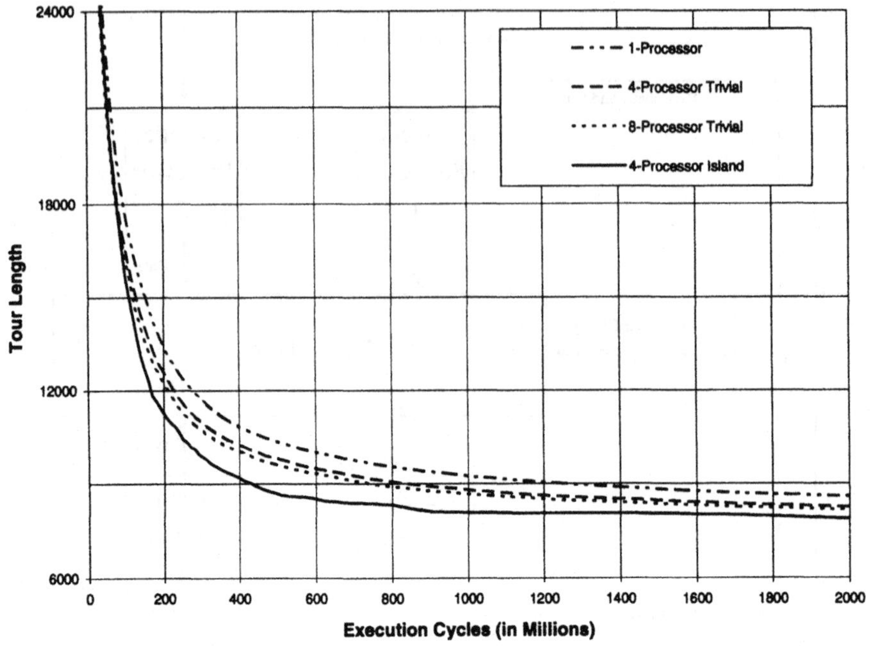

Fig. 4. GA Performance Comparison

processor island model has found a solution which requires 990 million cycles for the 8-processor trivially parallel model to match and 1.7 billion cycles for the single-processor to match.

6 Conclusions

We have described a hardware genetic algorithm on Splash 2. We believe the genetic algorithm described in this paper is a good match for reconfigurable computers for a number of reasons.

- We have shown that reasonably modest hardware resources (4 FPGAs and 4 memories) can significantly outperform state-of-the-art workstations on this algorithm. SPGA does this with a custom design that avoids operating system overhead, TLB and cache misses, and complex addressing calculations. In addition, it makes use of pipelining to achieve parallel execution.

- The individual data objects manipulated by the algorithm are small, mostly 8- or 16-bits with a few 24-bit values. This is a a good match for the computation and storage capabilities of today's FPGAs.
- The arithmetic requirèments of SPGA are modest, consisting of small-word additions, subtractions, and comparisons. This is also a good match for today's FPGAs.
- The additional work required to create parallel implementations of the algorithm is minimal. The parallel versions were able to find 'good' solutions in far less time than the single processor version.
- Other search methods often terminate at a specific solution; genetic algorithm search methods continue to find better solutions until terminated by the user. Custom hardware such as described herein may be required to take advantage of this for very long execution times.

A further conclusion of this project is that since few entire computations are purely systolic or SIMD, extra interconnections are often required for a complete implementation and may make the difference between poor and excellent performance. The breadth of computing models and FPGA interconnection paths supported by the Splash 2 proved most useful for SPGA. The linear pipeline was required for each processor's design and additionally the MIMD model was essential for supporting migration in the island-parallel version. All of the Splash 2 interconnect paths were eventually used in the design including: nearest-neighbor linear connections, the crossbar, and the X0 broadcast and handshake lines.

Areas for future work with SPGA include porting designs similar to SPGA to other reconfigurable hardware platforms, experimenting with different migration strategies, extending the system to allow migration between array boards, and using SPGA as a test bed for furthering parallel genetic algorithm research.

References

1. M. Gokhale, W. Holmes, A. Kopser, S. Lucas, R. Minnich, D. Sweely, and D. Lopresti. Building and using a highly parallel programmable logic array. *IEEE Computer*, 24(1):81–89, January 1991.
2. J. M. Arnold, D. A. Buell, and E. G. Davis. Splash 2. In *Proceedings of the 4th Annual ACM Symposium on Parallel Algorithms and Architectures*, pages 316–324, June 1992.
3. P. Bertin, D. Roncin, and J. Vuillemin. Introduction to programmable active memories. In J. McCanny, J. McWhirther, and E. Swartslander Jr., editors, *Systolic Array Processors*, pages 300–309. Prentice Hall, 1989.
4. D. P. Lopresti. Rapid implementation of a genetic sequence comparator using field-programmable gate arrays. In C. Sequin, editor, *Advanced Research in VLSI:*

Proceedings of the 1991 University of California/Santa Cruz Conference, pages 138–152, Santa Cruz, CA, March 1991.

5. P. Bertin, D. Roncin, and J. Vuillemin. Programmable active memories: a performance assessment. In G. Borriello and C. Ebeling, editors, *Research on Integrated Systems: Proceedings of the 1993 Symposium*, pages 88–102, 1993.
6. J. H. Holland. *Adaptation in Natural and Artificial Systems.* University of Michigan Press, 1975.
7. R. Vemuri, R.; Vemuri. Mcm layer assignment using genetic search. *Electronics Letters*, 30(20):1635–7, September 1994.
8. N.; Hong Ren Mou, E.S.H.; Ansari. A genetic algorithm for multiprocessor scheduling. *IEEE Transactions on Parallel and Distributed System*, 5(2):113–120, February 1994.
9. John E. Lansberry and L. Wozniak. Adaptive hydrogenerator governor tuning with a genetic algorithm. *IEEE Transactions on Energy Conversion*, 9(1):179–183, March 1994.
10. M. C. Leu; H. Wong; and Z. Ji. Planning of component placement/insertion sequence and feeder setup in pcb assembly using genetic algorithm. *Transactions of the ASME*, 115:424–432, December 1993.
11. M. Dorigo and V. Maniezzo. Parallel genetic algorithms: Introduction and overview of current research. In J. Stender, editor, *Parallel Genetic Algorithms: Theory and Applications*, pages 5–41. IO Press, Washington DC, 1993.
12. J. M. Arnold, D. A. Buell, and E. G. Davis. VHDL programming on Splash 2. In *More FPGAs: Proceedings of the 1993 International Workshop on Field-Programmable Logic and Applications*, pages 182–191, Oxford, England, September 1993.

Modular Architecture for Real-time Astronomical Image Processing with FPGA

Massimiliano Corba and Zoran Ninkov

Rochester Institute of Technology/Center for Imaging Science
54 Lomb Memorial Drive, Rochester, New York 14623, USA
e-mail mxcpci@borg.cis.rit.edu

Abstract. This paper describes the implementation of a two-dimensional, real-time centroiding algorithm using reconfigurable Field Programmable Gate Arrays (FPGA's). A centroiding computation is used to calculate the highest image light density location to sub-pixel resolution for bright point source events. The method is of interest mainly in astronomical applications. In order to achieve this computation in real-time, a solution which realizes high parallelism, pipelinability, and modularity is essential. Recent improvements in FPGA architecture allow for a new approach to realizing fast, wide and complex arithmetic functions thereby targeting these devices for real time imaging applications. A prototype has, therefore, been developed exploiting the high density, high speed and flexibility of FPGA's. The resulting prototype is able to perform centroiding on a 5x5 pixel array on data coming from an electronic imaging detector at a throughput rate of 10Mevents/s.

1 Introduction

This paper describes a flexible and modular implementation, using a systolic system design approach, of two-dimensional real-time centroiding. The filter herein described performs a weighted sum of the first moment around the central pixel based on a 5x5 pixel array. This computation is used to calculate the location (centroid) of the highest light density of the Point Spread Function, to sub-pixel resolution (1/16 of a pixel). In order to achieve the required centroiding spatial resolution, all single point source events on the detector are focused and constrained to be included in a 5x5 pixel array[1]. The spatial filter designed is intended to be mesh-connected into a large system dedicated primarily to image processing from a low-light level electronic imaging detector. The detector is used in astronomical applications, as for a star tracker or for photon counting in the ultra-violet region. For such applications, the capability of performing real-time imaging on a variable window size is essential [2]. Real time imaging is required to provide high spatial resolution, by applying event centroiding techniques, and high count rate capability. Flexible windowing, the capability of having variable window size, allows one to perform computations "on the fly" on the data flow that comes either from the full detector array or any variable portion of it (e.g.

10 rows x 20 columns). In this context, it is possible that arbitrary sub-arrays can be addressed at fast rates so high contrast images, bright point sources, can be read out quickly and then processed.

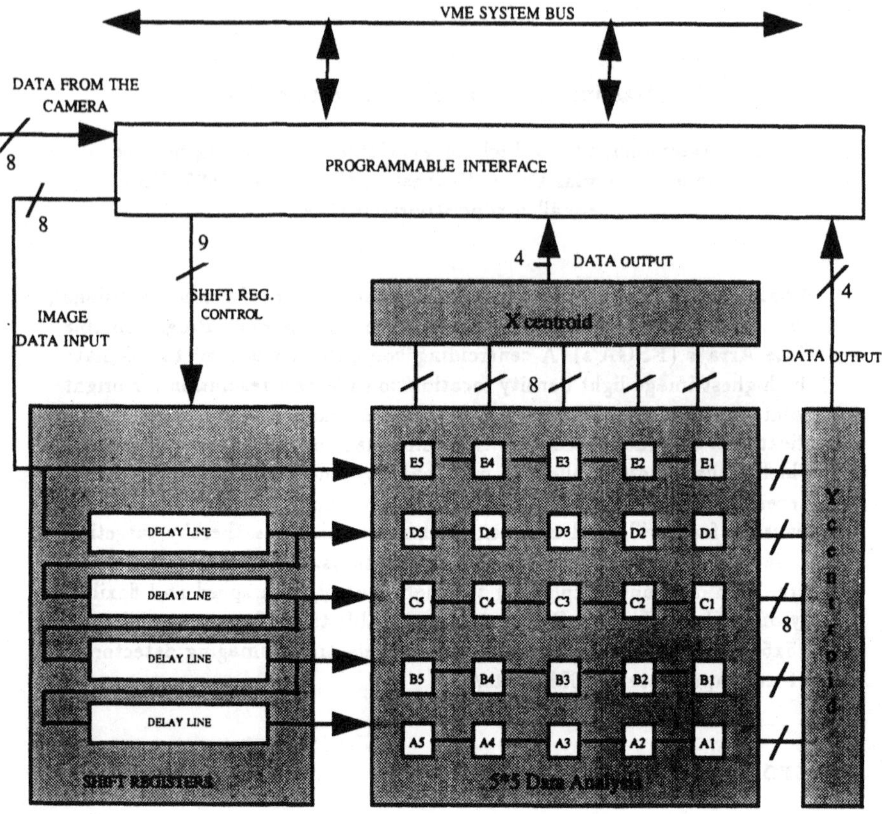

Fig. 1. Block diagram of the filter

As a general solution to the real-time requirement of signal processing, special purpose array processors are used to maximize the processing concurrency by pipeline processing or parallel processing or both [3]. Typical digital signal processing applications require attributes of parallelism, pipelinability, modularity and flexibility [4]. When extensive prototyping of custom made applications with these attributes is also necessary, the FPGA becomes an appealing choice of architecture. In fact, the Xilinx 4000 family provides the necessary attributes and furthermore, permits good arithmetic performance due to built-in special carry circuitry logic (carry ripple propagation at 600 picoseconds based on a two bits sum)(see [5]). It is thereby possible to develop fast, wide and complex arithmetic functions for image processing applications.

The paper discusses the spatial filter design implemented with the Xilinx 4000 family devices. Section 2 provides an overview of the filter design, the division architecture, the arithmetic precision determination and the resulting speed/area tradeoffs. Sections 3 provides a discussion of the centroider design considerations and its performance. Section 4 is a discussion of the flexibility available using reprogrammable FPGA technology.

2 Spatial filter design

The design of the spatial filter consists of four main functional blocks:
1) Interface to detector.
2) A programmable length shift-register.
3) A data analysis event condition.
4) The centroiding algorithm.
Figure 1 shows a simplified block diagram of the spatial filter.

The interface allows one to program, in real time, the size of the shift register (SR) to be equal to the size of the window of interest on the detector imager array. This process must be performed each time a different window size is required. The interface then clocks the image coming from the array into the SR, pixel by pixel. An operation to recognize valid events and to issue a valid event flag, initiating centroiding, is then performed (further described below). Upon completion of the event centroiding, each centroid address is combined with the row-column window address defining its position within the imager to form the final event address. This information is accumulated into a temporary FIFO buffer and then transmitted via VME bus interface to the PC computer memory for storage and later analysis.

The programmable length shift-register uses a structure that inputs vertically parallel (in the scanned direction) five 8-bit image data and delays them in the horizontal direction to produce two-dimensional pixel data. The use of a programmable length shift register increases the overall dynamic range since only the pixels within the window will be processed. The size can be programmed to be between 16 pixels and 512 pixels. A modern and efficient way to implement large shift-registers is to use a RAM-based approach that significantly reduces the overall size compared to the traditional cascade of data registers.

Digital comparators, not shown in figure 1, are connected to the last five registers of the five shift registers in the 5x5 data analysis section. The comparators determine if the valid event condition is met for the entire 5x5 pixel array in a single clock cycle (pixel). The valid event condition is that a multi-pixel event's accumulated charge is greater than a threshold level and that the central pixel of the event has the highest value. A valid event initiates event centroiding, transferring the last 5x5 pixels to the centroiding stages.

The centroiding algorithm is the core of the system. Basically, this algorithm (reduced to one dimension for simplicity) convolves the image with a linear filter,

$$t(x) = g(x) \otimes h(x), \tag{1}$$

where $g(x)$ is the image, $h(x) = -x$ is the linear filter, and $\otimes$ denotes convolution.

The position of the centroid is given by $t(0)$, where

$$t(0) = \frac{\sum_{x=o}^{m} g(x)x}{\sum_{x=o}^{m} g(x)}. \tag{2}$$

If filtering is applied on a 5 × 5 based pixel array, equation (2) reduces to:

$$x_c = \frac{2C_1 + C_2 - C_4 - 2C_5}{C_1 + C_2 + C_3 + C_4 + C_5}. \tag{3}$$

The complete centroiding involves the same computation in the Y direction, requiring two independent circuits for the hardware implementation. The multiplication operations in equation (3) are realized by simple shift operations. Particular attention is, therefore, required only on the division since the resulting polynomials for the numerator and denominator involve only a few terms.

2.1 Division Operation

The main obstacle to high-speed division has traditionally been that each bit of the quotient is determined by the carry-out bit from a full addition or subtraction performed on the n-bit divisor and partial reminder[6]. The choice of adding or subtracting the divisor, which corresponds to the calculation of one digit of the final quotient, is determined by the sign of the previously computed partial reminder. Hence the algorithms are sequential by nature. Care must be taken in the latency involved with the carry-propagate addition/subtraction when implementing the division. A choice of the algorithm to implement this must be made in order to optimize trade-offs among arithmetic precision, computational speed, and modular and bit expandable hardware implementation with Xilinx FPGA's. The algorithm considered is the well known Binary non Restoring Division (BRD), in which the result is:

$$Q = N/D$$

with the numerator N, and the divisor D, which are assumed to be fixed point two's complement numbers with wordlength W. In order for Q to be a fraction, N is required to be smaller in magnitude then D. The resulting quotient digits are denoted by $Q = (q_0, ...q_i, ..., q_{w-1})$, the partial reminder is denoted by x_i for the i-iteration.
Initialization:

$$x_0 = N, \tag{4}$$

$$q_0 = \begin{cases} +1 & if\, signs\, of\, N\, and\, D\, agree \\ -1 & if\, signs\, of\, N\, and\, D\, differ \end{cases} \tag{5}$$

Iterations:

$$x_{i+1} = x_i - q_i D 2^{-i};\ with\, i \in (0, ..., W-1), \tag{6}$$

$$q_{i+1} = \begin{cases} +1 & if sign(x_{i+1}) = sign(D) \\ -1 & otherwise \end{cases}. \tag{7}$$

Let N and D be positive, then the first partial reminder x_1 is computed by subtracting D from N, and, if the result is positive, setting $q_1 = 1$, otherwise $q_1 = -1$(which is represented by 0 in binary notation). From (6), the filter's output is derived from accumulating/subtracting partial products in a highly modular manner. This algorithm is best suited for implementation that uses ripple carry logic circuitry as that of the Xilinx 4000 family.

2.2 Arithmetic Precision Trade-offs

By careful selection of the bit width of addition/subtraction for overflow protection and on the bit width of the division for arithmetic precision, the design can be tailored to a specific applic tion in order to have high speed performance with low area cost. For photon counting applications[2] using 8 bit pixel inputs, all the addition/subtraction operations in equation (3) can be performed sufficiently on 11 bits including overflow protection. Also for photon counting applications, all of the bits of the quotient are not required. If W is the bit width of N and D, the area required to implement the division is, at least, on the order of $O(W^2)$. For example, if the spatial resolution for an application of the filter is 1/16 of a pixel unit, only four digits for the quotient must be generated[7]. The use of 4 bits of the quotient instead of the full W-bit quotient length results in a considerable reduction in the implementation area.

3 Centroider design considerations

The proposed design uses a combination of parallel and pipeline structures to realize the three independent parts of the equation: the numerator (N), the divisor (D), and the quotient (Q). From (3), it is shown that the spatial parallelism includes concurrent computations for N and D, while the temporal parallelism can be obtained by overlapping N and D with Q, as a pipeline of at least two stages. Since the numerator and divisor of equation (3) are produced concurrently, only two pipeline cycle times are necessary to complete the computation. Two non-overlapping phase signals which are generated from one clock input signal are used to latch the intermediate results of centroiding. Figure 2 shows the block diagram.

The cycle time is defined as the time span between two consecutive input data samples. It is determined by the longest operational time delay among the stages of the pipeline considered. Viewlogic timing simulations, for this application, have shown that two pipeline cycles are more than sufficient to balance the depth of logic gates between the pipelined latches. It is possible, however, to improve the time performance by inserting additional latches (increasing the number of pipeline levels). The maximum (upper) limit of this technique is fixed by the delay in obtaining a single addition/subtraction operation due to the propagation delay of the ripple carry from the least significant bit (LSB) to

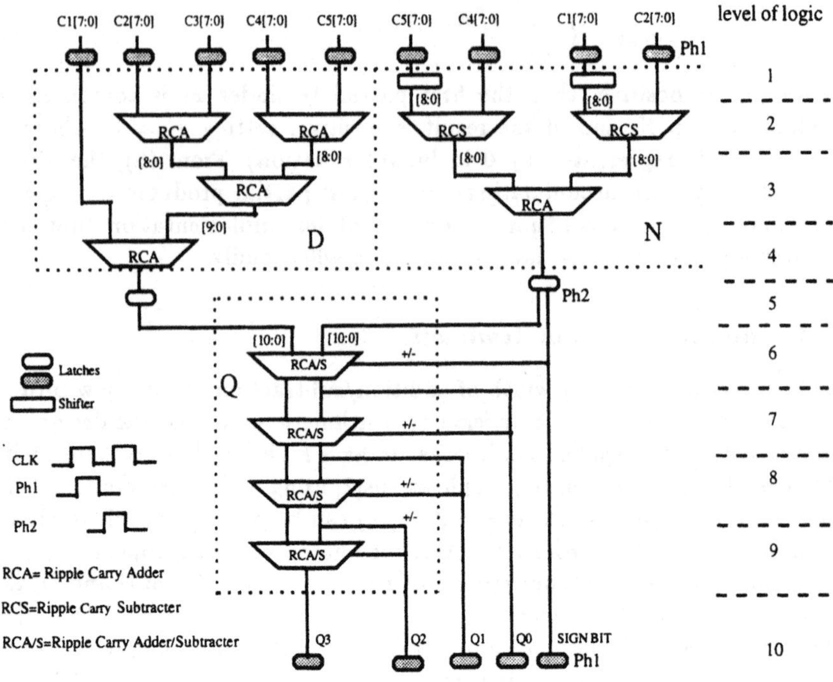

Fig. 2. Block diagram of the centroider module

most significant bit (MSB). Ultimately, this depends on the number of bits in the addition/subtraction.

3.1 Centroider performance

Viewlogic timing simulations for the centroiding filter have been summarized in Table 1. Results are based on the CLB density and routing resources of the 4000 family, speed "4" and worst case conditions. The implementation was optimized for timing performance. By balancing the implementation area and routing connections occupied by each single stage (RCA/S), it can be seen that the overall system time performance can be extrapolated as a multiple of a single 11-bit RCA/S stage propagation delay (15 ns). The only exception to this was the introduction of a latch stage among N, D, and Q to break the

		BEFORE OPTIMIZATION	
stage	N	2 level of logic (RCS/RCA)	30 Ns
stage	D	3 level of logic (RCS/RCA)	50 Ns
stage	Q	4 level of logic (RCS/RCA)	65 Ns
		AFTER OPTIMIZATION AND INSERTION	
		OF 1 LEVEL OF LATCH	
stage	N	2 level of logic (RCS/RCA)	30 Ns
stage	D	3 level of logic (RCS/RCA)	40 Ns
stage	Q	4 level of logic (RCS/RCA)	85 Ns

*Table*1.

computation into two pipeline stages. This imposes an additional propagation delay in the routing path to result in a total of 30 ns to produce the first RCA/S stage of the quotient. The reduced performance is clearly due to a different utilization of routing resources between the latch and RCA/S.

4 Flexibility using FPGA's

The design has been realized by using macro-function blocks of the XBLOX library. The use of XBLOX maintains modularity at the schematic level. XBLOX allows easy expansion of the schematic by changing only XBLOX attributes, in order to accommodate higher bit word lengths, for example. Control of the placement of these macros (i.e. floorplanning) can be used for time performance optimization. XBLOX used in combination with incremental design techniques [reference] facilitates modification of the filter coefficients and therefore, the filter configuration. This was used to generate new programming bitstreams for different filter configurations for easy prototyping.

A flexible mode of operation has been realized in which several different filter configurations are stored on an external ROM device. The unique intrinsic reprogrammability feature of the SRAM FPGA enables real-time programming of the filter by downloading any one of the available configurations as the application requires.

5 Conclusion

It has been demonstrated how a modular, flexible architecture for astronomical image processing can be implemented with SRAM FPGA to achieve real time performance. This modular structure is constructed using only a few simple macros with, simple, communication and control. The modularity of the algorithm and the flexibility of the FPGA architecture have been controlled to match system performance. The SRAM FPGA flexibility is shown to allow for fast prototyping capability.

A prototype for the application of photon counting was realized at NASA Goddard Space Flight Center. The resulting system is able to perform centroiding filtering on complete images (512 x 512 pixels) at the throughput rate of 10Mevents/sec based on 5x5 pixel windowing array, and clocking rate of 10MHz. Improvements over the 10Mevent/sec can be achieved by introducing additional latch stages. This system is now under test to determine the filtering performance on images coming from a CID (Charge Injection Device) camera detector. Specifically the "flat-field" is being tested to determine the fixed pattern introduced by the filter.

References

1. Kawakami, H, : The effect of event shape on centroiding in photon counting detectors. Nuclear Instruments and Methods in Physics Research, 1994.
2. Kimble, R, NRA 93-OSSA-7 Proposal: A photon counting intensified CID detector for UV space astronomy. May 1993, NASA/GSFC code 681.
3. Lampropoulos, G, A, and Fahmy, M, M: A new realization for 2-D digital filter. IEEE Trans. Acoust., Speech, Signal Processing, vol. 15, Jan. 1982.
4. Mertioz, B., G., : VLSI implementation of two-dimensional digital filters via two-dimensional chips. IEEE Trans. Circuits Syst., vol. CAS-33, No 6, Dec 1985.
5. Xilinx, The Programmable Logic Data Book. 1994 edition.
6. Hwang, K, : Computer Arithmetic. John Wiley Sons, 1979.
7. Scott, N., R., : Computer number Systems and arithmetic. Englewood Cliffs, Prentice Hall, 1988.

A Programmable I/O System for Real-Time AC Drive Control Applications

D.R. Woodward, I. Page [#], D.C. Levy [*], R.G. Harley

Department of Electrical Engineering, University of Natal, Private Bag X10, Dalbridge, Durban 4014, South Africa, Email: dwoodw@elaine.ee.und.ac.za.

[#] Programming Research Group, Oxford University Computing Laboratory, 11 Keble Road, Oxford England, OX1 3QD.

[*] Department of Electrical Engineering, University of Sydney, Sydney, New South Wales, Australia.

Abstract. This paper describes a programmable hardware structure capable of interfacing a conventional microprocessor to various sensors / actuators in a real-time AC drive control environment. The objective of the interface structure is to handle the low level data manipulation and supervisory control operations of the sensors / actuators and consequently free the resources of the host microprocessor for higher level tasks such as the real-time control algorithm of the application. The advent of high density Field Programmable Gate Arrays, executable specification and hardware compilation techniques provides an ideal path for the realisation of configurable hardware such as this interface structure. The paper illustrates the general structure of the interface and then concentrates on the design and implementation of an interface for a shaft position sensor commonly found in AC drive control applications, using the aforementioned techniques.

1 Introduction

Electrical drive systems have in the past been dominated by DC machines as a consequence of the simple requirements of their control structures. The mechanical commutator employed by a DC machine, however limits the speed and torque capability of the machine and requires periodic maintenance. More recently, due to advances in microprocessor technology, AC machines coupled with sophisticated control algorithms are capable of matching the dynamic performance of DC machines. The AC machine requires little maintenance and is generally smaller than its DC machine counterpart. The implementation of the control algorithms for the AC machine requires fast and complex signal processing which must be provided by the microprocessor based programmable controller [1].

A programmable controller capable of implementing complex AC drive control algorithms such as Field Oriented Control [2] with digital current control loops achieving sampling rates of approximately 5 - 10 kHz has several requirements. These requirements include high computational performance coupled to an efficient I/O system to enable the controller to sample the AC machine phase currents, shaft speed and position variables, and control the AC machine by generating pulse width modulation (PWM) inverter control signals to modify the magnitude, phase and frequency of the AC supply to the machine. Figure 1 shows the components of a typical Voltage Source Inverter (VSI) based AC drive system and clearly identifies the interfaces required to connect the various sensors and actuators in the AC drive environment to the microprocessor based controller. The example application shown in Fig. 1 cites only VSI based AC drive controllers; however, in practice many variations exist not only with regard to the type of inverter employed (eg. Current Source Inverter CSI, resonant link inverter) but also in the combination of other interfaces required for various applications.

Commercial AC drive control systems employ high performance embedded microprocessors coupled to interfaces based on Application Specific Integrated Circuits (ASICs), to provide the computational power required to execute the machine control algorithms and to interface to the numerous sensors and actuators within the AC machine environment. The use of a combination of microprocessor and ASIC devices in AC drive control systems, whilst providing acceptable performance, remains limited to high volume AC drive applications due to the high cost of development and limited user configurability of the ASIC devices. Lower volume and less general AC drive control systems thus have to rely on the facilities offered by standard embedded microprocessors, to handle the low level requirements of the sensor and actuator interfaces as well as the execution of the control algorithm, and consequently the performance of these control systems is limited.

To overcome these problems this paper describes a programmable interface architecture based on Field Programmable Gate Array (FPGA) technology, that is capable of interfacing a conventional microprocessor to the various sensors / actuators in a real-time AC drive control environment. The objective of the interface structure is to handle the low level data manipulation and supervisory control operations of the sensors / actuators and consequently free the resources of the host microprocessor for higher level tasks such as the real-time control algorithm of the application, whilst at the same time providing user programmability and configurability to the interface.

The recent availability of large Field Programmable Gate Arrays (FPGAs) containing tens of thousands of logic gates with the ability to be configured and reconfigured in a matter of milliseconds without the use of costly programming

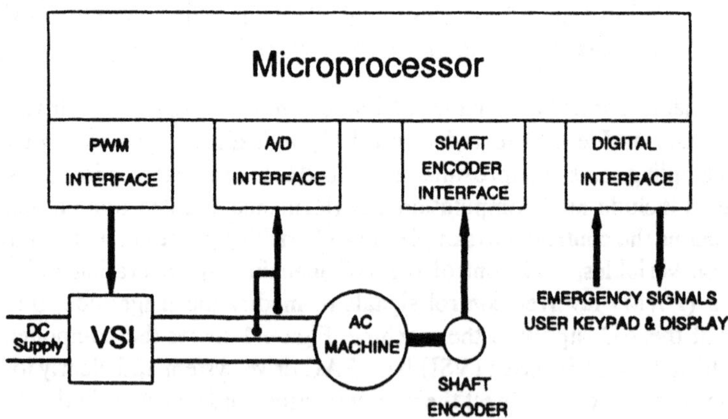

Fig. 1 A Typical AC Drive Configuration.

hardware, provides an excellent platform upon which the user configurability requirements of the interface can be implemented. To further address the requirements of user configurability the interface design adopts a slightly unconventional approach to generating the configuration data necessary to program the FPGA devices. Conventional methods of FPGA device configuration rely on the use of CAD software to layout gate level circuit diagrams or alternatively, Hardware Description Languages (such as VHDL) to define the required circuit configuration. Once this design entry phase is complete the resulting circuit netlist is processed by FPGA vendor supplied placement and routing software to generate the configuration information for the FPGA device/s. This level of design entry is, however, complex and requires extensive knowledge of digital logic and its timing requirements to enable a successful design to be implemented, and is thus likely to be beyond the capabilities of the average AC drives engineer.

The I/O Coprocessor design adopts the use of the high level algorithmic description method of design entry to describe and realise the architecture of the coprocessor. This level of design entry enables the user to describe the behaviour of a hardware circuit without actually having to define low level entities such as gates, registers and their interconnections. Typically the hardware algorithm is described using a special description language and then CAD software is utilised to convert the description into a netlist of gates and registers based on a fundamental structure and timing paradigm. The resulting netlist is then processed via placement and routing software to generate the FPGA configuration information as before. A number of researchers [3],[4],[5] have developed algorithmic description languages and their associated compilers, and demonstrated that functional hardware circuits can be reliably constructed using these methods whilst requiring significantly less effort on the part of the user. Of particular interest is the work conducted by Page et al [3], [6] of Oxford University who have defined an algorithmic description language Handel [7] and implemented CAD software capable of directly compiling the Handel language into a netlist form suitable for the configuration of Xilinx FPGA devices.

The Handel language is based on the concepts of Communicating Sequential Processes (CSP) [8] and consequently has a rich set of constructs that enable the simple description of several concurrently executing processes and a secure communication mechanism between them. Furthermore, the Handel language has a well defined set of formal semantics that allows the user to check a description based on the language for errors and treat the description as an "executable specification" which can be compiled like a conventional software language and used to simulate the behaviour of the intended hardware circuit. The Handel language and its associated compiler therefore provide an excellent algorithmic design entry platform through which to design and implement the I/O Coprocessor.

The remaining sections of the paper describe the overall structure of the I/O Coprocesor and then concentrate on the design of a shaft encoder interface based on the coprocessor structure and implemented using the techniques discussed in the preceding paragraphs.

2 The I/O Coprocessor Architecture

A general interface structure that is compatible with a wide range of I/O functions commonly found in AC drive applications has been developed. The interface structure takes the form of an I/O Coprocessor (see Fig. 2) which operates in conjunction with the host processor and communicates directly with it. The I/O Coprocessor is in turn composed of a number of concurrently operating subcomponents which implement the overall I/O task of the interface. The subcomponents of the coprocessor comprise a processor interface unit (PIU) and four independent I/O channel structures each containing an "I/O Processor" and an associated "I/O Port".

The structure of the I/O channels adopts the idea that the operations required to implement most I/O interface functions can be divided into two basic groups; word level operations and bit level operations. At the word level, complex operations such as scaling, offset, transformation, packaging etc. take place, whilst at the bit level simple operations such as Boolean manipulation, latching, counting, shifting etc. take place. Furthermore, word level operations generally occur at a relatively slow rate (in comparison to bit level operations) and the precise instant at which they occur is not critical, whilst at the bit level operations generally occur at high speed and the occurrence in time of the operation can be critical. Based on these observations each of the I/O channel subcomponents employ random logic circuit structures to perform the bit level operations of the I/O functions where speed is of the essence (I/O Port Units), and utilise microprocessor like structures to perform the word level operations that can tolerate lower response times (I/O Processor Units). This approach attempts to benefit from the speed advantages offered by random logic circuit structures whilst at the same time optimising the use of FPGA silicon resources by enabling "circuit structure reuse" where speed is not required.

The following sections describe the details of the coprocesssor subcomponents and the configurability of the overall I/O Coprocessor structure.

2.1 I/O Coprocessor Subcomponents

The Processor Interface Unit (PIU) :- The PIU combines the functions of a host processor interface and a command interpreter. The host processor interface connects the host microprocessor peripheral interface bus and the internal structures of the I/O Coprocessor as well as managing the flow of control signals (such as interrupt and error signals) between the I/O Coprocessor and the host processor. The command interpreter function is responsible for receiving and interpreting high level commands and data from the host microprocessor and distributing them to the relevant I/O channels. The algorithm executing within the host microprocessor may use these high level commands to pass data to or receive data / status from the I/O Coprocessor.

The I/O Processor Units :- The I/O Processor Units are based on conventional microprocessor architecture with built in memory structures and a highly optimised and restricted instruction set to handle the word level operations required by the I/O functions of each I/O channel. The architecture of the I/O Processor and the instruction word format is designed to facilitate the simple addition or removal of instructions from the overall instruction set to enable the instruction set to be optimised for particular applications without grossly affecting the basic structure of the processor. The "software routines" executing on the I/O Processors can be downloaded from the host processor to suit the I/O function required or the routine may be embedded in ROM like structures within the I/O Processor structure.

The I/O Port Units :- The I/O Port Units of the I/O Coprocessor are based on random logic circuits and implement the high speed, time critical, bit level operations required by the I/O functions of each channel. A simple interface links the I/O Port to the corresponding I/O Processor of each channel. The I/O Port Unit structures are responsible for ensuring that bit level operations occur in the correct sequence and at the correct instant in time and consequently require careful design and implementation to match the requirements of the sensors and actuators to which they are connected.

The basic subcomponents of the I/O Coprocessor are self contained and communicate with each other via unidirectional, synchronous point-to-point data channels. Consequently this modular structure is readily expandable to cater for multiple I/O functions and is partitionable over several FPGA devices to accommodate complex I/O functions requiring logic configurations which exceed the gate densities currently available in single FPGA devices.

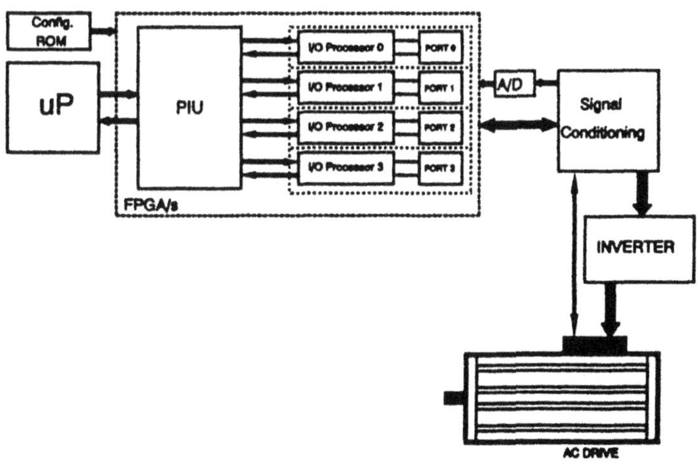

Fig. 2 The I/O Coprocessor Architecture.

2.2 I/O Coprocessor Configurability

To address the requirements of user configurability, the structure of the I/O Coprocessor provides three levels of programmable capability to the user.

At the lowest level where the FPGA configuration information is generated, sufficiently skilled users may generate different versions of the PIU to support alternative host processor types, the I/O Processors can be optimised to perform certain functions, and the bit level operations performed by the I/O Port Units can be modified to suit particular sensor / actuator requirements. The next level of programmable capability involves the generation of programs for each of the I/O Processors. Programs may be written and downloaded to each of the I/O Processors to suit a particular application without requiring the reconfiguration of the basic structure of the I/O Coprocessor. The third and final level of programmable capability possible with the I/O Coprocessor architecture is the execution of high level commands issued by application code executing on the host microprocessor and forming part of the overall control algorithm of the controller. These commands enable data to be passed to and from the I/O Coprocessor and the I/O Processors to be reset and booted with programs.

3 The Implementation of a Shaft Encoder Interface

This section describes the implementation of a shaft encoder interface for an AC drive application based on the I/O Coprocessor architecture. The diagram illustrated in Fig. 3 shows the overall process structure of an interface to connect a transputer processor to an incremental shaft encoder of the type commonly found in machine and robotics control applications. The interface decodes the quadrature squarewave signals produced by the shaft encoder sensor to generate angle and speed information which is then made available to higher level tasks executing as software processes on the host transputer processor. The I/O Coprocessor structure in this application only

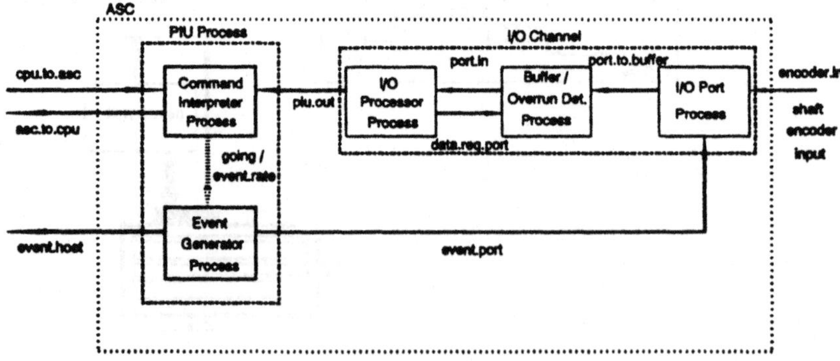

Fig. 3 The Shaft Encoder Interface Structure.

requires the use of a single I/O channel (I/O Processor + I/O Port) and thus consists of five interconnected parallel processes co-operating to achieve the required interface function. The details of each of these processes is explained in the following paragraphs.

I/O Port Process :- The I/O Port process accesses the two single bit shaft encoder signals via the "encoder.in" channel. Each squarewave signal is captured and passed through a simple digital filter to remove noise glitches before being passed onto a quadrature decoder. The decoder treats the shaft encoder as a simple four state finite state machine (FSM) with the states arranged as a loop (see Fig. 4) and determines the current state of the FSM by interpreting the filtered quadrature signals generated by the encoder. Using current state information and noting the previous transition from either the left or the right adjacent states, the decoder can determine the direction of rotation of the shaft encoder and increment or decrement a 16 bit position counter accordingly. The I/O Port Process thus generates a continually updated count value corresponding to the shaft position which is then available to be passed to the Buffer / Overrun Detection Process for onward transmission to the I/O Processor Process. The transfer of data between the I/O Port Process and the Buffer / Overrun Det. Process occurs at the sampling instants determined by the event signals generated by the Event Generator Process. Appendix A shows the Handel code listing for the I/O Port Process.

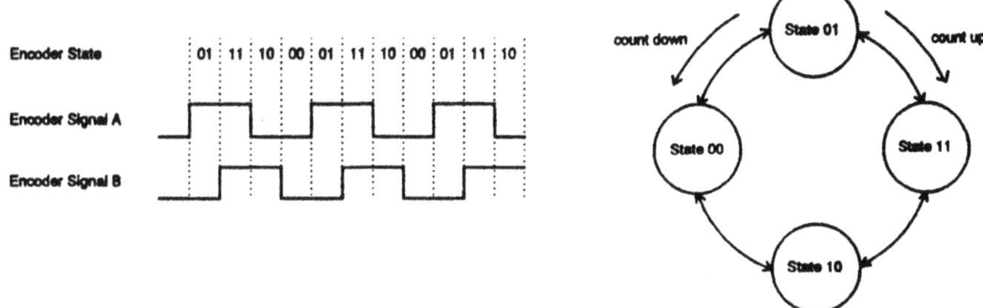

Fig. 4 Shaft Encoder State Diagram.

The Buffer / Overrun Detection Process :- The Buffer / Overrun Detection Process provides the interface with the means of detecting and informing the user of the occurrence of sampling overrun. Sampling overrun occurs when the event signal frequently is too high compared to the execution rates of the tasks within the I/O Processor or the host transputer based S/W. To remedy sampling overrun the user must either reduce the execution times of these tasks or lower the event (sampling) frequency).

The I/O Processor Process :- The I/O Processor process receives count data from the I/O Port process (via the Buffer / Overrun Det. Process) and manipulates this data to generate angle and speed estimates that are made available to the host transputer via the Processor Interface Unit (PIU) process. The I/O Processor Process implements a custom microprocessor with two general purpose 16 bit registers, a 4 bit instruction register, a 16 bit operand register, a 5 bit instruction pointer register, a single 16 bit temporary register, and a set of 18 instructions. In the interests of simplicity the program executing within the I/O Processor Process is stored in program memory implemented on the FPGA as ROM and is consequently fixed for this particular I/O Coprocessor implementation. However, a more general implementation would provide a facility to enable the I/O Processor program to be changed for different applications and interface functions.

The PIU Process :- The PIU process contains the interface to the host transputer processor and as can be seen from Fig. 3 consists of the Command Interpreter Process and the Event Generator Process. The Command Interpreter Process is linked to the host transputer via two channels "cpu.to.asc" and "asc.to.cpu", the former is used to pass high level commands to the I/O Coprocessor and the latter channel is used to return speed and angle data to the host transputer. The commands recognised by the Command Interpreter process are as follows; En / Dis-able event generation, Set event generator parameters, Get speed and angle data. The Event Generator Process generates a regular event signal which is passed to the I/O Port Process via the "event.port" channel and to the host transputer via the "event.host" channel. The operation of the Event Generator Process is controlled via the Command Interpreter Process by commands and parameters issued by the software processes executing on the host transputer.

The entire I/O Coprocessor structure for the shaft encoder interface has been described using the constructs available in the Handel language definition (due to space constraints only the I/O Port section of the description is shown in Appendix A). To determine the timing constraints of the interface system the next section discusses the timing behaviour of the critical I/O Port Process in detail.

3.1 Timing Constraints of the Shaft Encoder Interface

To ensure the correct operation of the interface it is instructive to investigate the timing behaviour of the critical I/O Port process of the I/O Coprocessor structure, the timing behaviour of the other processes is less critical and space does not permit their discussion. The Handel description of the I/O Port process shown in Appendix A can be seen to consist of a number of parallel (denoted by Par) subprocesses embedded within a continuous loop (While True). Close inspection of the digital filter process reveals that its operation requires that the input signals from the shaft encoder remain stable for at least three consecutive cycles of the While True loop, before a valid input is passed on to the decoder process. Whilst inspection of the decoder process reveals that at least a single cycle of the While True loop is required to detect a state transition and appropriately increment or decrement the counter value. Consequently the state time of the signals emanating from the shaft encoder must exceed the cycle time of the While True loop.

To get an accurate indication of the timing constraints of the interface it is therefore necessary to determine the cycle time of the While True loop in terms of the basic system clock of the I/O Port process. The Handel language compiler generates netlist descriptions of synchronously clocked logic to implement the constructs of the language. This mapping of the language constructs to hardware circuits enables a precise temporal behaviour for each of the language's constructs to be defined in clock cycles. Using this information it is possible to determine that a single cycle of the While True loop will take two system clock cycles to complete. This allows us to determine that the shaft encoder input signals must remain stable for at least six clock cycles of the system clock, whilst the state time of the shaft encoder signals must exceed at least two clock cycles of the system clock. Thus it is necessary to consider these constraints in addition to the normal constraints when selecting the system clock frequency for a particular shaft encoder application.

The entire shaft encoder interface has been described as a set of parallel processes using the Handel language and successfully simulated to verify correct operation. The netlist information has been generated and vendor supplied automatic place and route software is currently being used to generate FPGA configuration data to implement the interface. The results of this implementation were not available for inclusion in the paper and will thus be presented at the Poster session of the workshop.

4 Conclusion

A programmable interface structure based on Field Programmable Gate Arrays that is capable of interfacing a conventional microprocessor to the various sensors / actuators in a real-time AC drive control environment has been developed. The interface structure is based on a coprocessor like structure and was developed using the Handel algorithmic description language and its associated hardware compilation techniques. To illustrate the capabilities of the general programmable interface structure the paper discussed the design and implementation of an interface to connect a transputer processor to an incremental type shaft encoder of the type commonly found in machine and robotics control applications.

5 References

[1] W Leonhard, "20 Years Field Orientation, 10 Years Digital Signal Processing with Controlled AC Drives", invited paper, 6th conference on Power Electronics and Motion Control (PEMC), Budapest, Hungary, Oct 1990.

[2] F Blaschke, "The principle of field orientation as applied to the new TRANSVECTOR closed loop control systems for rotating field machines", Siemens Review, 1972.

[3] I Page, W Luk, "Compiling Occam into Field Programmable Gate Arrays", Research Report, Programming Research Group, Oxford University Computing Laboratory, Oxford, England.

[4] G De Micheli, D Ku, F Mailhot, T Truong, "The Olympus Synthesis System", IEEE Design & Test of Computers, Vol 7, No. 5, Oct 1990.

[5] A.S. Weban, J.W. O'Leary, G.M. Brown, "Codesign of Communications Protocols", IEEE Computer, vol 26, No. 12, Dec 1993.

[6] Many references to the Oxford work are available via the World Wide Web home page at URL http://www.comlab.ox.ac.uk/oucl/hwcomp.html.

[7] M Spivey, I Page, "How to Program in Handel", Technical Report, Oxford University Computing Laboratory, revised Dec 1994.

[8] C A R Hoare, "Communicating Sequential Processes", Prentice-Hall International, London, 1988.

Appendix A - I/O Port Process Handel Listing

```
WHILE TRUE
  PAR
    encoder_in ? encoder_data
    {{{ filter encoder signal a
    SIMUL
      t_minus_1_a := (encoder_data \\ 0) 1
      t_minus_2_a := t_minus_1_a
      t_minus_3_a := t_minus_2_a
    IF
      (t_minus_1_a AND t_minus_2_a) AND t_minus_3_a
        ch_a_fil := TRUE
      TRUE
        IF
          ((NOT t_minus_1_a) AND (NOT t_minus_2_a))
          AND (NOT t_minus_3_a)
            ch_a_fil := FALSE
          TRUE
            ch_a_fil := ch_a_fil
    }}}
    ... filter encoder signal b -- this is a repeat of the filter process above
    {{{ quadrature decoder
    IF
      (complex Boolean expression1 involving ch_a/b_fil, ch_a/b_fil_old)
        count := count + 1
      TRUE
        IF
          (complex Boolean expression2 involving ch_a/b_fil, ch_a/b_fil_old)
            count := count - 1
          TRUE
            DELAY
    ch_a_fil_old, ch_b_fil_old := ch_a_fil, ch_b_fil
    }}}
    {{{ output to buffer
    PRIALT
      event_0 ? event_any
        port_to_buffer ! count
      SKIP
        DELAY
    }}}
```

Reconfigurable Logic for Fault Tolerance

Rajani Cuddapah[1] and Massimiliano Corba[2]

[1] NASA/Goddard Space Flight Center Code 662, Greenbelt MD 20771, USA
e-mail rcc@gamma.gsfc.nasa.gov
[2] Code 681

Abstract. A novel application of SRAM-based FPGA technology is the development of fault tolerant systems in which reconfigurability is exploited in order to implement inherent redundancy. The approach is to use SRAM-based FPGA's in a mode where fault tolerance is achieved by detection of a fault and its location, and recovery from the fault via device reconfiguration. The scope of this paper is limited only to the demonstration of the flexibility of the SRAM-based FPGA architecture to tolerate faults which have been detected and located by means not described herein. Computer simulations of random faults and recovery from the faults has been performed. Results are described validating this technique and the success rate in terms of both routability and performance.

1 Introduction

Classically, redundancy or replicate circuitry has been used for fault tolerance in high integrity systems for use in hostile environments. A powerful alternate approach is to implement designs in reconfigurable logic and retain a set of unused cells for use as spares in case of the occurrence of a fault. These spares can then be used to repair the functionality of the circuit via device reconfiguration. SRAM based FPGA's are unique devices due to their flexibility. The FPGA is a two-dimensional array of logic blocks with routing channels distributed along columns and rows. The architecture of FPGA's is a logic cell array (LCA) that consists of the following elements: configurable logic blocks (CLB's), I/0 blocks (IOB's) and connection resources[1]. The user has control over each element of each LCA by way of a configuration program which is stored in on-chip memory. Designs may therefore be recoverable from faults simply by downloading a new configuration program, implementing allocated spares.
Historically fault tolerance has been studied using one of two strategies, in which fault replacement is carried out either by direct replacement or by shifted replacement[2],[3]. In this context, fault tolerance depends not only on spare allocation, but also on the organization of spare resources in the reconfigurable array[4][5]. The technology of SRAM-based FPGA's yields improved fault recovery due to increased flexibility. In this paper, we introduce several definitions that are used to establish an algorithm for evaluating the flexibility of SRAM

FPGA's to achieve the goal of fault tolerance. An experimental procedure involving computer simulation of and recovery from random faults is detailed and finally the results are documented.

2 Definitions

Routability:
Routability is defined as the total number of successfully routed connections in a design as a percentage of the total number of connections required for a successful implementation of the particular design. The flexibility of a FPGA interconnection resources is determined by the number and distribution of switches available in the FPGA architecture[6]. Although a large number of switches can yield high overall routability and flexibility, there is an associated performance and density cost because each switch introduces significant delay, due to parasitic capacitance, and occupies a significant amount of area.

Faults:
Two types of faults are considered in this study: CLB faults and network faults. First, the CLB fault is defined as a CLB that is no longer functional and requires complete replacement. Second, the network fault is defined as an connection fault which requires a new path assignment to realize the necessary interconnection. Study of recovery from IOB failures requires the consideration of peripheral devices as well as I/O pin redundancy and, therefore, are not considered in this paper. In general, since tristate outputs are available in some FPGAs' I/O facilities, it is possible to use the same redundancy techniques with these I/O's as with the CLB's.

3 Spare Allocation Models

There are two different spare allocation techniques that can be considered for fault recovery. One involves design redundancy and the other involves resource redundancy.

Design Replacement:
Design replacement is the case where a design is implemented or mapped into no more than half of a device thereby leaving at least half of the device for implementation of a redundant circuit. For this case, any defect detected in operation causes an alternate configuration file to be loaded which replaces the design into the redundant fraction of the chip. Note that speed performance is maintained in the redundant circuit since it is exactly replicated from the CLB and connection layout level, it is just located in different physical area of the chip.

Resource Replacement:
Resource replacement is the case where some percentage of resources are left

unused in a device for use as spares. For this case, any defective resources detected are replaced by spare resources. Timing performance is not necessarily retained in different configurations of the same functional design. For critical applications, effort must be made to ensure that net timing requirements are retained after reconfiguration.
Clearly, the fault coverage associated with the two approaches is very different. Design replacement allows for 100% fault coverage for only a single instance of fault occurrences that are confined to the area occupied by the initial design implementation. On the other hand, resource replacement allows for a smaller percentage of fault coverage, yet can conceivably be used repeatedly until all spares are exhausted.
For this study, the emphasis is on the investigation of resource replacement. It is the opinion of the authors that the feasibility of design replacement can be shown easily as it is simply based on the principles of redundancy. Resource replacement, conversely, can be made possible only by the unique flexibility of the SRAM FPGA. Note that both replacement options may be implemented using SRAM FPGA's, the choice of one over the other is strictly application dependent.

4 Experimental Procedure

The experimental procedure was to first implement a benchmark design[7] in a "step- and-repeat" fashion. The design was prepared for a Xilinx SRAM FPGA device in order to achieve maximum logic density using the APR (Automatic Place and Route) software utility provided by Xilinx. For each benchmark, the number of repeated circuits was different, depending on the "size" of the primary circuit. The manufacturer supplied APR software was used without any design constraints to yield an optimally routed design (ie., with the maximum number of benchmark circuit replications possible on the chip).
For the fault analysis simulations, the faults were forced into the design implementation by creating a constraints file which prohibits the APR from using the randomly chosen faults. The constraints file also contained information forcing the IOB information from the optimized circuit to be frozen into each subsequent iteration. This forced the APR to "work around" the faulty areas while maintaining the same I/O pin configuration.
Table 1 lists the name, the percentage of CLB's used, the percentage of IOB's used, and a brief function description of each optimized circuit. The table is based on implementation for an XC3020-50 which has 64 CLB's, 64 IOB's and a maximum clock speed of 50 MHz. The resulting optimized designs were used to study both CLB faults and network faults as described below.

CLB Faults
In order to study the feasibility of attaining fault tolerance for CLB faults, a C-language program was developed that randomly selected blocks as being faulty. To obtain the worst-case results, the number of faulty blocks selected was cho-

sen to be exactly equal to the number of spare CLB's remaining after the above described implementation for each benchmark circuit. These randomly selected blocks were then prohibited from usage by the APR utility. The C-program then invoked APR using the prohibited (randomly chosen) CLB's and the I/O block assignments from the optimized design as placement and routing constraints. This forced the usage of every single spare to replace CLB's used in the optimized design while maintaining the same I/O pin information as the optimized design.

Table 1. Circuit Descriptions

Benchmark	CLB's Used	IOB's Used	Description
#1	64%	67%	Datapath
#2	91%	45%	8 Bit Timer/ Counter
#3	94%	17%	Filler Circuit
#4	72%	56%	16 -Bit Accumulator

Network Faults

In order to study the case for the network faults, a "fault design" was created. This design was functionally arbitrary but deliberately dense in terms of interconnections. This design was used with benchmark circuit number 2 to simulate a faulty area which included both CLB's and interconnections. The number of CLB's used in the network fault circuit was exactly equal to the number of CLB faults (6 for Benchmark #2) used in the CLB fault study for a worst-case analysis. The manufacturer supplied APR software was used to place and route the fault design into a random area of a device LCA design, and then to include the benchmark design into the same device design, avoiding all resources occupied by the fault design. Again a constraints file was generated and used to prohibit APR usage of all resources occupied by the fault design and also to maintain the IOB information of the optimized design. The experimental procedure, illustrated in Figure 1, was performed over 200 times on the circuits described in Table 1 to study the CLB faults. The network fault analysis was performed

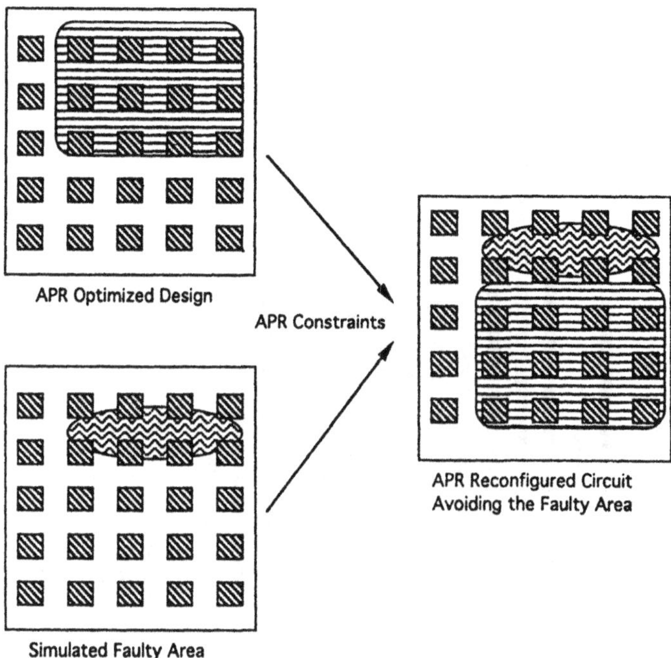

Fig. 1. Reconfiguration Illustration.

165 times following the same procedure on Benchmark #2 only. All of the LCA circuit designs were for a Xilinx XC3020-50 device.

5 Experimental Results

Assessment of the APR's avoidance of the randomly selected (simulated faulty) blocks and networks is based on both the routability and the performance characteristics of the reconfigured design. Both of these are documented by a report file which is automatically generated by the APR program. The report file provides a list of unrouted nets as well as a routed net timing analysis which can be used to compare the performance of the optimized placement and routing to the trials with the faults present.
A C-language analysis program was written to read the design report files generated by the APR utility. For evaluation purposes, this program was limited only to the analysis of routability and timing performance. It read out whether or not the benchmark circuits with faulty blocks or networks were successfully routable. It also supplied the longest net delays for user defined critical nets for comparison with the same nets in the optimized benchmark design.

CLB Fault Study Results
Table 2 summarizes the number of spares available in each benchmark design in units of CLB's and the routability results. All of the circuits tested were

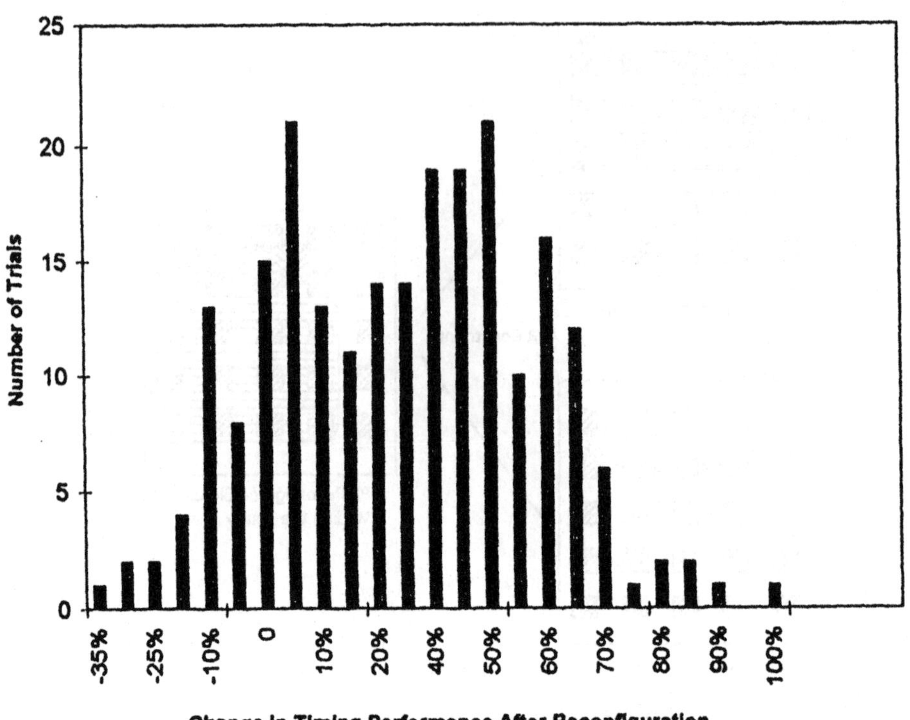

Fig. 2. Histogram of "CLK" net in Benchmark #1

Table 2. CLB Fault Routability Results

Benchmark	Spare CLB's	Routability with Number of Faults Equal Number of Spares
#1	23	100%
#2	6	100%
#3	4	100%
#4	18	100%

100% routable with the maximum number of faults coverable. The CLB fault recovery analysis was therefore 100% successful, in terms of routability, for 100% of the trials. Figure 2, shows a histogram of the "CLK", clock, net in Benchmark #1. The optimized routed design (without any faults) resulted in a maximum path (or propogation) delay in the "CLK" net of 22ns. This implies a maximum operating clock speed of 45.4MHz. After introducing 23 faulty CLB's into the device and recovering by reconfiguring the design to use the 23 allocated spare CLB's, the timing performance of this particular net changed as shown in the figure. It should be noted that in some of the trials the timing on this net was actually improved or made smaller. For these cases, however, longer path delays were suffered on other nets in the design. For more than 85% of the cases, however, increased path delays were noted, varying from 105% to 200% of the optimized path delay value.

Similar timing analysis was performed on all of the critical nets for all the four designs. In practical design procedures the criticality of nets must be assessed and their timing performance requirements must be established in order to define guidelines for acceptance of reconfigured designs. For example, one can establish a requirement that net delay increases of no greater than 10% for all critical nets throughout the design are acceptable. To determine the fault recovery rate, the total number of trials satisfying the requirement can be counted and divided by the total number of trials performed. Figure 3 shows the fault recovery rate for different values of delay requirements for all four of the benchmark circuits. A requirement of a 10% maximum increase in delay on all critical nets in Benchmark #1, for example, results in a fault recovery rate of 7.93%.

Network Fault Study Results

A similar analysis was performed for the network fault study on Benchmark #2. The routability with the fault circuit, using 6 faulty CLB's and associated networks, was 100% for 80.61% of the trials. A plot of the fault recovery rate for Benchmark #2 is shown in Figure 4. It is evident that faulty interconnections greatly affect not only the routability of the design, but the timing performance as well.

6 Conclusions

It has been shown that fault tolerance via the mechanism of reconfigurability is a viable technique. The merit of the technique has been evaluated in terms of both routability and circuit, timing performance for four different circuits. Designers for space-flight or other critical application hardware can use the herein described algorithm to determine the fault coverage of a design and the fault recovery rate for any given design implemented in SRAM-based FPGA's.
The work described herein was demonstrated by using only manufacturer supplied routing and placement tools. Note, in practical applications, constraints may be placed on critical nets, emphasizing their timing requirements. Placing specific timing requirements on critical nets can greatly improve the results

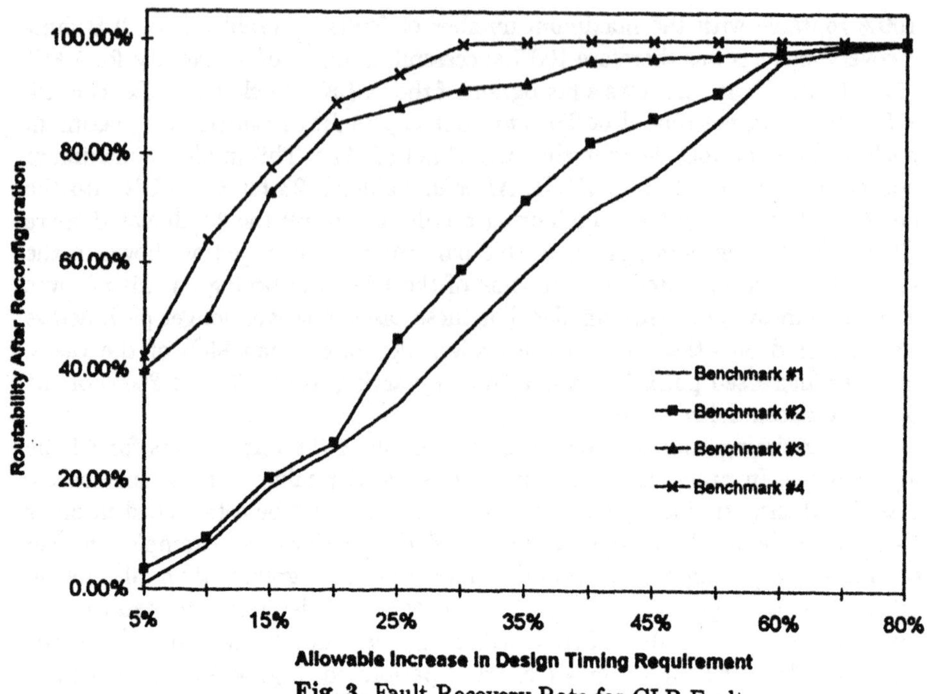

Fig. 3. Fault Recovery Rate for CLB Faults

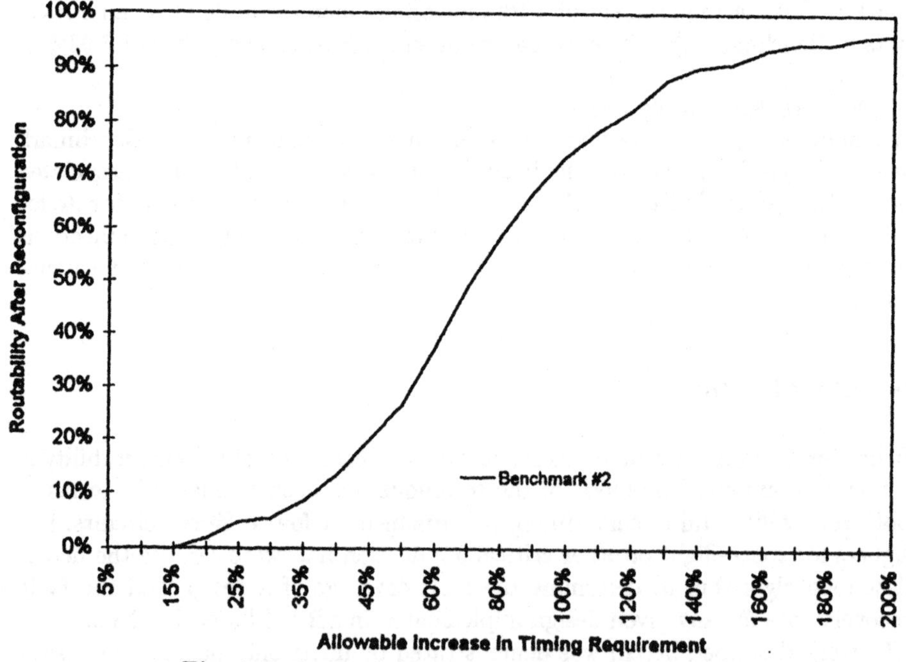

Fig. 4. Fault Recovery Rate for Net Faults for Benchmark #2

documented herein. The manufacturer supplied APR did not allow the same flexibility for defining constraints on particular interconnection paths as it did for discrete CLB's. As a consequence, the network fault study was limited.
For future work, a customized routing and placement tool could be investigated. Different router algorithms in which the selection for the network path can be constrained, similar to CLB constraints. These constraints would be considered in a detailed router's first stage when it enumerates the number of alternatives for the detailed route of each path corresponding to each net[8].

7 References

1. Xilinx, The Programmable Logic Data Book, 1994.
2. J. S. N., Jean, "Fault-tolerant Array Processors Using N-and-half-track switches, Conference Proc. on Application Specific Array Processors, pp. 426-437, 1990.
3. V.P. Roychowdhury, J. Bruck, and T. Kailath, "Efficient Algorithms for Reconfiguration in VLSI/WSI Arrays", IEEE Trans. on Computers, Vol 39, No. 4, pp. 480-488, April 1990.
4. S. Kuo, W. Fuchs, "Spare Allocation and Reconfiguration in Large Area VLSI", Conference Proc. 25th ACM/IEEE Design Automation Conference, pp. 608-612, 1988.
5. K. Sugihara, T. Kikuno, "On Fault Tolerance of Reconfigurable Arrays Using Spare Processors", Conference Proc. Pacific Rim International Symposium on Fault Tolerant Systems, pp. 10-15, September 26-27, 1991.
6. J. Rose, S. Brown, "Flexibility of Interconnection Structures for Field- Programmable Gate Arrays", IEEE J. Solid State Circuits, Vol. 26, No. 3, pp. 277-282, March 1991.
7. Programmable Electronics Performance Corporation Benchmark Suite 1, Version 1.2, May 28, 1993.
8. S. Brown, J. Rose, "A Detailed Router for Field-Programmable Gate Arrays", IEEE Trans. on Computer Aided Design, Vol 11, No. 5, pp 620-628, May 1992.

Supercomputing with Reconfigurable Architectures

Steven A. Guccione and Mario J. Gonzalez

Computer Engineering Research Center
Department of Electrical and Computer Engineering
University of Texas
Austin, Texas 78712

Abstract. Recently, several research and commercial systems based on reconfigurable logic have been implemented. These machines have demonstrated supercomputer levels of performance for a number of algorithms. While these demonstrations have been impressive, it is not clear that architectures based on reconfigurable logic will necessarily be suitable for algorithms commonly executed on supercomputers. This paper discusses the implementation of Livermore Fortran Kernels for a supercomputer class machine based on reconfigurable logic.

1 Introduction

Recently, a large number of systems based on reconfigurable logic have been designed and built [5]. Some of the largest of these machines have demonstrated supercomputer levels of performance on selected algorithms. As larger reconfigurable machines become available, it is widely believed that they will be used to implement algorithms typically found on existing supercomputers. It is not clear, however, that these architectures are well suited to these tasks. Most of the algorithms implemented to date are more typically found implemented in custom hardware rather than on large general purpose machines.

This study examines the feasibility of general purpose supercomputing using reconfigurable logic. Algorithms selected for the *Livermore FORTRAN Kernels (LFK)* [9] [2] are used to examine the performance of a reconfigurable logic based supercomputer. The LFK are chosen for several reasons.

First, the LFK suite is a widely used tool to measure CPU performance. This allows comparison to a wide range of existing architectures. Second, the LFK are composed of a number of tests which include a wide range of computational structures. While some of these structures are used to measure the peak performance of a system, others are constructed specifically to limit performance. This permits an examination of architectural weaknesses as well as strengths. Finally, the LFK are relatively compact and self contained. This allows their simulation on models of proposed hardware. Performance information gained from such simulations is valuable in guiding the design.

2 The Livermore Fortran Kernels

The LFK are a collection of 24 relatively small fragments of code. Each of these code fragments contains a CPU intensive loop, giving the test suite its informal name, "the Livermore Loops". The LFK were developed in 1970 to test the code generated by compilers. Over time, these codes have become a benchmarking tool for new supercomputer systems.

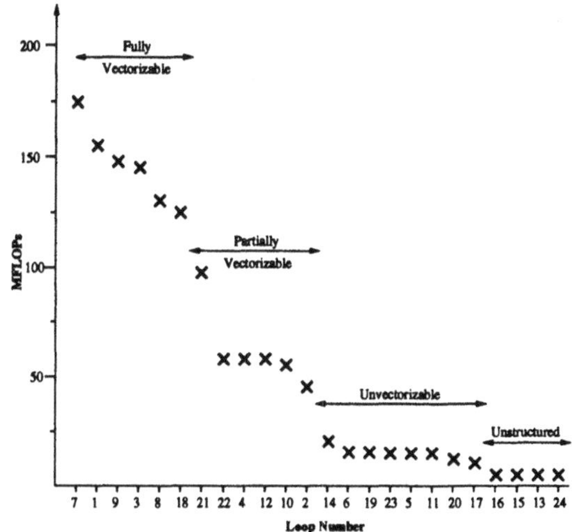

Fig. 1. Performance sorted my MFLOPs for a CRAY X-MP.

As the LFK have evolved into a benchmarking tool, new loops have been added to exercise specific features of both compilers and hardware. The number of loops has grown from the initial 12 to the current 24.

The kernels in this study were converted by hand from the original FORTRAN to a data parallel version of the *C* language. Simulations of circuits extracted from this data parallel code are compared against a standard C version of the LFK [3].

Finally, it should be noted that the LFK are specified for high accuracy floating point arithmetic. While some work is being done on the implementation of floating point arithmetic in reconfigurable logic [1], it is understood that using the technology available today, a very large RPU would be required to implement these functions. While numeric accuracy is important, it is the computational structures in the LFK which are of primary interest. It is these structures, not numeric accuracy, which has the greated impact of performance.

Figure 1 plots the performance of the 24 Livermore Fortran Kernels run on a *CRAY X-MP* using the *CFT77 3.0* compiler [10]. The numbers are listed in MFLOPs and are sorted by performance. From this sorted graph of the LFK, four distinct performance ranges can be identified. These are: *fully vectorizable*, *partially vectorizable*, *unvectorizable* and *unstructured.*

In general, kernels in each of these regions present different computational challenges. These will be discussed in more detail as the kernels are implemented. For brevity, only representative kernels from each region are discussed. Kernels were chosen primarily for their simplicity in illustrating the particular computational structures.

The reconfigurable hardware platform used for execution of these kernels is assumed to consist of a relatively large *Reconfigurable Processing Unit*, or *RPU*, dedicated memory tightly coupled to the RPU, and a host machine. The reconfigurable portion of the system is assumed to operate at 50 MHz. The highly pipelined circuits combined with the vector data accesses make this feasible.

3 Fully Vectorizable Loops

Kernels in the fully vectorizable category typically perform the highest on vector supercomputers. These kernels are characterized by being easily vectorized as well as providing enough work to occupy multiple functional units.

3.1 Loop 1: Hydrodynamic Code

Loop 1 is a fragment from a hydrodynamic simulation. The original FORTRAN code for this loop is shown in Fig. 2. This loop is easily vectorizable and can make concurrent use of several functional units.

```
    Do 1 k = 1,n
1     X(k) = Q + (Y(k) * ((R * Z(k+10)) + (T * Z(k+11))))
```

Fig. 2. The original FORTRAN code for Loop 1.

The translation of this algorithm to data parallel form is shown in Figure 3. Because of the structure and simplicity of this loop, the similarities between the FORTRAN code, the algorithm and the data parallel code are clear.

```
Z10 = delta(Z, 10);
Z11 = delta(Z10, 1);
X = q + (Y * ((r * Z10) + (t * Z11)));
```

Fig. 3. The data parallel code for Loop 1.

One feature of this implementation which may require further explanation is the use of the *delta()* function. This function is used to provide vectors whose indices are offset some small number of units. The *delta* is literally a delay. By providing a delayed version of the vector, data can be made available as it is required without having to re-access the memory system. The translation from an indexed style of coding is faily simple.

Figure 4 shows the RPU circuit extracted from the dataflow graph of this code. This is the circuit makes use of 5 functional units, and has a latency of 5 functional units.

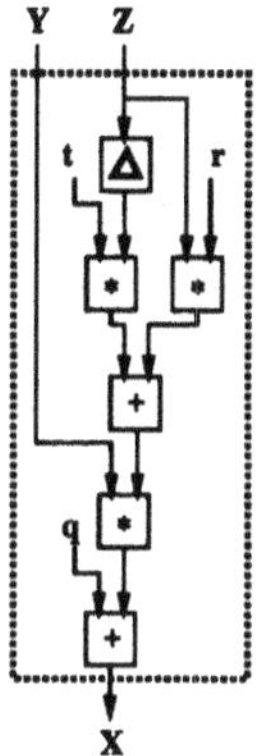

Fig. 4. The configured circuit for loop 1.

Estimating performance of this circuit is fairly simple. Assuming sufficient memory bandwidth and a clock speed of 50 MHz, the processor will produce one result per clock cycle, neglecting latency. Since all functional units are kept busy on each cycle, approximately 250 million operations per second are performed. If the functional units all perform floating point operations, this corresponds to 250 MFLOPs. Even at this modest clock speed, this exceeds the rate of computation of the CRAY X-MP.

3.2 Loop 3 - Inner Product

The second fully vectorizable loop is an inner product calculation. This is a multiply-accumulate function found in many applications, including the matrix arithmetic. Because of the widespread use of this type of calculation, most supercomputers are especially efficient at its execution. The data parallel version of the code is simply the *add-scan()* of a product, as shown in Figure 5.

The circuit extracted from this data parallel code is fairly simple, containing a multiplier and an add-scan functional unit. The memory bandwidth required is also fairly modest. Two vector inputs and a vector output are required.

```
Q = add-scan(Z * X);
```

Fig. 5. The data parallel code for Loop 3.

At a rate of 50 MHz, neglecting overhead, this circuit performs 100 million operation per second. This is somewhat less than the CRAY X-MP. The very small number of functional units indicates very little exploitable parallelism.

4 Partially Vectorizable Loops

The next group of kernels perform at a level somewhat below that of the fully vectorizable kernels. Here, these loops are referred to as *partially vectorizable loops*. In these kernels, the ability to fully use the vector units is reduced. While these loops do not have the performance levels of the fully vectorizable loops, their levels of performance are still substantial, but only a fraction of the peak performance achieved by the fully vectorizable loops.

4.1 Loop 12 - First Difference

Loop 12 is a first difference calculation. Despite its relatively low performance, this loop is fairly simple and is easily translated to data parallel code.

Figure 6 gives the translated data parallel code for the first difference calculation. The use of the *delta()* function provides the offset version of the vector Y, saving input bandwidth.

```
Y1 = delta(Y, 1);
X = Y - Y1;
```

Fig. 6. The data parallel code for Loop 12.

The circuit extracted from this code contains only a *delta* unit and a subtractor. Because this implementation requires only a single functional unit, the calculation proceeds at a rate of 50 million operation per second. This is similar to the rate achieved by the CRAY X-MP. The performance of this loop is reduced because of a lack of work to be performed on the data.

4.2 Loop 22 - Planckian Distribution

Loop 22 is from a Planckian distribution program. Here, the computation rate on traditional vector processors is slowed by conditional execution. Figure 7 gives the data parallel code for this loop.

```
if (U < (V * 20.0))
   Y = (U / V);
   else
      Y = 20.0;

W = X / (exp(Y) - 1.0);
```

Fig. 7. The data parallel code for Loop 22.

In this implementation, the conditional statement provides two alternate values for the elements in Y, depending on the result of the conditional statement. This permits a parallel computation of the two values, with the proper result being selected.

While this is a more complex loop, only 5 functional units, not including the comparison or the the multiplexer, are used. This assumes that the *exp()* function is counted as a single functional unit. Depending on the implementation, this operator may be composed of other simpler arithmetic and logical operations.

Assuming a clock speed of 50 MHz, this implementation achieves approximately 250 million operations per second. This is almost four times the rate of the CRAY X-MP reference machine. This increase is attributed to the ability to efficiently perform conditional operations.

5 Unvectorizable Loops

The kernels in this performance range are typically unvectorizable and are unable to make extensive use of vector hardware. Since they are not able to make use of the vector processing facilities that helped enhance performance in the previous loops, their performance is not only considerably lower, but also more uniform. These algorithms are typically forced to used the non-vector portion of the CPU, thus testing the performance of this portion of the architecture.

Most of these loops are unvectorizable because of data dependencies introduced by recurrence equations. While completely unvectorizable using traditional fixed instruction architectures, the use of structures such as parallel prefix *scan* operators open up new possibilities for these functions.

5.1 Loop 5 - Tridiagonal Elimination

Loop 5 is a fragment of code use in tridiagonal elimination. The original FORTRAN code contains a data dependency in X that prohibits vectorization. This kernel typifies a class of equations known as first order linear recurrence equations. Several approaches to parallelizing these equations have been proposed over the years [8] [7] [4].

The approach demonstrated here makes use of the fact that the computed values are actually independent if previously computed values are substituted into the subsequent equations. $X(5)$, for instance, can be written as in Equation 1.

$$X_5 = Z_5Y_5 - Z_5Z_4Y_4 + Z_5Z_4Z_3Y_3 - Z_5Z_4Z_3Z_2Y_2 + Z_5Z_4Z_3Z_2X_1 \tag{1}$$

This approach, while producing independent calculations, raises the computational complexity from $O(n)$ to $O(n^3)$. It is, however, possible to represent these equations using *scan* operators. While the actual construction of the *scan* based version of this code is beyond the scope of this paper, the parallel form of the equation contains easily discernible patterns amenable to *scan* operators. Figure 8 gives the data parallel code for this kernel.

```
X = mul-scan(-Z) * (add-scan((Z * Y) /
    mul-scan(-Z)) - x0)
```

Fig. 8. The data parallel code for Loop 5.

From this data parallel code, a pipelined circuit can be extracted. The ability to use non-standard operators such as *scans* has permitted a pipelined version of this kernel, greatly improving performance.

The circuit uses 7 functional units and two vector inputs and a single vector output. At 50 MHz, this circuit calculates 350 million operations per second. While the re-casting of the algorithm has added these extra functional units, thereby boosting the number of operations, the throughput of this circuit is still superior to other implementations, including those on supercomputers. While performance is increased, the new approach to this algorithm introduces some numerical stability problems not found in the original version.

5.2 Loop 11 - First Sum

Loop 11 is a first sum calculation. As in loop 5, a data dependency in the form of a recurrence is responsible for the low performance.

Unlike the recurrence equation in loop 5, the first sum in this kernel is very simple. It is essentially the definition of the *add-scan()* operator. The circuit extracted from this code is again, very simple. A single *add-scan* operator is used. A single vector input Y is used to produce a single vector output X.

While providing a vector solution for this algorithm, the first sum suffers a similar performance limitation to loop 12, the first difference kernel. Since only a single functional unit is used, the number of operations at 50 MHz is only 50 million per second. Even this low rate of calculation, however, still exceeds supercomputer levels of performance.

6 Unstructured Loops

These kernels are the lowest in performance on the CRAY X-MP reference machine. As with the unvectorizable loops, they are unable to take advantage of the special vector hardware. Additionally, these kernels contain structures that further reduce performance, even for the non-vector portion of the processor.

For lack of a better term, these loops will be referred to as *unstructured.* They are characterized primarily by the presence of unstructured control, usually in the form of *goto* statements, as well as complicated array indexing schemes.

Some of these loops actually exhibit a large amount of parallelism. It is often the way in which the algorithm is expressed, rather than any limitation in the underling algorithm, that reduces performance. For these reasons, some of these loops are better test of FORTRAN compiler optimizers than the underlying processor architecture.

6.1 Loop 24 - First Minimum

Loop 24 is selected as a representative of the unstructured loops because it has a deceptively simple implementation, while having the lowest performance of all 24 loops on a CRAY X-MP. The original FORTRAN code for this kernel is given in Figure 9.

```
      max24 = 1
      Do 24 k = 2,n
24       if (X(k) .lt. X(max24)) max24 = k
```

Fig. 9. The original FORTRAN code for Loop 24.

An attempt to translate this algorithm into data parallel code reveals some of its limitations. First, this code uses a conditional operator, which interferes with vectorization. Next, it performs an operation involving only two scalar quantities. Finally, X is indexed by a scalar quantity which changes unpredictably. All of these factors combine to dramatically reduce the performance of this kernel.

The goal of this loop. however, is to find the location of the minimum value in the vector X. Constructing a data parallel solution will require more than a simple translation from the original FORTRAN specification of the algorithm.

Figure 10 gives the data parallel code for this algorithm. In this implementation, the *min-scan()* operator is used to determine the minimum value. The conditional operator is then used to select the index for the minimum value. Since a looping index is not directly available, the *add-scan(1)* statement is used to generate these index values.

```
Min = min-scan(X);
Min1 = delta(Min, 1);
Diff = Min1 < Min;

Index = add-scan(1);
M = max-scan(Index * Diff);
```

Fig. 10. The data parallel code for Loop 24.

While a substantial modification of the original algorithm, this version is fully pipelinable and executes at approximately 200 million operations per second. While much of this figure is due to the additional functional units, this implementation still produces the desired result in approximately N clock cycles for a vector of length N.

A final note on unstructured algorithms. Many may not be suitable for reconfigurable machines. Unstructured access to vector data is a problem. Vector indexing of the form $X[Y[n]]$ is particularly difficult. Without special hardware support in the memory system, this type of calculation will almost certainly involve the host.

7 Conclusions

The table in Figure 11 gives an analysis of the performance of the seven kernels implemented. Perhaps not surprisingly, the algorithms which fared well on supercomputers also fared well on reconfigurable logic based machines. What is more surprising is the high levels of performance achieved by some of the loops which performed poorly on the CRAY X-MP, the supercomputer reference machine.

Loop Number	Vector Inputs	Vector Outputs	Latency	Functional Units	Estimated MFLOPs	CRAY X-MP MFLOPs
1	2	1	5	5	250	160
3	2	1	2	2	100	138
12	1	1	2	1	50	63
22	3	1	5	5	250	68
5	2	1	6	7	450	14
11	1	1	1	1	50	14
24	1	1	5	5	200	3

Fig. 11. The LFK performance parameters.

While many of the algorithms are easily implementable and exhibit very high performance, some structures are still problematic. First, simple recurrences can be implemented efficiently using scan circuits. Currently, however, no simple algorithm exists for translating more complex recurrences into these circuits. Secondly, unstructured algorithms, particularly those which make use of indirect array indexing, are not well suited to reconfigurable logic. Using the host to vectorize these types of array accesses before they are submitted to the RLU may be a solution for some algorithms.

This study was performed primarily to examine the feasibility of general purpose supercomputing using reconfigurable logic. While not a solution to all problems, the results for a large class of popular computational structures is promising. Furthermore, it should be noted that the algorithms in the LFK are taken from real applications, written for traditional architectures. It is possible that new classes of algorithms which exploit the unique features of reconfigurable logic will provide even higher levels of performance for a larger class of problems.

References

1. Barry Fagin and Cyril Renard. Field programmable gate arrays and floating point arithmetic. *IEEE Transactions on Very Large Scale Integrated (VLSI) Systems*, 2(3):365–367, September 1994.
2. John T. Feo. An analysis of the computational and parallel complexity of the livermore loops. *Parallel Computing*, 7(2):163–185, June 1988.
3. Martin Fouts. The Livermore Loops in C. NASA Ames Research Center memo, 1994.
4. Daniel D. Gajski. An algorithm for solving linear recurrence systems on parallel and pipelined machines. *IEEE Transactions on Computers*, C-30:190–206, March 1981.
5. Steven A. Guccione. List of FPGA-based computing machines. World Wide Web page http://www.utexas.edu/~ guccione/HW_list.html, 1994.
6. Steven A. Guccione and Mario J. Gonzalez. A data-parallel programming model for reconfigurable architectures. In Duncan A. Buell and Kenneth L. Pocek, editors, *IEEE Workshop on FPGAs for Custom Computing Machines*, pages 79–87, Los Alamitos, CA, April 1993. IEEE Computer Society Press.
7. Peter M. Kogge. Parallel solution of recurrence problems. *IBM Journal of Research and Development*, 18(2):138–148, March 1974.
8. Peter M. Kogge and Harold S. Stone. A parallel algorithm for the efficient solution of a general class of recurrence equations. *IEEE Transactions on Computers*, C-22(8):786–793, August 1973.
9. Frank H. McMahon. The Livermore Fortran kernels: A computer test of the numerical performance range. Technical Report UCRL-53745, Lawrence Livermore National Laboratory, December 1986.
10. Wayne Pfeiffer, Arnold Alagar, Anke Kamrath, Robert H. Leary, and Jack Rogers. Benchmarking and optimization of scientific codes on the CRAY X-MP, CRAY-2 and SCS-40 vector computers. *The Journal of Supercomputing*, 4(2):131–152, June 1990.

Automatic Synthesis of Parallel Programs Targeted to Dynamically Reconfigurable Logic Arrays

Maya Gokhale and Aaron Marks

David Sarnoff Research Center, CN5300, Princeton NJ 08543, USA

Abstract. Dynamically reconfigurable Field Programmable Gate Arrays (FPGAs) offer virtually unlimited numbers of gates to an application. This technology makes feasible large applications which can be temporally partitioned, with each phase being rapidly loaded onto the chip as required. We demonstrate in this paper an automatic technique to temporally partition a parallel program. Our technique partitions along a data parallel C program's function scopes. A configuration bit stream is generated for each function, and the host control program is generated which automatically loads the function's configuration file as the function is entered during execution. Preliminary results show that this partitioning makes it possible to

- run larger problem sets,
- run programs which would not otherwise fit on the chip, and
- include program-specific debug code without performance penalty.

Our compiler targets the NAPA accelerator board, a PCI bus based parallel system whose processors consist of Multi Chip Modules composed of National Semiconductor CLAy^{TM} FPGAs.

1 Introduction

The emergence of reconfigurable logic arrays has made it possible to create on the same physical hardware platform many different virtual computing architectures. In contrast to the conventional approach of adapting an application to an existing architecture, we can now mold the architecture to conform to each application's unique requirements. Typically the hardware description of an application is loaded into the reconfigurable logic array, and then the application is run to completion. A different application can then be loaded and run.

New FPGAs being designed and built such as the National Semiconductor CLAy devices carry this flexibility even further by allowing the architecture to change dynamically as an application runs. With a partial reconfiguration capability combined with high speed configuration memory close to the chip, an unlimited number of gates becomes available.

Using dynamic reconfigurability effectively requires sophisticated synthesis tools. We are developing high level procedural language as a preferable alternative to hardware description language for programming FPGAs. This paper describes research in progress to apply and extend the compiler technology developed in conjunction with the Splash 2 project ([JA92], [GS95b]) to dynamically reconfigurable systems. Our research addresses the problem of automatically partitioning a parallel program to exploit dynamic reconfiguration.

Our compilation strategy in mapping the high level data-parallel bit C (dbC) ([SG93], [GS95a]) to the Splash system has been to create a customized instruction set for each dbC program. Due to limitations of static reconfiguration, we accumulate all instructions used throughout the program into a single virtual processor. This virtual processor is then replicated on an FPGA chip to the resource limit of the chip.

Dynamic reconfigurability allows us to create an instruction set at a smaller granularity than the entire program. In particular, we partition the program on subroutine boundaries, and create a customized instruction set for each subroutine. This partitioning was chosen for two reasons. First, a function scope is the logical partitioning unit in the programming language. Thus (in a well-written program) logically related code can be found grouped within a subroutine. Second, the programmer can use the function scope to control partitioning.

Dynamic reconfigurability at the function level is particularly synergistic to dbC's data parallel model in which a collection of processors, one or more to a physical chip, execute in synchrony the instructions broadcast from the host: since each processor can get a new instruction set upon function entry, the amount of space occupied by a virtual processor on the chip can be smaller. This, in turn, allows more virtual processors to be packed on a chip, and thus larger problems can be run on the FPGA-based system than with a purely static configuration.

The remainder of the paper is organized as follows. The next section describes related work. Next we discuss the dbC language and compiler, followed by an overview of the NAPA board. The next section illustrates our function-level partitioning strategy with an application. We compare dynamic and static reconfiguration with regards to the application, a systolic DNA sequence match program. We end with conclusions and future work.

2 Related Work

Another project which is concerned with dynamic instruction set creation and management is the BYU DISC system ([WH95]). With this system, the logic designer creates via conventional CAD tools an "instruction" which conforms to the requirements of the DISC support library. The DISC environment loads instructions on demand, managing the difficult problems of placement and partial reconfiguration. Our research, working from the high level programming language, is synergistic to DISC, and we are currently exploring with BYU the feasibility of combining our front-end technology with the DISC back end.

Many of the same concerns that motivate our work are evidenced in [LM93], which implements a genetic data base search utilizing run-time reconfiguration on a single Xilinx 3090. Our work differs in that we have developed a general automatic compilation system. The genome problem is merely one example of its use. Also, in contrast to the Xilinx chip, our CLAy platform directly supports dynamic and partial reconfiguration.

The PRISM-II team [WAL+93] focus on the problem of synthesizing efficient logic for indefinite iteration while-loops and other control constructs. Our system is concerned with execution unit synthesis, with the control flow being directed by the general purpose host processor.

3 data-parallel bit C (dbC)

dbC is a high level parallel programming language which has been ported to a diverse range of machines ([GS95a]) including bit-serial Single Instruction Multiple Data (SIMD) arrays, workstation clusters, and FPGA-based processor arrays. dbC is *not* a C-like hardware description language. It is a true procedural language offering the programmer the computational model of an abstract SIMD processor array (see Figure 1).

In this model, a host processor issues a single instruction stream to an array of Execution Units (EU) which perform each instruction on local instances of data in lock step. Sequence control such as stepping to the next instruction, (conditional) "jump," and function call is performed by the host processor. The EUs merely execute the instruction issued by the host. The EUs are "virtual" processors in the sense that on many of the dbC platforms multiple EUs are mapped onto a single physical node.

Data-dependent conditional execution is accomplished by giving each EU a mask or "context bit." A masked instruction is executed by all processors having the context bit on, while the remaining processors wait. Unmasked instructions are executed by all processors regardless of the state of the context bit. In addition to local computation, the EUs can cooperatively perform certain global computations which reflect the state of the entire processor array. In these combining operations, commonly referred to as *Reduce* operations, each EU contributes data which is accumulated into a single result and returned to the host. The host can also read and write data to an individual processor or to the entire array. The processors are connected in a fixed interconnection topology, and can communicate data to their adjacent neighbors. In this work, we assume a linear interconnection topology.

These aspects of the data parallel model are reflected in dbC. The programmer configures a virtual processor array by initializing pre-defined variables `DBC_net` (number of dimensions) and `DBC_net_shape[DBC_net]` (size of each dimension). As Figure 1 illustrates, the programmer allocates data on the EUs by adding the `poly` modifier to the normal C data declaration. This allocates one instance of the variable on each EU. Normal C expression syntax is used to compute with poly variables. The context bit is controlled indirectly by use of the block structured `where` construct, which is similar in syntax to the conventional C `if`. Use of the `where` is illustrated in Figure 5. The `all` construct is used for unmasked execution. Infix operators such as <? = are overloaded to indicate reduction operations (in this case, the operator is "min", and the

expression <? $= p$ returns to the host the minimum value of p among all active EUs). Compiler intrinsics are used to initiate host-processor communication and interprocessor communication.

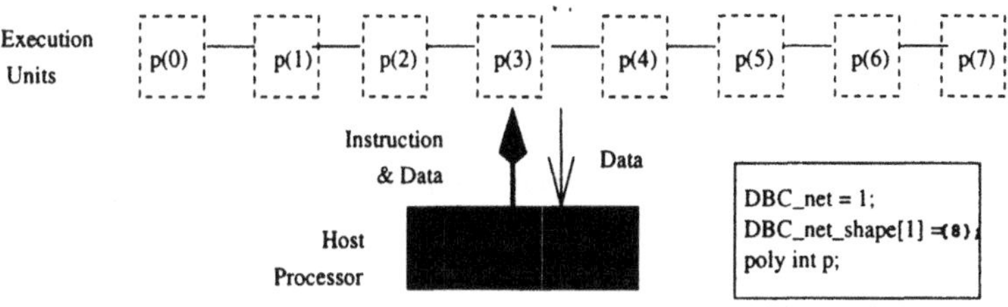

Fig. 1. A SIMD Array

4 dbC Compiler

Figure 2 shows the flow of compiler phases to target statically reconfigurable logic arrays. The dbC source is processed by the compiler front end, which generates compiler Intermediate Format (IF). The other dbC platforms (conventional SIMD array and workstation cluster) directly interpret the IF. To target reconfigurable logic arrays, we process the IF through additional phases. The compiler backend generates a VHDL description of the Execution Units. In addition it generates a C program which runs on the host processor. The C program sequences the EUs and performs any other sequential processing specified in the dbC program. The VHDL is then processed by commercial CAD tools to ultimately generate a configuration bit stream specific to the FPGA technology of the processor array.

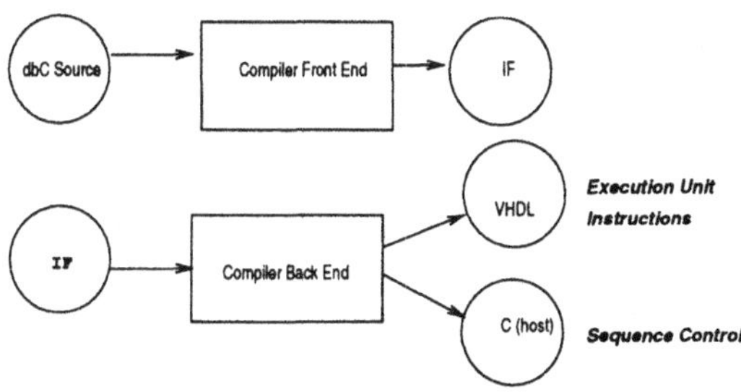

Fig. 2. Compiler Phases for Statically Configured Processor Array

Figure 3 shows the compiler flow for dynamically reconfigurable logic arrays. Here, the compiler emits a separate VHDL source module for each subroutine of the dbC program and a file listing the global variables of the program. Each VHDL module is processed by the CAD tools to generate a self-contained configuration bit stream. The CAD tools use the global symbol information to ensure that those registers are placed in the same fixed locations on the chip. Thus when, for example, F1 is swapped out and F2 is swapped in, the global variables as seen by F1 are in the same locations as those seen by F2. This preserves the global variables' state is across reconfiguration.

In addition to the individual VHDL source and the global symbol list, the compiler generates commands in the host C program to reconfigure the processor array when control reaches a new function.

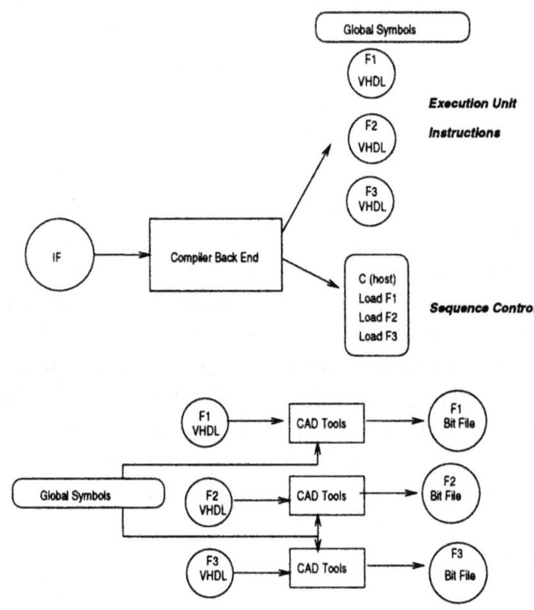

Fig. 3. Compiler Phases for Dynamically Configured Processor Array

5 The NAPA Architecture

Our static compiler targets the Splash 2 processor array. To explore the potential of dynamic reconfiguration, we have re-targeted the static compiler to National's Adaptive Processor Architecture (NAPA). The NAPA processor array (see Figure 4) contains up to 4 ring-connected (64-bit interconnect) Adaptive

Processing Units (APU). Each APU consists of a Field Configurable Multi Chip Module (FCM) and 4 banks of 128×16 SRAMs. The processor array controller, the Adaptive Control Unit (ACU), is composed of an FCM and a single bank of SRAM. Dedicated interconnect (8-bit) from the ACU to each APU is used for reconfiguration and is available to the application. Global resources include 32MB of DRAM and a 64-bit global bus shared by the ACU and APUs. The processor array is connected to the host via a PCI bus interface.

Each FCM contains four National Semiconductor CLAy^{TM} 31 "tiles" arranged in a 2×2 array. The FCM features over 20,000 useable gates and 432 user I/O. It is packaged in a 625 pin Ball Grid Array with 448 inter-module connections per device, greatly reducing inter-tile routing delay.

The ACU is reconfigured from the host via the PCI bus. An APU may be reconfigured from the host, from the DRAM under ACU control, or from the ACU SRAM, again under ACU control. 8-bit parallel or 1-bit serial reconfiguration modes are available.

We map the dbC abstract programming model onto NAPA by packing Execution Units onto each FCM. Thus if the dbC program requires 64 EUs, we allocate 16 per FCM. Each EU contains registers to hold the dbC poly variables and function units to perform the dbC program operations on the poly variables.

If the dbC program contains global operations, the logic to perform the global operation is distributed between the array FCMs and the control FCM. For example, for a reduce min operation, each APU does a local reduce min over its EUs and sends the local min to the ACU via the dedicated ACU–APU interconnect. The ACU, then does a global min over the four APUs and returns the result over the PCI bus to the host.

Nearest neighbor communication can be accomplished over the APU-to-APU interconnection ring. The Global Bus is used for Host-to-EU communication.

6 Exploiting Reconfigurability in the Genome Sequence Match Problem

Our compiler strategy to automatically exploit dynamic reconfigurability is to put each dbC function into a separate configuration. We have tested this methodology with an application which was previously compiled for the Splash 2 processor array. In this section, we briefly describe the algorithm, and then compare the static Splash 2 version with the dynamically reconfigured NAPA version.

Sequence Match A well-known Grand Challenge problem, the genome sequence match application searches for a match in a large data base between known sequences and a newly generated sequence. Our implementation uses a systolic dynamic programming algorithm, which is described in detail in [Hoa93]. The heart of the algorithm is the computation of *edit distance*, a measure of the similarity of the target streams (from the genome data base) to the source stream (spread across the EUs). The dbC source code for the edit distance computation is shown in Figure 5. Each EU performs this computation on its own instances

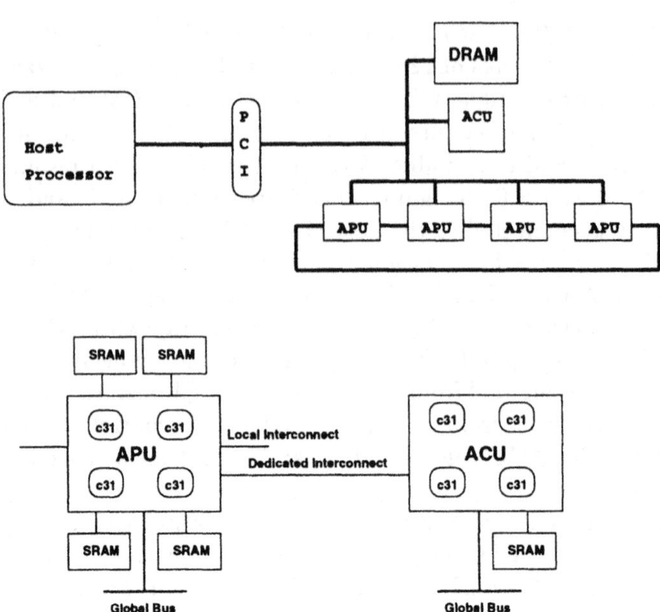

Fig. 4. National's Adaptive Processor Architecture (NAPA)

of the DNA characters `source_stream` and `left_char`, and updates its instance of the edit distance counter `distances`.

Reconfigurability The sequence match program is logically partitioned into three phases. The first phase loads the processor array with the new sequence, one DNA character per Execution Unit. The second compute phase streams the target stream through the linearly connected EUs and does the edit distance computation of Figure 5. The final phase reads data back from the EUs. It is called after the entire data base has been streamed through the EUs to recover the distance counters. This final print phase also functions as a debug routine which prints out intermediate values. It can be called from within the dbC program or explicitly through a conventional debugger running on the host. In any case it is called once at the end of the program.

The compiler flow for this program is shown in Figure 6. Each of the 3 functions is compiled down to a separate bit file[1]. The host C program loads each configuration before it calls the function. `LOAD_CONFIG` is simply a macro which loads the configuration *if it is not already loaded.* If the configuration is already loaded, no action is required. Thus the `Compute_Distances` configuration is loaded only once in the program. An obvious compiler optimization which we

[1] In order to simplify the presentation, we do not show the ACU configuration files. In reality, each APU configuration has its matching ACU configuration. Both must get loaded before the function can begin.

```
    where (source_stream & left_char){
      where (distances +1 < prev_distances)
        distances++;
      else
        distances=prev_distances;
    } else {
      where (distances>prev_distances+1)
        distances=prev_distances+2;
      else
        distances++;
    }
    where (distances>left_dist+1)
      distances=left_dist+1;
}
```

Fig. 5. dbC Source Code Computing Edit Distance

will include in the future is to move the call `LOAD_CONFIG Compute_Distances` (a loop invariant step) outside the while loop, so that configuration check at every iteration is eliminated.

Results On Splash, we were able to achieve a maximum of four Execution Units per Xilinx chip. The EUs contained load and compute functions, but not the `Print_Array` due to lack of space on the chip. Thus we used the print function only during simulation[2]. State readback was used to recover the counter values.

To compile for NAPA, we put each function into a different configuration. The limiting factor is the amount of logic required for the compute function. Due to the unavailability of the NAPA board and the FCM partitioning tools (they are still under construction), our experiments have been done in the Splash 2 environment. Compiling just the compute function gives us 5 EUs per chip, an increase of 25%. This improvement is above and beyond the benefit of being able to run the print function on the hardware, which we cannot do in a statically reconfigured system due to lack of space. The cost at runtime for reconfiguration is negligible in this example, as only three reconfigurations are done, each taking on the order of microseconds.

7 Conclusions

We have built a system which maps a high level data parallel program onto dynamically reconfigurable logic arrays. The compiler partitions the program along function scopes, creating a different configuration bit file for each function. Our

[2] The dbC system includes a simulator which runs on workstations, so that a program can be debugged independently of the parallel machine.

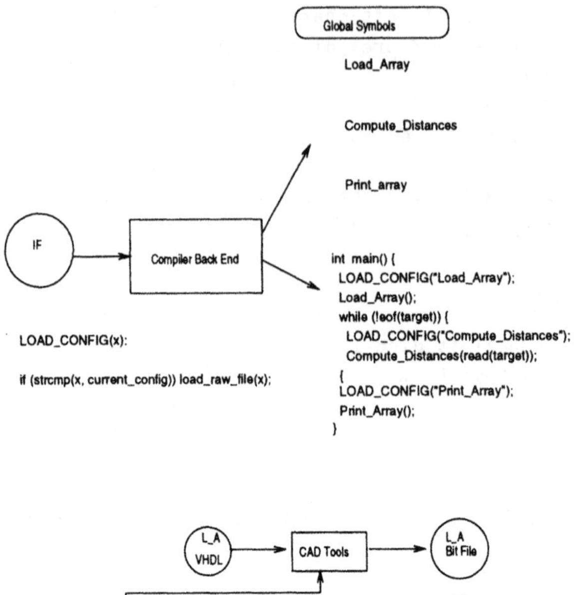

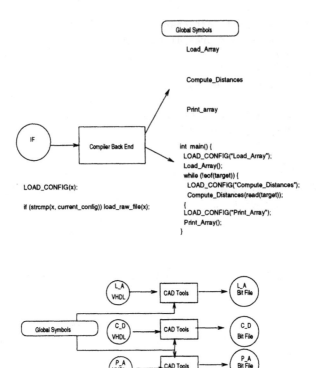

Fig. 6. Compiler Flow for Genome Sequence Match

preliminary experiments show that this approach to dynamic reconfigurability yields some significant advantages. In particular, when we partition temporally disjoint sections of the program into separate configurations,

- we can run larger problem sets (in the case of the genome sequence match, a longer source stream);
- we can run more phases of the problem (in the case of the genome sequence match, the print phase, which did not fit in Splash 2 version);
- by encapsulating debug code in a separate subroutine, we can benefit from debug functionality without performance penalty. The debug configuration is simply loaded when needed.

Our immediate work is to complete the port of the dbC run time environment, including the global reduce operations, to the NAPA board, whose final architecture has only recently been finalized. Longer term goals include more experience with applications and further compiler optimizations. We are also investigating extending our technique of creating customized instruction sets to other programming models.

Acknowledgements We gratefully acknowledge the support of National Semiconductor under ARPA contract DABT63-94-C-0085. We thank the Center for Computing Sciences (formerly Supercomputing Research Center) for the use of the Splash 2 programming environment and associated tools. Dzung Hoang wrote the dbC version of the DNA pattern match application. Brian Schott wrote the dbC runtime system.

References

[GS95a] M. Gokhale and J. Schlesinger. A Data Parallel C and its Platforms. *Frontiers '95*, February 1995.

[GS95b] M. Gokhale and B. Schott. Data Parallel C on a Reconfigurable Logic Array. *Journal of Supercomputing*, September 1995.

[Hoa93] D. T. Hoang. Searching genetic databases on Splash 2. In D. A. Buell and K. L. Pocek, editors, *Proceedings of IEEE Workshop on FPGAs for Custom Computing Machines*, pages 185–191, Napa, CA, April 1993.

[JA92] E. Davis J. Arnold, D. Buell. Splash-2. *Proceedings, 4th Annual ACM Symposium on Parallel Algorithms and Architectures(SPAA '92)*, 1992.

[LM93] E. Lemoine and David Merceron. Run time reconfiguration of fpga for scanning genomic databases. *FCCM '93*, April 1993.

[SG93] Judith Schlesinger and Maya Gokhale. dbC Reference Manual. Technical Report TR-92-068, Revision 2, Supercomputing Research Center, 1993.

[WAL+93] M. Wazlowski, L. Agarwal, T. Lee, A. Smith, E. Lam, P. Athanas, H. Silverman, and S. Ghosh. PRISM-II compiler and architecture. In D. A. Buell and K. L. Pocek, editors, *Proceedings of IEEE Workshop on FPGAs for Custom Computing Machines*, pages 9–16, Napa, California, April 1993.

[WH95] M. J. Wirthlin and B. L. Hutchings. A dynamic instruction set computer. *Proceedings of IEEE Workshop on FPGAs for Custom Computing Machines*, 1995.

Prototyping Environment for Dynamically Reconfigurable Logic

Patrick Lysaght, Hugh Dick, Gordon McGregor, David McConnell and Jonathan Stockwood

Dept. Electrical and Electronic Engineering
University of Strathclyde
204 George Street
Glasgow G1 1XW

email: p.lysaght@eee.strath.ac.uk

Abstract

A prototyping environment specifically designed for research into dynamically reconfigurable logic is presented. The system provides for the rapid prototyping of designs which exploit the intrinsic dynamic reconfigurability of certain types of FPGAs. The development environment is universal in that it can accommodate a range of current and future FPGA architectures with only minor modification and also provides a flexible interface to host computer platforms so that it can be used with all of the major CAD tools.

1. Introduction

Dynamically reconfigurable systems represent one of the newer and more innovative areas of current FPGA research. Such systems are characterised by their ability to reconfigure subsets of their logic and routing resources at run-time while continuing to operate normally. This paper presents the design and development of a new, universal prototyping system for the investigation of dynamic reconfiguration at the device level. Dynamic reconfiguration at this level is defined as the selective updating of a subset of a single FPGA's programmable logic and routing resources while the remainder of the device's programmable resources continue to function without disruption. The class of FPGAs that have this capability is called dynamically reconfigurable logic.

Dynamic reconfiguration at device level is effectively the combination of partial and run-time reconfigurability within a single FPGA. The term partial reconfigurabilty is sometimes used as a substitute for dynamic reconfigurability [Hadley95]. In this paper partial reconfigurability is used to describe the ability to reconfigure subsets of a device while device operation is suspended. Reconfiguration latency may be defined as the delay between a request for reconfiguration and the successful completion of that reconfiguration. An FPGA that is not partially reconfigurable has a constant reconfiguration latency. Since the latency is typically a function of device capacity, as FPGAs improve the latency becomes proportionately longer. The introduction of partial reconfigurability makes the latency approximately proportional to the size of the circuitry that is being changed. The operation of all of the circuitry on the FPGA, however, must still be suspended during the reconfiguration intervals. With larger devices, this implies that greater amounts of circuitry are inactive during reconfiguration, despite the reduction in latency. This effect is compounded by the fact that the opportunities to deploy reconfiguration typically increase with the size of the application. The advantages of device level dynamic reconfiguration will thus be more significant as FPGAs become larger, since only those circuits that are to be changed will be affected by dynamic reconfiguration.

Previously reported development systems, such as Splash 2 [Arnold93] or Virtual Computer [Casselman93], typically provide environments for implementing reconfigurable applications at system level and are designed around particular FPGA device families. The system reported in this paper, in contrast, is universal because it has been designed to support the investigation of any family of FPGAs that is dynamically reconfigurable. Much emphasis has been placed on providing a reusable tool to enhance the life-time of the prototyping environment. The system has been designed to be easy to adapt and extend to ensure that is capable of meeting future requirements.

The remainder of the paper is organised as follows: section two reviews the motivation for the development of the prototyping system and also describes the essential requirements that it has to fulfil. The third section introduces the overall system architecture. This is followed in the next section by a detailed examination of its two principal subsections, the host and target interfaces. Conclusions and potential improvements to the system, resulting from more than a year's experience of using it, are presented in the final section.

2. Prototyping Environment Requirements

The motivation for the design and development of the prototyping environment was to facilitate research into dynamically reconfigurable logic. In particular, the aim of the work is to identify where dynamic reconfiguration may be used to advantage in digital circuit synthesis. The prototyping environment is used to test new circuit designs and to evaluate their performance. This empirical approach, in tandem with other related work, forms a larger programme of research whose aim is to investigate all aspects of dynamically reconfigurable logic. Among the topics being studied are: algorithms and applications that are amenable to dynamic reconfiguration [Lysaght94a], [Lysaght94b], new FPGA and system architectures, system analysis and performance measurement, design methodologies, and design automation [Stockwood95].

It was necessary to create a new system because no satisfactory alternative was found. Although there are a wide variety of development systems available, none was suitable for our purposes. Most of the development systems marketed by FPGA vendors have limited functionality and are more suitable for demonstration purposes and the implementation of relatively simple applications. Very few boards using dynamically reconfigurable FPGAs have been reported and virtually all of these have been produced by the FPGA manufacturers themselves. This lack of choice is not altogether surprising since, at present, dynamically reconfigurable FPGAs command very little market share. Systems designed for reconfigurable computing applications, such as Splash 2 and Virtual Computer, are not appropriate for our requirements since they are reconfigurable *systems* and typically do not use dynamically reconfigurable FPGAs. Both of these examples use Xilinx Logic Cell Arrays which are not dynamically reconfigurable. (Note that, in general, a dynamically reconfigurable system does not require the use of dynamically reconfigurable FPGAs.)

The two main platforms for running FPGA CAD software are Personal Computers (PCs) and UNIX workstations. An early requirement of the development system was that it could be interfaced easily to either type of computer. Thus if the choice of a specific software tool dictated the use of a particular machine, it would be easy to connect the new host to the development system. A bi-directional interface was specified to allow configuration and control data to be sent from the host to the target and to allow status information and computation results to be transferred back from the target to the host.

A further requirement was that the target FPGA system could be operated either under the control of a local microprocessor or in stand-alone mode. The local microprocessor would be embedded in the target system where it would provide considerable flexibility for rapid prototyping when experimenting with new applications of dynamically reconfigurable logic. Its inclusion created a need for suitable software tools to develop programs for it and also an efficient means of downloading software into it.

As dynamic reconfiguration is an area of active research, it was accepted that the requirements for the system were likely to change with time and with increased experience of its use. Moreover, new families of dynamically reconfigurable FPGAs are anticipated in the future. Both these considerations demand a flexible and extensible prototyping platform that can be adapted to a rapidly evolving set of hardware requirements. Much emphasis was placed on providing a reusable tool to maximise the life-time of the prototyping environment.

3. System Overview

In this section the architecture of the current prototyping environment is presented. The following section addresses how this choice of architecture meets the specific design requirements which have just been described. Figure 1 shows a block diagram of the major components of the prototyping system. The host computer, which by default is a PC, is shown on the left and the target system is shown on the right.

The interface between the host and the target parts takes the form of a serial link between two INMOS Transputers. A B008 motherboard, incorporating a T805 Transputer, occupies one ISA bus interface slot in the PC. At the target end, a B430 prototyping Transputer Module (TRAM) with a T222 Transputer and 32K bytes of RAM, functions as the embedded microprocessor.

The T805 Transputer and the B430 TRAM communicate via the bi–directional INMOS serial link, which can support data transfer rates of up to 20 Mbps. The reliability of host to target communication is enhanced by the presence of Transputer devices at both ends of the link, allowing test software to be used at any time to verify the link's integrity. An additional benefit of the use of the Transputer link is that the prototyping system can be situated at a maximum distance of approximately one metre from the host computer. This allows the designer to arrange his computer, development system, power supplies and other tools, such as a logic analyser, more easily on the workbench.

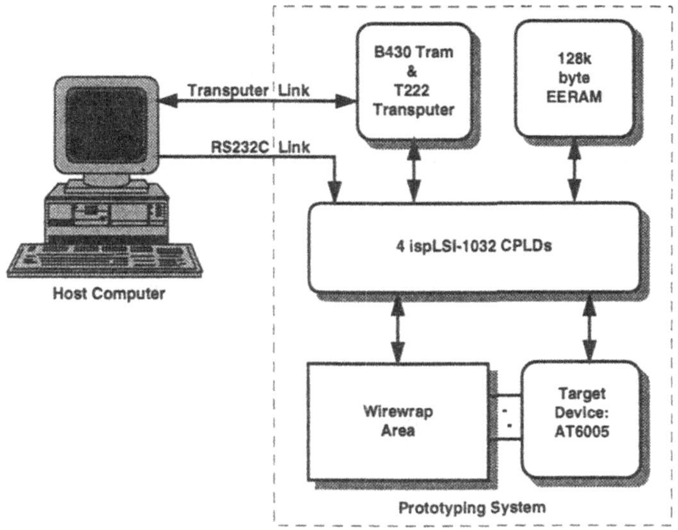

Figure 1 - Block Diagram of Prototyping System

The system supports a range of FPGAs and configuration modes through the use of four Lattice ispLSI–1032 Complex Programmable Logic Devices (CPLDs). The architecture of these devices is very suitable for implementing the logic to control the FPGA's configuration interface. The CPLDs are non–volatile, electrically erasable integrated circuits and are configured in–circuit via the PC's RS232C serial interface. The Lattice devices are customised to provide the interface for each new FPGA. The current board supports a single Atmel AT6005.

To enable stand-alone operation of the target FPGA, a byte wide, 128K non-volatile RAM is provided for storing configuration bit-streams. The CPLDs provide the interface and memory access control between the target device and the memory, and between the memory and the host platform via the Transputer link.

The actual prototyping board was manufactured as a 28 cm by 25.5 cm, four-layer printed circuit board and was assembled in-house. The physical layout of the board is shown in Figure 2. It does not include the B430 TRAM, which is connected directly to it via a short parallel cable. The original intention was to connect the two parts via the standard TRAM mounting. However a need for parallel data transfer and memory-mapped access to the embedded system meant that the limited number of outputs and the serial data path available with the standard interface were inadequate.

Four ispLSI-1032 CPLDs were selected even though devices with higher logic capacities were available during the design phase. These larger devices were rejected because they would have required the use of surface mount technology. Their use would also have resulted in a loss of flexibility due to the reduction in the overall number of input/output pins.

A wirewrap area of 14.5 cm by 7 cm is provided on the PCB to allow more target FPGAs and support logic to be integrated into the system. This logic can be interfaced with the CPLDs and the other components of the system, using wirewrap

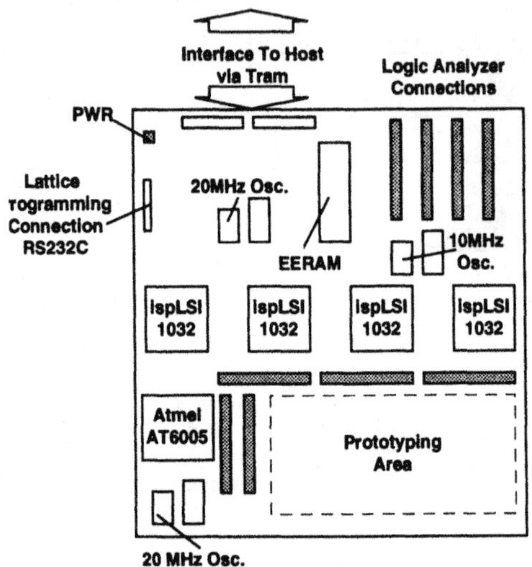

Figure 2 - Physical Board Layout

connections. The board also makes special provision for debugging via dedicated connections for logic analyser pods. Any signal may be monitored by routing it to these observation points either by programming the CPLDs or by wirewrapping the necessary connections.

4. System Architecture

In contrast to existing FPGA development systems, which are typically designed for one family of FPGA devices, the prototyping environment is *universal*, providing support for potentially any interesting FPGA. The design of the board was influenced by the design of the "user-definable emulators" offered with Hewlett Packard's universal development systems for microprocessors. An emulator consists of a stable host interface and an interface to the target that is customised for a particular microprocessor architecture. Since the responsibility for customisation of the emulator lies with the user, the emulators are called user definable. The prototyping system is similarly separated into two parts; a consistent host-system interface and a universal interface to the target FPGAs. Both of these interfaces are now explored.

4.1. Host Interface

The host interface is provided by the two Transputers within the system. Transputers can be connected together to form a fast and expandable parallel computing resource, allowing multiple prototyping boards to be connected together. However, the primary reason for this choice of processor was the intrinsic support, in hardware and software, for high–speed serial link management. Existing FPGA development systems that are physically separate from the host computer, are often coupled to the host with non-standard, custom–designed interfaces. Other systems, that are housed inside the host computer, are tied to a particular architecture by the use of a system-specific bus interface.

In contrast, the system reported in this paper employs host and target Transputers connected via a standard INMOS serial link to provide a proven and reliable "building block" solution to the interface problem. This approach effectively minimises hardware and software development time and thus accelerates progress towards more important research objectives. Plug-in Transputer cards are available for most popular computers, allowing the system to be used with all of the common CAD tools. As different FPGAs are used, this independence has the benefit of allowing the best tools to be used for a given device.

The main disadvantage of this approach is the additional cost introduced by the use of Transputers. It has been assumed that if a new computer platform is needed, the cost of purchasing a new Transputer motherboard and software for that particular computer will be borne by the user. The cost of motherboards for workstations, in particular, can be a significant overhead. There is an alternative approach, however, that overcomes this potential problem. If the PC that houses the Transputer motherboard is equipped with a network interface card to access a Local Area Network (LAN) and a PC version of the Network File System (NFS) software, the design files can be shared effectively via the virtual file system. The user having created his files on the workstation can access them from the PC and download them into the development system.

This arrangement can be further improved by running an enhanced version of the Transputer server software [Murphy94]. The free software effectively allows the server to be controlled remotely over the network. The user can control the prototyping system directly from the workstation environment. The PC thus becomes a dedicated server for the FPGA prototyping system. This is an ideal application for older PCs that are no longer useful for day-to-day computing applications. In many situations, this solution would be more affordable than the alternative of purchasing new Transputer equipment. Whichever option is chosen, the development of software to control the downloading of configuration data across the serial link has proved to be a relatively simple task using the comprehensive set of ANSI C development tools, supplied by INMOS. Thus the Transputer meets all of the system requirements identified previously; that of a reliable link, host independence, local microprocessor support and standard programming language support.

4.2. A Universal FPGA Interface - Hardware Device Drivers

Perhaps the most distinguishing feature of the development system is the fact that it is a universal prototyping platform which supports experimentation with any FPGA architecture. The requirement for universality implies that the interface must be modified to support each new target device. These changes occur infrequently and are easily accomplished because of the high degree of similarity between FPGA interfaces. The CPLDs implementing the interface are non-volatile parts that can be programmed up to 1,000 times reliably, which is an acceptable upper limit in this application. Their non-volatility greatly simplifies the boot sequence of the prototyping board, since it avoids the additional complexity of having to load the CPLD configurations prior to initialising the target FPGA(s).

In the microprocessor–controlled reconfiguration modes (see Figure 3a), the main purpose of the CPLDs is to function as a *hardware device driver.* The hardware device driver is similar in concept to its software equivalent, as found in many operating systems. Its purpose is to conceal any FPGA device–specific interfacing details, such as configuration control signal generation and routing of data pins, from the software part of the prototyping system. The encapsulation of the majority of the device–specific functions in this way allows the interface to the host computer to be standardised and kept independent of the particular target FPGA. Despite the use of the hardware device driver, a small amount of manual customisation is inevitable for power connections and to accommodate different device packages.

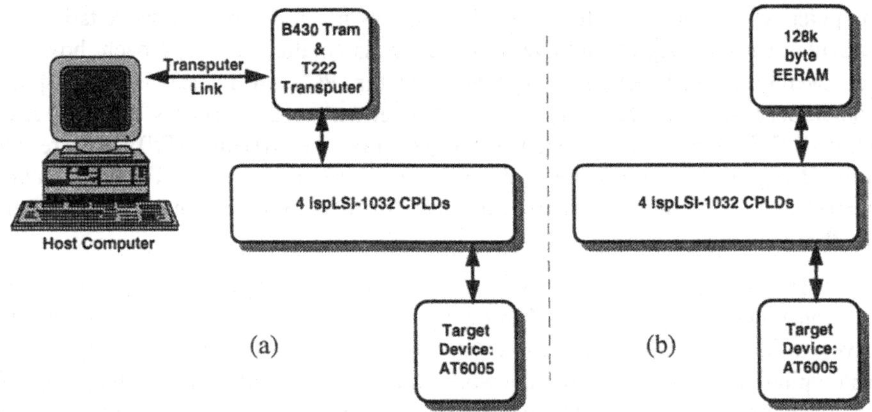

Figure 3: Implementation of a) microprocessor–controlled and b) stand–alone configuration modes

The decision to adopt a universal design was influenced by the current state of the FPGA market. At the time of writing this paper, the AT6000 series from Atmel Inc. are unique in being the only commercially available FPGAs that are dynamically reconfigurable. The present CAD tools supplied with these devices are first generation FPGA CAD tools, although new products are imminent. When combined with the absence of alternative, high quality tools from third-party vendors, experience has shown that application development is quite time consuming. The ability to rapidly provide prototyping support for newer and more advanced dynamically reconfigurable FPGA architectures, as they become available, is thus an obvious benefit.

One solution to shortening the design cycle for dynamically reconfigurable circuits has been to partition the logic such that the majority of the static sections are automatically synthesised on an FPGA with more mature CAD tool support, such as the Xilinx XC4000 family. Only those portions of the design that require to be dynamically reconfigured are then manually placed on the Atmel device [Lysaght94b]. Partitioning in this way results in an unusual hierarchy of three levels of reconfigurable devices consisting of the non-volatile CPLDs, the reconfigurable Xilinx parts and the dynamically reconfigurable AT6005.

At present, drivers have been produced for the Atmel AT6005, and the Motorola and Toshiba versions of the Pilkington Micro-electronics Dynamically Programmable Logic Device (DPLD) [Jones94].

4.3. Configuration Modes

The prototyping system supports a wide variety of configuration modes, enabled by the hardware device driver logic. For most applications the board operates as a slave to the host environment during the development and debugging stages. In this mode data can be retrieved from the system and stored or analysed by the host. However, there are restrictions on the maximum reconfiguration speed when using the AT6000 FPGAs in this way. Stand-alone operation is supported to allow hardware controlled reconfiguration, to achieve the best performance for advanced applications of dynamic reconfigurability.

When the board is used in stand–alone configuration mode, the CPLDs implement a fast DMA mechanism which allows the target FPGA to be programmed directly from the onboard EERAM (Figure 3b). Initially, the configuration data must be downloaded from the host to the RAM, via the Transputer link and CPLDs. Since the RAM is non–volatile, the prototyping system can subsequently operate without any microprocessor intervention. The main benefit of this mode is faster reconfiguration of the target FPGA, but the mode can also be used for demonstration purposes, without the need for the PC. Moreover, since the target Transputer is removed from the configuration control path, it can be used as a dedicated local processor.

5. Conclusions and Future Work

The development system has now been in use for over a year and has been interfaced to three different FPGAs. The first major revision of the board based on this experience is currently under development. Two principal enhancements to the system are planned. The first of these is the integration of a more powerful Transputer with greater system memory onto the prototyping board. The second is to absorb the programming interface for the CPLDs into the new, embedded Transputer. These improvements will remove the need for the separate TRAM and the RS232C connection.

One disadvantage of the present system is that the limited bandwidth of the Transputer link has proved unsuitable for some data-intensive applications that required post-processing by the PC. The need for a higher bandwidth link to the host will be significantly reduced by the inclusion of the more powerful local Transputer with increased local memory. As more powerful FPGA components become available it is likely that more of the processing will be carried out within the system, without reference to the host computer, which will further reduce the problem.

The provision for clocks derived from the system's crystal oscillators will also be investigated. The inclusion of a programmable clock generator, similar to that reported in the HARP development system [Page94], is being evaluated.

The use of several different reconfigurable FPGAs in the system brings with it the need for multiple CAD packages. Alternative, third-party synthesis tools, such as the Exemplar Logic's VHDL compiler, are being investigated to provide a single mechanism for design entry. The use of libraries for the different FPGA families will allow designs to be targeted at particular devices from the VHDL specification. This approach cannot be extended to the dynamically reconfigurable parts of the design because manual synthesis is currently the only option for circuits of this type [Lysaght94b]. A further observation is that current FPGA simulation tools provide limited support for designs that incorporate different families of FPGAs.

The system has now become stable enough to be used for undergraduate projects in addition to its normal use in postgraduate research. This new use has underlined the need for more comprehensive documentation, a more intuitive, graphical user interface and an on-line help facility. Finally, the system is quite complex to debug in the event of failure, so a self-test suite will be added.

The wider use of the system demonstrates its potential as a flexible tool for introducing people to dynamic reconfiguration, as well as being a platform for research. The progress to date indicates that the objectives in developing the prototyping environment are being achieved.

6. Acknowledgements

The authors gratefully acknowledge the EPSRC and the Defence Research Agency who support made this work possible. We also wish to express out thanks to Pilkington Micro-electronics Ltd. and Atmel Inc. for their assistance.

We are grateful to Dr. Eric Pirie for suggesting the term "hardware device driver" to describe the operation of the universal interface.

7. References

[Arnold93] J. M. Arnold, *The Splash 2 Software Environment,* IEEE Workshop on FPGAs for Custom Computing Machines, Napa, California, Apr. 93

[Casselman93] S. Casselman, *Virtual Computing and The Virtual Computer*, IEEE Workshop on FPGAs for Custom Computing Machines, Napa, California, Apr. 1993

[Hadley95] J. Hadley & B. Hutchins, *Design Methodologies for Partially Reconfigured Systems,* IEEE Workshop on FPGAs for Custom Computing Machines, Napa, California, Apr. 1995

[Jones94] G. Jones & D. Wedgwood, *An Effective Hardware/Software Solution for Fine Grained Architectures*, 2nd International ACM/SIGDA Workshop on Field Programmable Gate Arrays, California, 1994

[Lysaght94a] P. Lysaght, J. Stockwood, J. Law and D. Girma, *Artificial Neural Network Implementation on a Fine-Grained FPGA*, In, R. W. Hartenstein, M. Z. Servit (eds.) Field-Programmable Logic, Springer Verlag, Germany, 1994, pp. 421-431

[Lysaght94b] P. Lysaght & H. Dick, *Implementation of Adaptive Signal Processing Architectures based on Dynamically Reconfigurable FPGAs*, EUSIPCO, Edinburgh, UK, Sept. 1994

[Murphy94] B. Murphy, *An Enhanced Transputer Server Architecture*, 1994

file://unix.hensa.ac.uk/parallel/transputer/software/iservers/etsa/etsa.tar.Z

[Page94] I. Page, *The HARP Reconfigurable Computing System*, Oxford University Hardware Compilation Group, Oct. 1994

file://ftp.comlab.ox.ac.uk/pub/Documents/techpapers/Ian.Page/harp_desc.ps.gz

[Stockwood95] J. Stockwood & P. Lysaght, *A Simulation Tool for Dynamically Reconfigurable Field Programmable Gate Arrays*, ASIC, Austin, Texas, Sept. 1995

Implementation Approaches for Reconfigurable Logic Applications*

Brad L. Hutchings and Michael J. Wirthlin

Brigham Young University, Dept. of Electrical and Computer Engineering, 459 CB, Provo, UT 84602

Abstract. Reconfigurable FPGAs provide designers with new implementation approaches for designing high-performance applications. This paper discusses two basic implementation approaches with FPGAs: compile-time reconfiguration and run-time reconfiguration. Compile-time reconfiguration is a static implementation strategy where each application consists of one configuration. Run-time reconfiguration is a dynamic implementation strategy where each application consists of multiple cooperating configurations. This paper introduces these strategies and discusses the implementation approaches for each strategy. Existing applications for each strategy are also discussed.

1 Overview

Reconfigurable logic is an emerging branch of computer architecture that seeks to build *flexible* computing systems that can achieve very high levels of performance -much higher performance than is possible with the highest performance microprocessors, or in many cases, even supercomputers. At the heart of these computing systems is the SRAM-based field-programmable gate array (FPGA), an integrated circuit that consists of a large, uncommitted array of programmable logic and programmable interconnect that can be easily configured *and reconfigured* by the end user to implement a wide range of digital circuits. These FPGA-based systems achieve high levels of performance by using FPGAs to implement custom algorithm-specific circuits that accelerate overall algorithm execution. Yet, unlike other custom VLSI approaches, these systems remain flexible because the same custom circuitry for one algorithm can be *reused* as the custom circuitry for a completely different and unrelated algorithm. It is this ability to remain flexible while also being able to implement custom, algorithm-specific circuitry that is fueling interest in reconfigurable logic as a new paradigm for high-performance system design.

A number of significant applications have demonstrated promising performance improvements by replacing software computing with customized logic [1]. Most of these applications are developed as a static hardware configuration that remains on the FPGA for the duration of the application. This implementation

* This work was supported by ARPA/CSTO under contract DABT63-94-C-0085 under a subcontract from National Semiconductor.

approach, referred to as Compile Time Reconfiguration (CTR), involves the development of discrete hardware images for each application on the reconfigurable resource. Because hardware resources remain static for the life application, conventional design tools provide adequate support for application development.

To further take advantage of the flexibility and performance gained by using FPGAs, some applications reconfigure hardware resources *during* application execution. By doing do, applications can optimize hardware resources by replacing idle, unneeded logic with usable performance enhancing modules. This implementation approach, called Run-Time Reconfiguration (RTR), provides a *dynamic* hardware allocation model. With RTR, applications may allocate hardware resources as run-time conditions dictate. Allowing dynamic hardware allocation, however, introduces a number of additional design problems that current design automation tools do not address. The lack of sufficient design tools and a well-defined design methodology prevents the wide-spread use of this technique. This paper will discuss the design issues involved with both CTR and RTR applications and provide application examples of each.

2 Compile-Time Reconfiguration (CTR)

Compile-time reconfiguration is the simplest and most commonly-used approach for implementing applications with reconfigurable logic. The distinctive feature of CTR applications is that they consist of a *single* system-wide configuration. Prior to commencing operation, the FPGAs comprising the reconfigurable resource are loaded with their respective configurations. Once operation commences, the FPGAs will remain in this configuration throughout the operation of the application as seen in Figure 1.

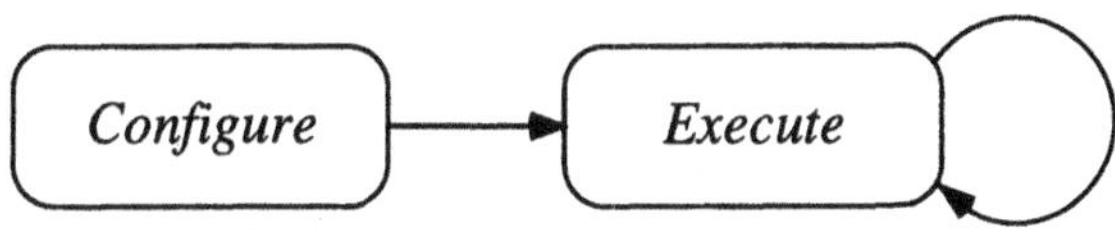

Fig. 1. Compile-Time Reconfiguration

This approach is very similar to using an ASIC for application acceleration. From the application point of view, it would matter little whether the hardware used to accelerate the application were an FPGA or a custom ASIC because the hardware remains constant throughout its operation. However, FPGAs provide an important cost advantage because the same silicon can be reused over and over again to implement any number of applications. This economy of silicon allows one, in effect, to implement a different ASIC for every application in a cost effective manner.

2.1 Implementation

Many of the steps used in developing CTR applications are similar to those used in designing ASICs. Designs are entered as either schematics using schematic capture, or as text when using a synthesis tool. Once the design is captured and verified, the CAD tool generates an intermediate vendor-specific netlist. FPGA-specific CAD tools process the netlist to create the final physical placement and interconnect of the circuit. The final step in this basic design process is to generate the configuration bitstream that will be downloaded into the FPGA.

The key similarity between conventional digital design and CTR application design is the static hardware allocation strategy. In both cases, hardware is assumed *static* for the life of the design (or application). Although FPGA technology certainly allows a more dynamic approach, CTR applications hold to the static approach for design simplicity. Because of the similarity with conventional design approaches, almost all reconfigurable logic applications are compile-time configured.

2.2 CTR Application Examples

With reconfigurable computing machines now becoming available, successful CTR applications have been developed and have demonstrated the viability of reconfigurable computing. Because of the widespread availability of FPGAs, many general and special purpose systems have implemented custom applications. This section will give a short review of some applications developed on two well-known reconfigurable systems: PAM [2] and Splash [3].

A number of the most successful applications on PAM utilize the ability of custom hardware to perform efficient long integer operations. Specifically, long multiplication was implemented for 2K-bit coefficients at 66 Mbits/s making it faster than any machine available to the developers (at least 16 times faster than the Cray II or Cyber 170/750) [4].

A genetic database search has been implemented on Splash-II with impressive results [5]. This application implements a well-known dynamic programming algorithm to compute the edit distance, a measure of similarity, between genetic sequences. The Princeton Nucleic Acid Comparator(PNAC) VLSI chip for this algorithm has been ported to Splash-I [6]. By pre-storing the source sequence into the system before operation, a new implementation written in VHDL improved PE utilization of the algorithm to near 100%. Packing 14 PEs into a single FPGA, a complete Splash system with 16 boards contains 3608 PEs. Peak performance of the system is approximately 43,000M cells updated per second (CUPS) for an increase of 1300 times over a MP-1, 7000 times over a CM-2 and 64,000 times over a 486DX-55.

3 Run-Time Reconfiguration (RTR)

Whereas CTR applications statically allocate logic for the duration of an application, RTR applications uses a dynamic allocation scheme that re-allocates hardware at run-time. Each application consists of *multiple* configurations per FPGA with each configuration implementing some fraction of the application. Thus whereas CTR applications configure the FPGAs once before execution, RTR applications typically reconfigure them many times during the normal operation of a single application as seen in Figure 2.

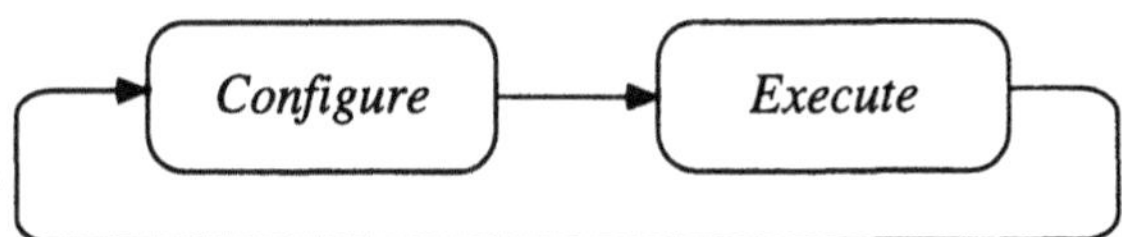

Fig. 2. Run-Time Reconfiguration

The dynamic nature of the hardware introduces two new design problems. The first is dividing the algorithm into *time-exclusive* segments that do not need to (or cannot) run concurrently. This process of dividing an algorithm into time-exclusive segments is referred to as *temporal partitioning.* Similar to structural partitioning where circuit elements are organized into strongly connected partitions, temporal partitioning is a process that organize circuit elements into strongly *concurrent* partitions, i.e., groups of circuit elements that must or tend to operate concurrently. Typically, an algorithm is temporally partitioned by breaking it down into distinct phases or operational modes. Each of these phases or modes is then designed as a distinct circuit module (FPGA configuration). These configuration are then downloaded into the FPGAs as required by the application.

The configurations that make up an RTR application should be organized such that they can remain resident for a reasonable amount of time. In addition, they should perform their task relatively independent of other configurations. Because all digital design tools assume a static hardware model, no tools support this partitioning step. Temporal partitioning currently requires tedious and error-prone user involvement.

The second design problem introduced by RTR systems involves coordinating the behavior between configurations of RTR applications. This behavior is manifested in the form of inter-configuration communication, or the transmission of intermediate results from one configuration to the next. This occurs during the normal progression of the application as each configuration will typically process data and then produce some intermediate result that serves as the input to a succeeding configuration. This has a tremendous effect on the design process because all of the configurations must be carefully designed such that they step through the various phases of an application and communicate intermediate results with other configurations.

There are two basic approaches that can be used to implement RTR applications: *global* and *local.* Both techniques use multiple configurations for a single application and both techniques reconfigure FPGAs during execution of the application. The principle difference between these two techniques is the way that the dynamic hardware is allocated. These differences and their impact on the design approach will be discussed in the sections that follow.

4 Global RTR

Global RTR allocates *all* (FPGA) hardware resources in each configuration step. More specifically, global RTR applications are divided into distinct temporal phases, where each phase is implemented as a single *system-wide* configuration that occupies all system FPGA resources. At run-time, the application would step through each phase by loading all of the system FPGAs with the appropriate configurations associated with a given phase. As an example, Figure 3 shows how a single application is made from three configurations: A, B, and C. The application executes by sequentially configuring and executing the three temporally disjoint hardware configurations.

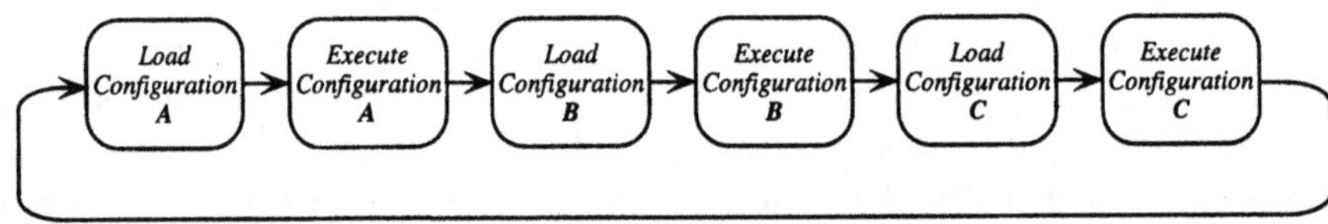

Fig. 3. Global Run-Time Reconfiguration

4.1 Implementation Issues

The designer's primary task when implementing global RTR applications is to temporally divide the application into roughly equal-sized partitions to efficiently use reconfigurable resources. It is important to realize, however, that the temporal partitioning step is iterative as it is impossible to know the physical size of the partition until it has been placed and routed. The designer must make an initial partition and then iterate both on the design and the partition until the sizes of the partitions are approximately equal. Simulation also presents a unique challenge as none of the currently available CAD tools can directly simulate the transition from one configuration to the next. The main disadvantage of this approach is the need for equally-sized temporal partitions. If it is not possible to evenly partition the application, inefficient use of FPGA resources will result.

Once an application has been temporally partitioned, a means must be provided to support inter-configuration communication. Because each phase occupies all reconfigurable resources, interfaces between configurations are fixed and all circuit modules can be designed under the same general context. Typically, applications are organized such that one phase of the application computes some intermediate result that is used by the next phase. During operation, the current configuration will terminate by placing its results in some fixed-function system resource, e.g., a memory or register. The next configuration, when loaded, will commence operation by reading this intermediate result and then proceed to compute the next result. This process of "configure and execute" continues until the application has terminated.

The main advantage of global RTR is its relative simplicity. Because the application partitions are coarse grained, conventional CAD tools (schematic capture, VHDL synthesis) can be used successfully once the manual temporal partitioning step has been completed. Each partition can be designed and implemented as one independent circuit module and the CAD tool is free to perform any global optimizations as necessary.

4.2 Extant Global RTR Systems

This section will describe two applications. Both applications are implemented as a series of sequentially-executed stages. Each stage is implemented as an overlay that reconfigures all FPGA resources each time they are loaded.

Hardware Acceleration of Image Processing In order to implement image processing systems without the costs and time of development associated with custom ASICs, the *im*puter (**im**aging - com**puter**) was developed based on FPGA technology [7]. The unique feature of this system is the ability to rapidly reconfigure the FPGA resources with distinct image processing algorithms during execution to compute successive algorithms in real-time. Because image processing algorithms can be partitioned into several well defined steps, the hardware implementation for each step can be independently specialized.

The imputer has implemented the following functions on its platform with significant speedup over competing systems: interframe difference(15x speed-up), Sobel filter(28x speed-up), and a threshold filter(9x speed-up). Even with reconfiguration times at 10% of total processing time, the imputer produces significant performance improvements.

RRANN RRANN was designed as a proof-of-concept system to demonstrate the effectiveness of using RTR in neural networks. It implements the popular backpropagation training algorithm as three time-exclusive FPGA configurations: feed-forward, backpropagation and update. System operation consists of sequencing through each of these three configurations *at run-time*, one configuration at a time. As one circuit module finished (indicating the completion of the corresponding stage), all FPGA hardware was reconfigured with the next stage's circuit module. RRANN demonstrated that RTR can increase the functional density of a neural network by 500% when compared to FPGA-based implementations that do not use RTR [8]. This density enhancement was obtained by eliminating idle circuitry from each stage and then implementing five additional neurons with the reclaimed FPGA resources. Additionally, once the neural network has completed the training process, the update and backpropagation configurations no longer need to be loaded and the FPGAs can remain in the feed-forward configuration. This eliminates the need to reconfigure and further enhances performance while maintaining the original density enhancements.

5 Local RTR

Local RTR takes an even more flexible approach to reconfiguration than does global RTR. As the name implies, these applications *locally* (or selectively) reconfigure subsets of the logic as the application executes. Local RTR applications may configure any percentage of the reconfigurable resources at any time; individual FPGAs may be configured or even single FPGA devices may themselves be partially reconfigured on demand. This flexibility allows hardware resources to be tailored to the run-time profile of the application with finer granularity than possible with global RTR. Whereas global RTR approaches implement the execution process by loading relatively large, global application partitions, local RTR applications need load only the necessary functionality at each point in time. This can reduce the amount of time spent down-loading configurations (smaller configurations take less time to download), and can lead to a more efficient run-time allocation of hardware. Figure 4 provides an example of a local RTR application. Between each execution stage, only a *sub-set* of the reconfigurable hardware is configured.

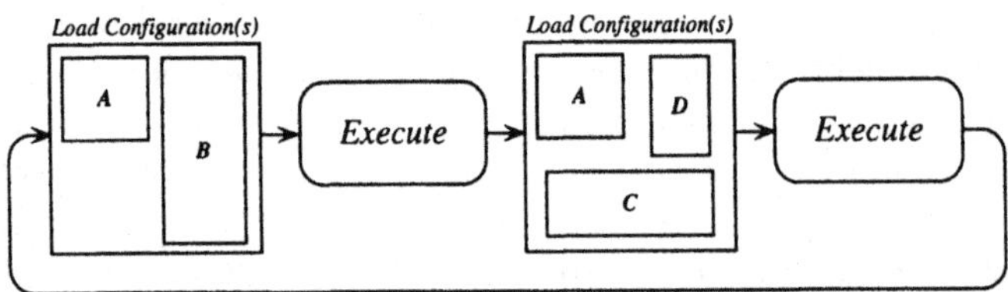

Fig. 4. Local Run-Time Reconfiguration

The organization of local RTR applications is based more on a *functional* division of labor than the phased partitioning used by global RTR applications. Local RTR applications are typically implemented by functionally partitioning an application into a set of fine-grained application operations. These operations need not be entirely temporally exclusive as many of them may be active at any time. This is in direct contrast to global RTR where only one configuration (per FPGA) may be active at any given time. Still, it is important to organize the operations such that idle circuitry is eliminated or greatly reduced. Each of these operations is implemented as a distinct circuit module and these circuit modules are then downloaded to the FPGAs as necessary during the operation of the application. Note that, unlike global RTR, several of these operations may be loaded simultaneously and each operation may consume any portion of the system FPGA resources.

5.1 Implementation Issues

The main advantage that local RTR provides over global RTR is the ability to create fine-grained functional operators that make more efficient use of FPGA resources. This is important for applications that are not easily divided into equal-sized temporally exclusive circuit partitions. However, this advantage can cause a very high design penalty because of the increased flexibility and complexity of the system. Unlike global RTR where circuit interfaces remain fixed between configurations, local RTR allows these interfaces to change with each configuration. For example, when circuit configurations become small enough for multiple configurations to fit in a single device, the designer is now forced to ensure that all configurations will *interface* correctly, one with another. Moreover, the designer may also have to ensure not only structural compliance but *physical* compliance as well. That is, when the designer creates circuit configurations that do not occupy an entire FPGA, they will have to ensure that the physical footprint of a configuration is compatible with other configurations that may currently be loaded. In addition, the designer must ensure that any physical circuitry is placed and routed such that inter-configuration communication is supported properly.

Currently available CAD tools are a very poor match for local RTR implementations. They do not allow the designer to simulate the transition between configurations nor do they allow designers to conveniently control the routing and placement of configurations in the context of other configurations. Of those available, only structural design (schematic capture) tools combined with physical place and route editors are appropriate for creating configurations that must interact with each other within a single FPGA.

5.2 Extant Local RTR Applications

The only known local RTR systems produced to date have been in the areas of neural networks and programmable processors. Lysaught et al. [9] reported on a design that emulated a large neural network by reconfiguring the network between layers. This allowed the Atmel FPGA to emulate a larger neural network than if it had been statically configured. Hadley and Hutchings [10] report on RRANN-2, a follow-on of the RRANN project that uses local RTR to implement the backpropagation algorithm. Whereas Lysaught's design reconfigured between layers, RRANN-2 reconfigures between algorithm stages in order to eliminate idle circuitry and allow additional concurrency. The main difference between RRANN-2 and the earlier RRANN project is the introduction of partial reconfiguration to reduce reconfiguration time to increase performance for smaller networks.

The Dynamic Instruction Set Computer (DISC) is a general processor architecture that uses local RTR to offset the limited resources available on an FPGA and allow an essentially infinite application specific instruction set [11]. Of particular interest is the way that DISC combines conventional software programming approaches with hardware design. Designers are free to implement arbitrary special-purpose instructions and then "glue" them together with other instructions (both general-purpose and special-purpose) as assembly and 'C' programs.

Implemented on a single FPGA, the processor is divided into two distinct subsections - the static global control module and the dynamic instruction space. The static control unit remains on the FPGA at all times to manages global resources such as memory, monitor communication channels and to store global state. The dynamic instruction space is reserved for dynamically allocated processor instruction modules that are paged into the processor as required by the control flow of a particular program.

Before a DISC application begins, the global control unit is loaded into the hardware and the dynamic instruction space is cleared. During program execution, instruction modules are *paged* onto the hardware space as dictated by control flow. When the hardware space has been filled, idle instruction modules are removed to allocate additional hardware for newer instruction modules. To reduce run-time configuration overhead, the FPGA is partially configured to load or remove only those instruction modules that are necessary.

A unique feature of DISC is that instruction modules can be placed *anywhere* along a one dimensional area in the dynamic instruction space. The ability to place a module anywhere in the dynamic hardware space provides the capability of maximized hardware utilization. This feature, known as relocatable hardware, allows physical placement of hardware resources to be determined at *run-time*.

A set of specialized image processing instructions was designed to show the performance advantages of replacing a series of general purpose instructions with a single high performance instruction. Specifically, a custom mean-filter instruction provided 80x speedup over the general-purpose instructions used to implement the same algorithm.

6 Conclusion

Several reconfigurable systems and applications have been developed to demonstrate the significant performance improvements possible with reconfigurable logic. To date most applications have been compile-time reconfigured where FPGAs are configured once for the duration of the application. Very few run-time reconfigured applications have reported. This is likely due to the additional difficulty required to design and implement such applications. In addition, the benefits of RTR have only recently been demonstrated.

Although significant progress has been reported, the field of reconfigurable logic appears to still be in its infancy. Most FPGAs are not completely suited to the process of run-time reconfiguration; configuration time is excessive and all internal state is typically lost when devices are configured. In addition, the only available commercial CAD tools have been developed for designing ASICs and do not help the designer to exploit the inherent flexibility of FPGAs. Additional work on CAD tools, system design, and FPGA-devices will need to commence before the advantages of reconfigurable logic can be fully realized in conventional digital systems. In addition, promising application areas still need to be developed.

References

1. S. Trimberger. A reprogrammable gate array and applications. *Proceedings of the IEEE*, pages 1030–1041, July 1993.
2. P. Bertin, D. Roncin, and J. Vuillemin. Programmable active memories: a performance assessment. In G. Borriello and C. Ebeling, editors, *Research on Integrated Systems: Proceedings of the 1993 Symposium*, pages 88–102, 1993.
3. J. M. Arnold, D. A. Buell, and E. G. Davis. Splash 2. In *Proceedings of the 4th Annual ACM Symposium on Parallel Algorithms and Architectures*, pages 316–324, June 1992.
4. M. Shand, P. Bertin, and J. Vuillemin. Hardware speedups in long integer multiplication. *Computer Architecture News*, 19(1):106–114, 1991.
5. D. T. Hoang. Searching genetic databases on Splash 2. In D. A. Buell and K. L. Pocek, editors, *Proceedings of IEEE Workshop on FPGAs for Custom Computing Machines*, pages 185–191, Napa, CA, April 1993.
6. D. P. Lopresti. Rapid implementation of a genetic sequence comparator using field-programmable gate arrays. In C. Sequin, editor, *Advanced Research in VLSI: Proceedings of the 1991 University of California/Santa Cruz Conference*, pages 138–152, Santa Cruz, CA, March 1991.
7. D. Ross, O. Vellacott, and M. Turner. An FPGA-based hardware accelerator for image processing. In W. Moore and W. Luk, editors, *More FPGAs: Proceedings of the 1993 International workshop on field-programmable logic and applications*, pages 299–306, Oxford, England, September 1993.
8. J. G. Eldredge and B. L. Hutchings. Density enhancement of a neural network using FPGAs and run-time reconfiguration. In D. A. Buell and K. L. Pocek, editors, *Proceedings of IEEE Workshop on FPGAs for Custom Computing Machines*, pages 180–188, Napa, CA, April 1994.
9. P. Lysaght and J. Dunlop. Dynamic reconfiguration of FPGAs. In W. Moore and W. Luk, editors, *More FPGAs: Proceedings of the 1993 International workshop on field-programmable logic and applications*, pages 82–94, Oxford, England, September 1993.
10. J. D. Hadley and B. L. Hutchings. Design methodologies for partially reconfigured systems. To be published in *Proceedings of IEEE Workshop on FPGAs for Custom Computing Machines*, 1995.
11. M. J. Wirthlin and B. L. Hutchings. A dynamic instruction set computer. To be published in *Proceedings of IEEE Workshop on FPGAs for Custom Computing Machines*, 1995.

Use of Reconfigurability in Variable-length Code Detection at Video Rates

Gordon Brebner[1] and John Gray[2]

[1] Department of Computer Science
University of Edinburgh
Mayfield Road
Edinburgh EH9 3JZ
Scotland

[2] Xilinx Development Corporation
52 Mortonhall Gate
Edinburgh EH16 6TJ
Scotland

Abstract. This paper shows how properties of the data produced by a particular compression scheme lead to an elegant paged logic implementation of a decoder. The implementation uses the new Xilinx XC6200 FPGA family, which supports dynamic use of programmable logic with special hardware. The work illuminates an eternal verity of computing that, when a resource is limited, its use can be extended by paging. In this case, the resource is logic.

1 The Group Three Fax Compression Scheme

In this work, properties of the data produced by a particular compression scheme are used to derive a paged logic implementation of a decoder. The compression scheme considered is the one used in the ITU-T (formerly CCITT) T.4 standard for group 3 fax transmission [3]. This combines run-length coding with Huffman coding [1]. The essence of Huffman coding is that it encodes symbols from an alphabet into binary codewords, with frequently occurring symbols having short codewords and less frequently occurring codewords having longer codewords. Given this, the encoding of a sequence of symbols is a binary string consisting of concatenated codewords. In the case of the ITU-T T.4 fax standard, the 'symbols' are the different lengths of alternating white and black runs in scan lines, and the binary codewords range in length from 2 to 13 bits. By this means, scan lines are encoded into bit streams that can be transmitted by fax apparatus.

The decoding problem is that codewords must be identified within a binary string. This is made possible by a key property of Huffman codes: no codeword is an initial substring of another. Thus, decoding can be performed by extracting codewords, in sequence, using a scan of the binary string from beginning to end. One method of detecting codewords in hardware is to allow the successive bits of the input to 'steer' a signal through a binary decoding tree, from the root to a leaf. The leaves of the tree represent codewords, and the arrival of a steered signal indicates the detection of a codeword in the input stream. Figure 1 shows a simple decoding tree for a code with the codewords **00**, **1000**, **1001** and **101**. Since decoding trees grow exponentially with the maximum size of codeword, there is usually a severe limitation to the size of a tree that can be built in fixed hardware or using PAL technology. However, if the hardware can be reprogrammed dynamically, it is possible to extend the steered signal through an arbitrarily large tree.

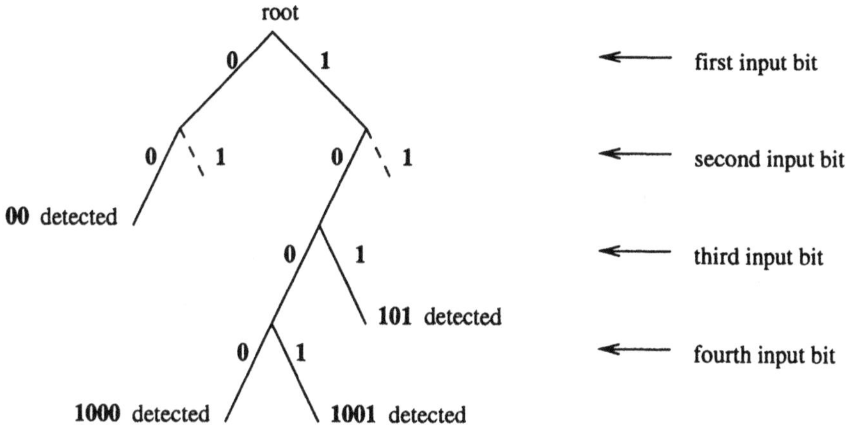

Fig. 1. Simple Decoding Tree

The systems context for designs of this type is usually a memory to memory transfer of data as compressed data is expanded. In this example, compressed fax data is being expanded for printing or bitmap manipulation. Binary strings from memory are decoded to give lengths of white and black runs in scan lines, which are then used to construct a bitmap.

2 Properties of Input Data

Group 3 fax apparatus operates at a normal resolution of 200 dots per inch. To accommodate paper of A4 width, a horizontal scan line contains 1728 pixels. Each scan line is usually handled independently, and consists of alternating runs of white and black pixels. The transmitted data for a scan line is an encoding of the sequence of run lengths. Since certain run lengths occur far more frequently than others, the T.4 fax standard defines two sets of variable-length codewords for the integers between 0 and 1728 (the possible run lengths), one set for white and a different one for black.

The codewords chosen in the T.4 standard were based on an analysis of 'typical documents'. Frequently observed run lengths were allocated short binary codewords, whereas less frequent lengths were given longer binary codewords. The allocation procedure used Huffman's algorithm to select codewords based on occurrence frequencies, in a way that ensures no codeword is an initial substring of any other. To illustrate the code, Table 1 shows the encodings of the first few run lengths, from 0 up to 10.

White run length	Codeword	Black run length	Codeword
0	00110101	0	0000110111
1	000111	1	010
2	0111	2	11
3	1000	3	10
4	1011	4	011
5	1100	5	0011
6	1110	6	0010
7	1111	7	00011
8	10011	8	000101
9	10100	9	000100
10	00111	10	0000100

Table 1. Codewords for Run Lengths 0 to 10

Since the data used by ITU-T to determine the choice of codewords was not readily available, experiments were performed to measure the frequency of different run lengths. Two types of 'typical document' were chosen: pages taken from a technical report, and pages taken from letters. The former had a higher density of black, and a more regular distribution of black. The latter differed from this because of more spacing out and the presence of logos etc. It was found that the distribution of run lengths for both white and black, for both types of document, was a reasonable match to the distribution expected from examining the code chosen by ITU-T. Table 2 summarises the distribution of *codeword lengths* that resulted when the test pages were encoded.

Codeword	White		Black	
length	Reports	Letters	Reports	Letters
2	—	—	54.8%	48.4%
3	—	—	17.4%	20.1%
4	59.1%	41.8%	11.8%	10.9%
5	13.0%	15.9%	3.0%	3.0%
6	13.3%	14.8%	4.7%	4.8%
7	4.7%	7.9%	5.2%	4.8%
8	8.1%	14.4%	1.3%	2.5%
> 8	1.8%	5.2%	1.8%	5.5%

Table 2. Distribution of Codeword Lengths for Test Data

From the table, it can be seen that the pages taken from the technical report have a shorter expected codeword length for both white and black. At least 85% of codewords have length at most six, and at least 90% of codewords have length at most seven. The same statistics for the pages taken from letters are not so impressive, being 72% and 80% respectively. This reflects the fact that these pages are probably ‘less typical’ (i.e., have a more random structure) than the pages used to derive the Huffman code. It is clear that there can be no single best code for all fax data. However, the codeword length frequencies obtained experimentally here give an indication of what is achievable in a realistic setting.

3 Architecture of Design

3.1 Systems Level

The overall architecture of the decoding system is shown in Figure 2. The idea is that there is a standard unit tree that can detect codewords of size up to a particular length. One or more ‘fixed trees’ are permanently resident in the FPGA, and other ‘swap trees’ are paged into the FPGA from primary memory as necessary, under the control of the processor.

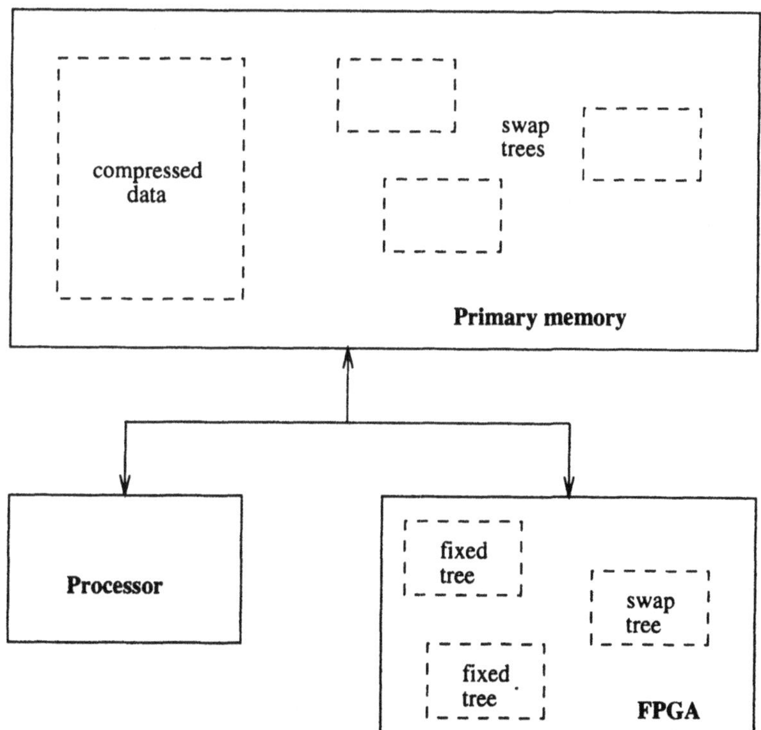

Fig. 2. Decoding System Architecture

When decoding is in progress, one of the resident trees is active. If it completes the detection of a codeword, a success indication is given to the processor. Otherwise, an "incomplete" indication is given to the processor. Then, the processor may either complete the codeword detection by using a table stored in memory, or swap in the next (lower) sub-tree of the decoding tree.

There are two key design decisions at this level. The first concerns the optimum size of a unit tree. The statistics included in the previous section point to possible answers for this question. For example, for textual documents, a six-bit decoding tree would detect approximately 85% of codewords directly. In fact, this tree size was used for the design described here, since it represented a reasonable compromise between having a high chance of successfully decoding within a single unit tree, and having moderate overheads when swapping trees. A further advantage is that the end-of-line code, which is a guaranteed feature of each scan line, is 12 bits long, and so can be detected using exactly two six-level trees.

The second decision concerns the number and arrangement of fixed trees included in a complete system. Two fixed trees — one for white, the other for black — are likely to be present, otherwise swapping is necessary for each run of white or black. In the case of black runs, all except four codewords begin with two zero bits. If the four special cases are detected separately, then a fixed six-bit tree can be used to detect all codewords of length up to eight bits. In the case of white runs, the decoding tree is significantly more balanced, so this approach is not fruitful. However, an alternative possibility is to include a pair of fixed trees, one for each of the two top-level sub-trees of the decoding tree. This idea could be further extended to having four six-bit sub-trees, which would allow detection of all codewords of length up to eight bits.

The final design decisions are driven by the performance specification of the whole system, and the resources available on the FPGA. Here, the prototype design was targeted at the XC6215 part, which offers a 64×64 cell array on one chip. Since a six-bit unit tree would fit into a 32×21 cell array, this allowed four unit trees to be present simultaneously on a single chip, together with any necessary glue logic. Therefore, using the above ideas, three fixed trees for decoding all codewords of length up to seven bits, together with one swap tree for longer lengths, could be comfortably accommodated. With a tighter design, it is likely that five, or even six, unit trees could be accommodated, depending on the amount of extra logic needed for interfacing.

3.2 Logic Level

To accommodate word-size transfers between the processor and the FPGA, a back to back arrangement of two five-bit decoding trees was used to construct a six-bit unit tree. Each five-bit tree is one of the two top-level sub-trees of the unit tree.

The XC6200 family allows direct register access to the FPGA. Every cell may be configured to contain both a function unit and a flip-flop, if required. The output of any cell's function unit can be read, and the flip-flop within any cell can be written. Registers can be configured within each column of the array, with the cells forming an 8-bit, 16-bit or 32-bit register selected using a map register for the column. To exploit this capability in the design here, a 32-bit shiftable I/O register was inserted between the five-bit decoding trees to supply the input bit stream and to collect codeword detection information. This register eliminates the need for wiring to the I/O pins of the array, at no performance penalty. The general layout is illustrated in Figure 3.

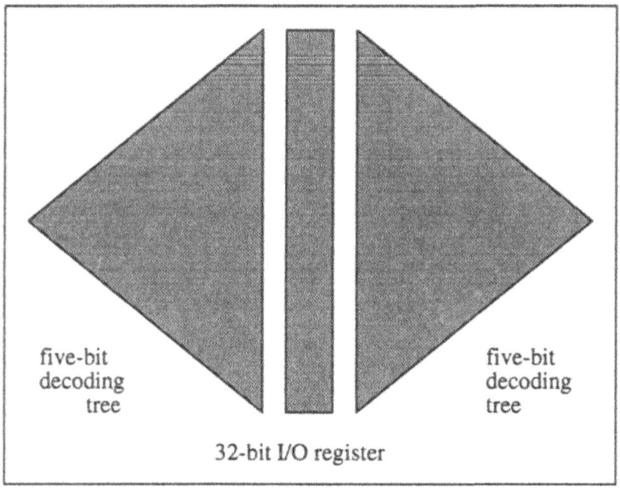

Fig. 3. Six-bit Unit Tree Layout

The input bit stream is loaded, one byte at a time, into 24 bits of the I/O register. The decoding sub-trees are driven from this register, the choice of sub-tree being made by the first bit, and then the steering at each of the five levels of the chosen sub-tree being driven by the next five bits. The decoding sub-trees have a simple B-tree form that interconnects detector cells. A dense layout is possible for an XC6200 series FPGA, given the assistance of the associated design tools, in particular the Camelot mapping, placement and routing software. The minimum size layout for a unit tree occupies a 32×15 cell area; a layout with better performance occupies a 32×21 cell area.

The detector cells represent nodes in the tree, and embody structural information: (a) whether the incoming tree edge is labelled **0** or **1**; and (b) whether the node is a leaf. This information may be 'soft-wired' for minimum area and maximum detection performance (but lower swap performance). Alternatively, the information may be register written, to maximise swap performance, at some space and detection performance overhead. A schematic diagram of a detector cell is shown in Figure 4. When a steering signal reaches a detector cell, and the detector's input bit matches its incoming tree edge label, the cell is activated. If it is a leaf cell, a 'codeword detect' signal is output; detect signals at each level of the tree are collected together using an OR chain. Otherwise, a steering signal is output to the next tree level, if there is one.

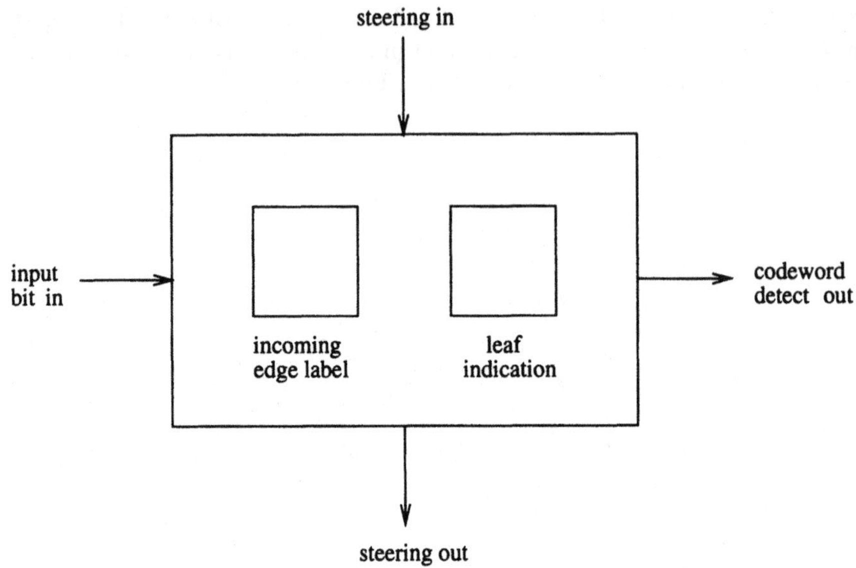

Fig. 4. Detector Cell Schematic

The I/O register contains a six-bit output from the codeword detection circuitry, each bit used to indicate a codeword detection at one of the six levels of the tree. If all six bits are zero, then the detection is incomplete. In this case, the first six bits in the input part of the I/O register indicate the appropriate next stage for the decoding process, whether this is an index of a swap tree or a pointer to a lookup table. When a codeword (or part of a codeword) has been detected and the output read by the processor, the input part of the I/O register is shifted by the appropriate amount using additional circuitry, and is replenished with further data bytes by the processor when required.

The current design does not detect invalid codes. However, this feature could be obtained easily by adding the generation of a 'code error' signal output to existing leaf cells which correspond to the detection of bit strings that are not prefixes of valid codewords.

3.3 Performance Considerations

With this architecture, performance when detecting a codeword of length at most six bits is determined by two factors. First, the number of gate delays before detecting the codeword, which will be in the range one to six for a single unit. Second, the time required to 'carry' a detect signal out of a level of the decoding tree, which will be 1 to either 15 or 31 gate delays, depending on wiring strategy. Therefore, codeword detection times would be in the range 20 to 70 ns for typical gate delays.

For codewords with length greater than six bits, one of 64 sub-trees must be swapped in. For the XC6200 FPGA family, register writes to the FPGA are executed at SRAM speeds. If detection cells are hard-wired to give a minimum space solution, 30 writes are required but, if detection cells are register written, only 10 writes are needed. At first sight, these figures suggest that the paged tree solution might be slower than a straightforward software solution using a lookup table. However, the fact to remember is that the paging scheme has good locality. For example, if 90% of codewords are detected by a fixed tree, then the cost of swapping in a sub-tree can be amortised: 10 writes incurred for a page fault would average out to one write per codeword.

Given the above remarks, a reasonable estimate for the expected detection time for a codeword is 150 ns. Since each codeword represents a run length, the final output rate of the overall decoding system, measured in bits per second, depends on the length. The experimental data, for report pages and letter pages, had average white run lengths of 46 and 121 respectively, and average black run lengths of 4.5 and 5.3 respectively. Thus, expected bit output rates would be 30 Mbps for black, and at least an order of magnitude more for white. These are above video rates, and are substantially better than the rates obtainable from finely-tuned software implementations,

4 Exploitation of Reconfigurability

The discussion in the previous section has shown that there is significant scope to exploit configurability in order to achieve the best performance from the fax decoding system. This can be summarised under three headings:

Logic Level

At the logic level, there is the option of configuring detection cells either by 'soft wiring' or by register writing. The former saves area, and gives improved performance. The latter gives improved swap performance.

Systems Level

At the system level, there are various options for choosing the total number of unit trees that are simultaneously resident, and then for choosing the balance between fixed trees and swap trees. A further consideration is the choice of unit tree size. These options involve trade-offs between area and paging behaviour.

Processor Level

At the overall system level, there is a basic trade-off between using table lookup and using the detection circuitry. At one extreme, there is the conventional software decoding approach; at the other extreme, there is an approach that uses the circuitry exclusively. The middle option is the use of the detection circuitry as a first stage, then the use of lookup tables to avoid logic paging.

The previous section gives timing estimates derived under the assumptions of a 2 ns gate delay, and a register write time of 40 ns per transfer. The latter figure depends on the efficiency of the FPGA control store performance. With the imminent availability of XC6200 parts, it will be possible to test these timings experimentally. This will also allow investigation of the various logic level and systems level trade-offs.

The fundamental question arises from the processor level trade-off. That is, whether logic paging can be more efficient than just handling infrequent cases in software. As the previous section indicates, the cost of handling a page fault can be as little as 10 memory writes. The circuitry can then complete the lookup in the time that would be needed for only one or two memory operations. This total time should be reasonable compared with a software solution, when instruction fetches, bit-twiddling and table lookups are taken into account. However, the full benefit might only be derived if paging did not require the intervention of the processor, by using DMA transfers. Whether or not this is done, some processor time is freed for other, more general purpose, work. Only experimentation, with the circuitry embedded in a realistic overall system will allow verification of these estimates. However, the expectation is that higher continuous rates of throughput will be possible, certainly compared with total software implementations, and also compared with hybrid implementations.

5 Conclusions

The work described in this paper indicates that reconfigurability can be exploited in several ways for an application with suitable properties. In this case, the major properties are that the detection circuitry has a regular (tree-like) structure, and that probabilities can be derived for the use made of different parts of the circuitry. This forms the classic basis for effective paging. Some of the trade-offs possible are exactly those found in the management of virtual memory by operating systems. Other, new, trade-offs arise from the fact that the hardware itself can be changed.

The issues raised in this paper apply in other circumstances. The total approach is clearly applicable to any Huffman decoding circuitry, and so would be of use for applications where very high speed operation is more significant than it is between group 3 fax apparatus. One particularly interesting application is to dynamic Huffman codes, where individual compressed files (or other units of data) carry details of their tailored Huffman code as extra information. Assuming that the tailored code guarantees short codewords with high probability (and the Huffman coding algorithm does guarantee the smallest expected codeword length possible), then the configurable hardware will perform well.

In general, the approach is applicable to any tree-structured circuit with predictable localised behaviour. By allowing customisation of the function in each tree cell, it is possible to envisage a general-purpose paged tree system. Further, the ideas are applicable to other types of planar graph. The most obvious such type of graph is the two-dimensional array, where rectangles form a natural paging unit. Two examples of this style of reconfiguration, used for artificial neural network implementation, can be found in [2] and [5]. An example of gross paging of an entire array, to give a multifunction progammable logic device, can be found in [4].

The interaction between the configurable hardware and the processor is also very important, since it takes place only via registers that are read and written by the processor. No additional wiring is required to interface the hardware, which is a major advantage in simplifying the management of FPGA real estate. The overall scheme brings configurable hardware into the mainstream of computing, and points the way to effective and flexible acceleration of software in the future.

References

1. Cormen, Leiserson and Rivest, *Introduction to Algorithms,* MIT Press 1990, pp.337–344.
2. Eldredge and Hutchings, "RRANN: A HArdware Implementation of the Back-propagation Algorithm Using Reconfigurable FPGAs," Proc. IEEE International Conference on Neural Networks, 1994.
3. International Telecommunication Union, "ITU-T Recommendation T.4: Standardization of Group 3 Facsimile Apparatus for Document Transmission," ITU Geneva 1993.
4. Ling and Amano, "WASMII: a Data Driven Computer on a Virtual Hardware," Proc. IEEE Workshop on FPGAs for Custom Computing Machines, 1993.
5. Lysaght, Stockwood, Law and Girma, "Artificial Neural Network Implementation on a Fine-grained FPGA," Proc. 4th International Workshop on Field-Programmable Logic and Applications, 1994.

Classification and Performance of Reconfigurable Architectures

Steven A. Guccione and Mario J. Gonzalez

Computer Engineering Research Center
Department of Electrical and Computer Engineering
University of Texas at Austin
Austin, Texas 78712

Abstract. Recently, several systems have been designed which use reconfigurable logic to perform general purpose computation. While the number of these systems being constructed continues to increase, their relationship to conventional architectures is not clear. This paper proposes a model which unifies traditional instruction set architectures with reconfigurable architectures. From this model, four major architectural categories of reconfigurable machines are given. From this classification, issues of performance, programmability and scalability are addressed.

1 Introduction

A promising new approach to computing is currently being explored by researchers. This approach uses reconfigurable logic devices to perform computations previously reserved for either traditional instruction set computers or custom hardware.

Starting from a small handful of research projects in the late 1980s, over 40 systems based on reconfigurable logic have been constructed to date [11]. This rate of growth appears to be increasing rapidly.

Machines based on this technology have taken several diverse architectural approaches. It is the goal of this paper to first define the general features of these machines which make them unique and then to place these machines in a framework which permits comparison to other architectures.

From this general framework, four architectural categories of reconfigurable machines are defined and examined in closer detail. Finally, performance issues concerning these machines are examined. Particular attention is paid to performance limitations, rather than peak performance potential of these systems.

2 A Reconfigurable Model of Computing

With the commercial availability of relatively large reconfigurable logic devices and the evolution of computer aided design tools, it has become feasible to build fairly large and powerful systems based on reconfigurable logic. While it is clear that large gains in performance can be achieved with this approach, little has been reported on architectural issues concerning these systems. Unfortunately, these machines appear on the surface to be sufficiently different from existing approaches to computation that direct comparison to traditional architectures is difficult and often confusing.

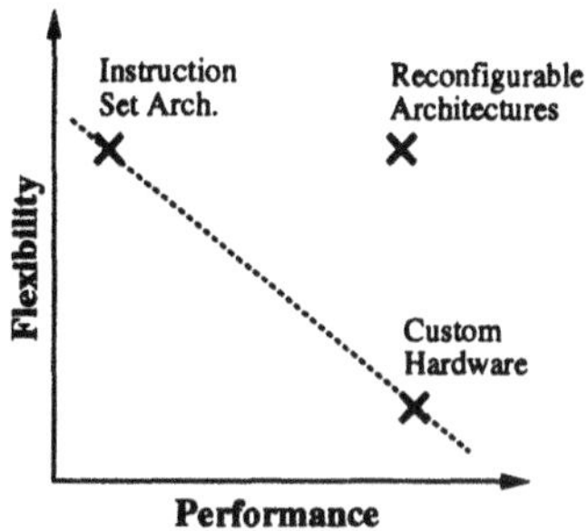

Fig. 1. The traditional tradeoff of flexibility and performance, and the potential of reconfigurable architectures.

Figure 1 gives a general diagram of traditional approaches to computation and their relation to reconfigurable architectures. Until the recent large scale use of reconfigurable logic to perform calculation, there was a generally accepted tradeoff between flexibility and performance in computing systems. In general, the more flexible a machine was, the simpler the programming, but the lower the performance. At one extreme, instruction set architectures can easily implement a wide variety of algorithms, but at only moderate levels of performance. On the other extreme is custom logic, which typically performs a single task very efficiently, but other tasks poorly or not at all. In between are various domain specific architectures which trade performance for flexibility.

The use of reconfigurable logic appears to offer the flexibility of instruction set architectures with the potentially high performance of fully custom hardware. This unique combination has led to difficulties in analyzing reconfigurable machines. Performance comparisons to both custom hardware and to instruction set machines can be found in the literature. While these comparisons are useful for benchmarking, they provide little insight into how performance gains are achieved and what levels of performance can be expected for other algorithms.

2.1 A Hardware Model

On selected algorithms, the performance of reconfigurable machines approaches that of custom logic. This level of performance is typically two to three orders of magnitude greater than that of implementations on instruction set architectures. For this reason, it is tempting to make performance comparisons to a custom hardware reference.

It should, however, be a foregone conclusion that any custom hardware implementation of an algorithm can also be similarly implemented on a suitably large reconfigurable machine. The custom hardware implementation may be used to determine the maximum achievable performance for a given algorithm, but this is only useful in the cases where a comparable custom hardware solution exists.

Viewing reconfigurable systems as a form of custom logic does nothing to aid in predicting performance for algorithms for which no custom hardware reference platform exists. Neither is it clear that comparing a highly programmable machine to a fixed one is appropriate. Despite the similarities in performance, it appears that comparing reconfigurable machines to fixed custom logic implementations of algorithms can only be useful in providing a rough expectation of performance levels.

2.2 A Software Model

The ability to dynamically reconfigure hardware is often seen as the unique feature of machines based on reconfigurable logic. However, for any hardware to be used for more than a single purpose, some level of reconfigurability is necessary.

A traditional instruction set processor may be viewed as a reconfigurable processor. At the heart of the system, the arithmetic and logic unit, or *ALU* can be viewed as a *reconfigurable processing unit*, or *RPU*. A dedicated path is provided to the ALU for rapid reconfiguration. Depending on the data sent to this port, the ALU performs various different logical operations on the inputs. At a higher level, this reconfiguration data is viewed as the operation codes which partially define the behavior of the system. Figure 2 shows and ALU with instruction operation codes being used to reconfigure the ALU.

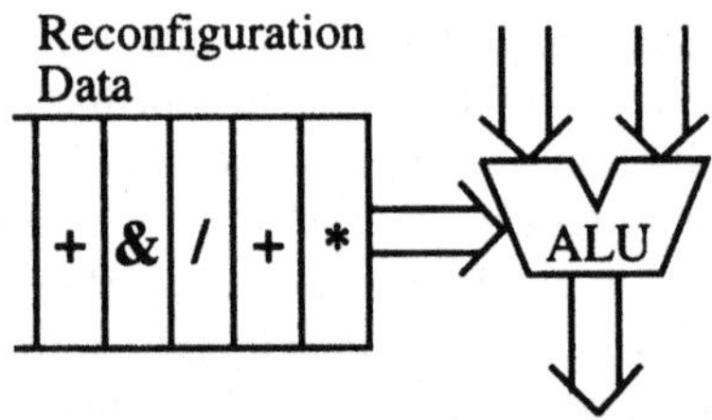

Fig. 2. A reconfigurable model of computing.

From this perspective, the traditional ALU is actually a specific class of RPU. The ALU is characterized by:

- A dedicated port for reconfiguration data
- Few possible configurations
- Rapid but frequent reconfiguration

The number of bits used to configure a typical ALU is less than 10, with 8 being a representative number. This permits at most 2^8 unique configurations. These configurations define the operations available to the machine.

ALU reconfiguration typically takes place on the order of once per clock cycle. While the number of possible functions is limited, sequential combinations of these operations permits a large number of useful functions to be performed. While not particularly efficient for any single task, this approach provides a fairly constant level of performance for a wide variety of algorithms. The primary drawback to this scheme is the bandwidth consumed by constant reconfiguration. Since this bandwidth is limited, operations must be performed in a more or less serial manner. In spite of these limitations, the inherent flexibility of this approach has been extremely successful.

By contrast, RPUs based on reconfigurable logic devices make large scale use of reconfiguration. Instead of the roughly 8 bits used to reconfigure a traditional ALU, thousands of bits are used to reconfigure an RPU. This very large number of bits permits a very large number of possible functions. It is the writing of these bits to the control port of the RPU which is is one of the major limiting factors in the use of larger RPUs.

3 An Architectural Classification

A model has been proposed that considers all general purpose machines to be reconfigurable. What differs is the way in which reconfiguration is managed. As noted, the traditional instruction set architecture opts for frequent reconfiguration with a small number of possible operations. Machines based on reconfigurable logic use a much more flexible RPU, but at the cost of requiring a high reconfiguration overhead. Based on the way in which reconfiguration is managed and utilized to perform computation, these machine can be further classified.

Two relatively independent architectural parameters can be used to subdivide reconfigurable machines into four general categories. These parameters are:

- RPU size
- Dedicated local memory

The first parameter, the RPU size, is the amount of reconfigurable logic used to implement the RPU. This value can be measured more or less by the number of equivalent logic gates in the RPU. This will determine the complexity of the functions which can be implemented by the RPU.

The second parameter is dedicated local memory. This is the memory directly accessible to the RPU. The absence or presence of dedicated memory will effect the system at several levels. Architecturally, dedicated memory implies that the reconfigurable portion of the system may operate independently from the host. From a software perspective, a programming model which supports an independent processor and memory space is indicated. Finally, at the application level, dedicated memory will effect the types of algorithms that can benefit effectively from the use of reconfigurable processing.

Based on these two parameters, reconfigurable machines can be divided into four major categories. These are *Application Specific Architectures (ASA)*, *Reconfigurable Logic Coprocessors (RLC)*, *Custom Instruction Set Architectures (CISA)* and *Reconfigurable Supercomputers (RS)*. Figure 3 shows the four types of reconfigurable architectures and their RPU sizes and presence or absence of dedicated local memory.

	No Local Memory	Local Memory
Small RPU	CISA	RLC
Large RPU	ASA	RS

Fig. 3. An architectural classification of reconfigurable machines.

In this table, a small RPU is defined to be less than 10^5 equivalent gates and a large RPU is assumed to be greater 10^6 equivalent gates. This boundary is somewhat arbitrary and leaves a "grey area" for machines between 10^5–10^6 equivalent gates. Machines which have RPUs whose gate count is somewhere in this region may have features of two classes of machines.

3.1 Application Specific Architectures

The first class of reconfigurable systems are *Application Specific Architectures (ASA)*. These machines were some of the earliest to exploit the advantages of reconfigurable logic. They have no dedicated memory and have relatively large RPUs. These machines are primarily characterized by a very narrow area of application.

One popular use of such application specific architectures is in the acceleration of logic simulation. Here, reconfigurable logic is used to prototype custom hardware. This approach has resulted in dramatic speedups over more traditional software simulations. A good overview of this area can be found in [15].

Another example of an application specific machine is *GANGLION* [6]. This machine was used to implement a fixed size three-layer neural network. This system made use of reconfiguration to provide a dramatic speedup over established software techniques. This hardware was, however, only useful to simulate a single neural network configuration. Modifying the number of neurons in the system was not possible.

While perhaps the earliest large-scale use of reconfigurable logic, these machines function much like custom hardware. While they may take advantage of reconfiguration to accomplish their tasks, they are typically used for a single application. In this sense these machines are more closely related to traditional fixed custom hardware than more general purpose reconfigurable machines.

3.2 Reconfigurable Logic Coprocessors

The second class of machines based on reconfigurable logic are called *Reconfigurable Logic Coprocessors (RLC)*. These machines are relatively small, with only a few thousand equivalent gates in the RPU. They contain dedicated memory directly coupled to the RPU. Since the RPU is relatively small, the memory on these systems is similarly limited. Typically on the order of 1 megabyte or less is provided.

Figure 4 gives a high-level diagram of the RLC approach. Some examples of this approach to reconfigurable computing are the *Algotronix 2x4* [1] [14], the *AnyBoard* system [8], the Xputer [12] and the *BORG* system [5].

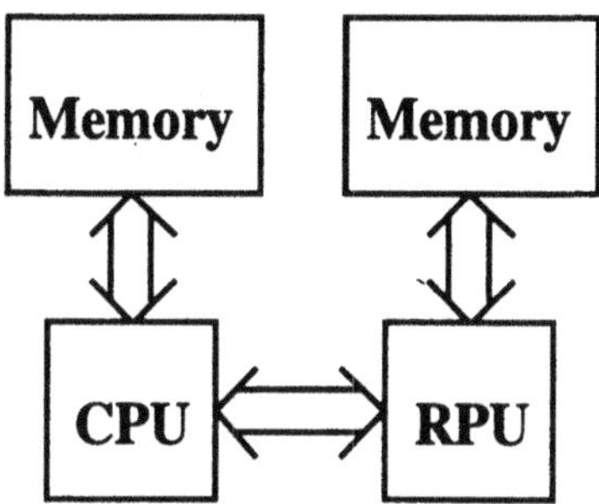

Fig. 4. The coprocessor approach.

Because of the relatively small RPU, these systems are used primarily as small custom logic prototyping systems and are programmed using circuit design tools and methodologies. They may be used effectively to perform tasks of low computational complexity requiring high throughput. Digital signal processing is one fertile area of application for this class of machine.

3.3 Custom Instruction Set Architectures

The third type of reconfigurable system is the *Custom Instruction Set Architectures*, or *CISA*. These machines trace their roots to earlier custom microcode machines. They attempt to increase performance by providing customized instructions typically unavailable in traditional instruction set architectures.

Figure 5 gives a diagram of this approach to reconfigurable computing. These machines differ from reconfigurable logic coprocessors in that they are typically more tightly coupled to the host CPU and have no dedicated memory. Some examples of CISA machines are the *PRISM* systems [2] [17]. the *flexible processor* [18], *Spyder* [13], the *ArMen* machine [16] and the *CM-2X* [7].

CISA machines typically offer a more traditional programming environment than other types of systems. This is primarily because the architecture is based on the instruction set model of computation. This shared programming model

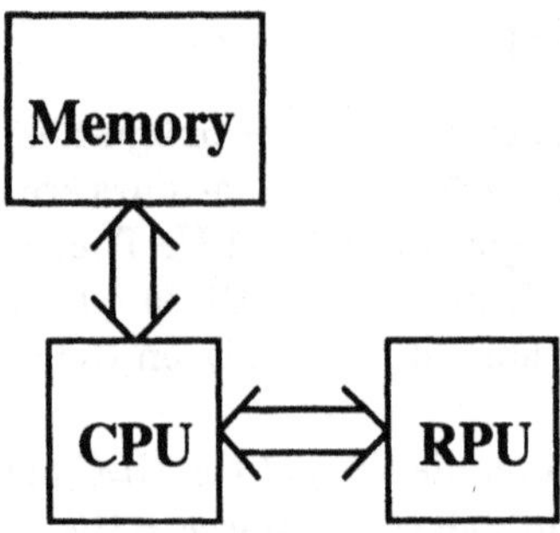

Fig. 5. The CISA approach.

permits the host and RPU to cooperate closely. The function configured into the RPU is viewed by the host as another instruction available to the processor. This permits a simple interface for existing tools and languages. It is likely that these systems will continue to be used for research in high level language programming of reconfigurable machines.

While these machines tend to be easier to program, the use of the instruction set model of computing makes this approach more or less serial. While the RPU can effectively implement complex bit operations not found in traditional architectures, it is not possible to further exploit parallelism within the RPU via pipelining.

Additionally, scaling to larger RPUs permit more complex functions, but the increase in the amount of logic will tend to slow the speed of the RPU. Depending on the coupling to the host processor, this may require a decrease in the system clock speed.

Except for special cases, this approach to reconfigurable computing offers relatively modest gains in performance. It is interesting to note that of the five machines cited above, three are multiprocessor systems. This multiprocessor approach should further boost performance by exploiting data parallelism, but at the cost of replicated hardware.

3.4 Reconfigurable Supercomputers

The final class of reconfigurable machines are *Reconfigurable Supercomputers (RS)*. These machines have large RPUs, on the order of one million equivalent gates. They typically have large amounts of dedicated memory and a high bandwidth link to a powerful host processor.

Architecturally, reconfigurable supercomputers resemble reconfigurable logic coprocessors. The difference is primarily one of scale. Reconfigurable supercomputers are several times larger, both in RPU and memory size, than reconfigurable logic computers. This permits these machines to perform larger and more complex algorithms.

Some examples of reconfigurable supercomputers are the *PAM* systems [3], the *Splash* systems [9] [10] and the *Virtual Computer* [4]. All of these systems tend to be on the low end of the scale, with none having an RPU with one million equivalent logic gates. These systems are perhaps better referred to as reconfigurable mini-supercomputers. They are, however, distinguished by their large memory and I/O bandwidth, as well as their fairly powerful host machines.

Like the smaller reconfigurable logic coprocessors, these systems currently tend to be programmed using hardware design tools and methodologies. Because of their larger size, however, they can be used to implement larger and more complex algorithms, often involving more general arithmetic operations. Several applications have been implemented on these machines achieving speeds surpassing that of large vector supercomputers.

4 Performance Issues

Based on the reconfigurable model of computation, all reconfigurable machines can be viewed as coprocessors which implement custom instructions. A portion of the increase in performance comes from replacing a sequence of instructions normally executed on the host with a single complex instruction implemented in the RPU.

If N cycles of processing on a traditional machine are replaced by reconfigurable logic and R cycles are used to configure the hardware, the reconfigurable logic must be used at least M times to amortize the overhead of reconfiguration. Assuming a single cycle operation for the reconfigurable logic processor, M is given by Equation 1. This represents the *break even* point where reconfigurable logic may be profitably used.

$$M = \frac{R}{N-1} \qquad (1)$$

In order to make effective use of the reconfigurable logic, M operations must be performed to cover the cost of reconfiguration. More operation using the reconfigurable logic will server to save $(N-1)$ cycles in the overall execution of the algorithm.

This simple analysis indicates that in order to minimize the break even point, reconfiguration should be rapid and the number of instructions replaced large. Unfortunately, these parameters are not independent. Typically, large numbers of instructions will require large amounts of logic to implement, this will, in turn, require longer reconfiguration. While the exact value of M will vary, it should be closely tied to the type of reconfigurable device used to implement the hardware.

This implies that more complex functions will require more repetitive operations to be competitive.

5 Conclusions

A model of computing which bridges the existing gap between traditional instruction set architectures and evolving reconfigurable architectures has been described. Rather than attempting to define reconfigurable computers in terms of instruction set machines, instruction set machines have been demonstrated to be a special type of reconfigurable machine.

Reconfigurable machines can be grouped into four categories based on the size of their reconfigurable logic units and the presence or absence of dedicated memory. Ignoring application specific architectures, machines with no dedicated memory may be more easily programmed using the traditional instruction set model of computation. This model, while simplifying programming, is inherently serial and will limit performance.

Systems with dedicated memory promise higher performance, but require a more complex programming model. This model must manage control and communication between the host and the RPU as well as recognize and make effective use of the dedicated memory.

Finally, reconfigurable supercomputers with over one million equivalent logic gates are on the horizon. Indications are that these machines will be competitive with existing parallel and vector supercomputers, with orders of magnitude less hardware. The ability to configure and interconnect several complex arithmetic and logic blocks is likely to make the use of high level languages a practicality, if not a necessity.

References

1. Algotronix, Ltd. *CAL1024 Datasheet*, 1990.
2. Peter M. Athanas and Harvey F. Silverman. Processor reconfiguration through instruction-set metamorphosis. *IEEE Computer*, 26(3):11–18, March 1993.
3. Patrice Bertin, Didier Roncin, and Jean Vuillemin. Introduction to programmable active memories. Technical Report 3, DEC Paris Research Laboratory, 1989.
4. Steven Casselman. Virtual computing and the virtual computer. In Duncan A. Buell and Kenneth L. Pocek, editors, *IEEE Workshop on FPGAs for Custom Computing Machines*, pages 43–48, Los Alamitos, CA, April 1993. IEEE Computer Society Press.
5. Pak K. Chan, Martine D. F. Schlag, and Marcelo Martin. BORG: A reconfigurable prototyping board using field-programmable gate arrays. In *First International ACM/SIGDA Workshop on Field Programmable Gate Arrays*, pages 47–51, 1992.
6. Charles E. Cox and W. Ekkehard Blanz. GANGLION – a fast field-programmable gate array implementation of a connectionist classifier. *IEEE Journal of Solid-State Circuits*, 27(3):288–299, March 1992.
7. Steven A. Cuccaro and Craig F. Reese. The CM-2X: A hybrid CM-2 / xilinx prototype. In Duncan A. Buell and Kenneth L. Pocek, editors, *IEEE Workshop on FPGAs for Custom Computing Machines*, pages 121–130, Los Alamitos, CA, April 1993. IEEE Computer Society Press.
8. David E. Van den Bout. The anyboard: Programming and enhancements. In Duncan A. Buell and Kenneth L. Pocek, editors, *IEEE Workshop on FPGAs for Custom Computing Machines*, pages 68–77, Los Alamitos, CA, April 1993. IEEE Computer Society Press.
9. Maya Gokhale, William Holmes, Andrew Kosper, Dick Kunze, Dan Lopresti, Sara Lucas, Ronald Minnich, and Peter Olsen. SPLASH: A reconfigurable linear logic array. In *International Conference on Parallel Processing*, pages I–526–I–532, 1990.

10. Maya Gokhale, William Holmes, Andrew Kosper, Sara Lucas, Ronald Minnich, and Douglas Sweely. Building and using a highly parallel programmable logic array. *IEEE Computer*, pages 81–89, January 1991.
11. Steven A. Guccione. List of FPGA-based computing machines. World Wide Web page http://www.utexas.edu/~ guccione/HW_list.html, 1994.
12. Reiner W. Hartenstein, Alexander G. Hirschbiel, Michael Reidmüller, Karin Schmidt, and Michael Weber. A novel ASIC design approach based on a new machine paradigm. *IEEE Journal of Solid-State Circuits*, 26(7):975–989, July 1991.
13. Christian Iseli and Edwardo Sanchez. Spyder: A reconfigurable VLIW processor using FPGAs. In Duncan A. Buell and Kenneth L. Pocek, editors, *IEEE Workshop on FPGAs for Custom Computing Machines*, pages 17–24, Los Alamitos, CA, April 1993. IEEE Computer Society Press.
14. Thomas Andrew Kean. *Configurable Logic: A Dynamically Programmable Cellular Architecture and its VLSI Implementation.* PhD thesis, University of Edinburgh, Department of Computer Science, January 1989.
15. Henry L. Owen, Ubaid R. Khan, and Joseph L. A. Hughes. FPGA-based emulator architectures. In Will Moore and Wayne Luk, editors, *More FPGAs*, pages 398–409. Abingdon EE&CS Books, Abingdon, England, 1993.
16. F. Raimbault, D. Lavenier, S. Rubini, and B. Pottier. Fine grain parallelism on a MIMD machine using FPGAs. In Duncan A. Buell and Kenneth L. Pocek, editors, *IEEE Workshop on FPGAs for Custom Computing Machines*, pages 2–8, Los Alamitos, CA, April 1993. IEEE Computer Society Press.
17. M. Wazlowski, L. Agarwal, T. Lee, A. Smith, E. Lam, P. Athanas, H. Silverman, and S. Ghosh. PRISM-II compiler and architecture. In Duncan A. Buell and Kenneth L. Pocek, editors, *IEEE Workshop on FPGAs for Custom Computing Machines*, pages 9–16, Los Alamitos, CA, April 1993. IEEE Computer Society Press.
18. Andrew Wolfe and John P. Shen. Flexible processors: A promising application-specific processor design approach. In *Proceedings of the 21st Annual Workshop on Microprogramming and Microarchitecture*, pages 30–39. IEEE Press, 1988.

Author Index

Lecture Notes in Computer Science

For information about Vols. 1–903

please contact your bookseller or Springer-Verlag

Vol. 904: P. Vitányi (Ed.), Computational Learning Theory. EuroCOLT'95. Proceedings, 1995. XVII, 415 pages. 1995. (Subseries LNAI).

Vol. 905: N. Ayache (Ed.), Computer Vision, Virtual Reality and Robotics in Medicine. Proceedings, 1995. XIV,

Vol. 906: E. Astesiano, G. Reggio, A. Tarlecki (Eds.), Recent Trends in Data Type Specification. Proceedings, 1995. VIII, 523 pages. 1995.

Vol. 907: T. Ito, A. Yonezawa (Eds.), Theory and Practice of Parallel Programming. Proceedings, 1995. VIII, 485 pages. 1995.

Vol. 908: J. R. Rao Extensions of the UNITY Methodology: Compositionality, Fairness and Probability in Parallelism. XI, 178 pages. 1995.

Vol. 909: H. Comon, J.-P. Jouannaud (Eds.), Term Rewriting. Proceedings, 1993. VIII, 221 pages. 1995.

Vol. 910: A. Podelski (Ed.), Constraint Programming: Basics and Trends. Proceedings, 1995. XI, 315 pages. 1995.

Vol. 911: R. Baeza-Yates, E. Goles, P. V. Poblete (Eds.), LATIN '95: Theoretical Informatics. Proceedings, 1995. IX, 525 pages. 1995.

Vol. 912: N. Lavrac, S. Wrobel (Eds.), Machine Learning: ECML – 95. Proceedings, 1995. XI, 370 pages. 1995. (Subseries LNAI).

Vol. 913: W. Schäfer (Ed.), Software Process Technology. Proceedings, 1995. IX, 261 pages. 1995.

Vol. 914: J. Hsiang (Ed.), Rewriting Techniques and Applications. Proceedings, 1995. XII, 473 pages. 1995.

Vol. 915: P. D. Mosses, M. Nielsen, M. I. Schwartzbach (Eds.), TAPSOFT '95: Theory and Practice of Software Development. Proceedings, 1995. XV, 810 pages. 1995.

Vol. 916: N. R. Adam, B. K. Bhargava, Y. Yesha (Eds.), Digital Libraries. Proceedings, 1994. XIII, 321 pages. 1995.

Vol. 917: J. Pieprzyk, R. Safavi-Naini (Eds.), Advances in Cryptology - ASIACRYPT '94. Proceedings, 1994. XII, 431 pages. 1995.

Vol. 918: P. Baumgartner, R. Hähnle, J. Posegga (Eds.), Theorem Proving with Analytic Tableaux and Related Methods. Proceedings, 1995. X, 352 pages. 1995. (Subseries LNAI).

Vol. 919: B. Hertzberger, G. Serazzi (Eds.), High-Performance Computing and Networking. Proceedings, 1995. XXIV, 957 pages. 1995.

Vol. 920: E. Balas, J. Clausen (Eds.), Integer Programming and Combinatorial Optimization. Proceedings, 1995. IX, 436 pages. 1995.

Vol. 921: L. C. Guillou, J.-J. Quisquater (Eds.), Advances in Cryptology – EUROCRYPT '95. Proceedings, 1995. XIV, 417 pages. 1995.

Vol. 922: H. Dörr, Efficient Graph Rewriting and Its Implementation. IX, 266 pages. 1995.

Vol. 923: M. Meyer (Ed.), Constraint Processing. IV, 289 pages. 1995.

Vol. 924: P. Ciancarini, O. Nierstrasz, A. Yonezawa (Eds.), Object-Based Models and Languages for Concurrent Systems. Proceedings, 1994. VII, 193 pages. 1995.

Vol. 925: J. Jeuring, E. Meijer (Eds.), Advanced Functional Programming. Proceedings, 1995. VII, 331 pages. 1995.

Vol. 926: P. Nesi (Ed.), Objective Software Quality. Proceedings, 1995. VIII, 249 pages. 1995.

Vol. 927: J. Dix, L. Moniz Pereira, T. C. Przymusinski (Eds.), Non-Monotonic Extensions of Logic Programming. Proceedings, 1994. IX, 229 pages. 1995. (Subseries LNAI).

Vol. 928: V.W. Marek, A. Nerode, M. Truszczynski (Eds.), Logic Programming and Nonmonotonic Reasoning. Proceedings, 1995. VIII, 417 pages. 1995. (Subseries LNAI).

Vol. 929: F. Morán, A. Moreno, J.J. Merelo, P. Chacón (Eds.), Advances in Artificial Life. Proceedings, 1995. XIII, 960 pages. 1995 (Subseries LNAI).

Vol. 930: J. Mira, F. Sandoval (Eds.), From Natural to Artificial Neural Computation. Proceedings, 1995. XVIII, 1150 pages. 1995.

Vol. 931: P.J. Braspenning, F. Thuijsman, A.J.M.M. Weijters (Eds.), Artificial Neural Networks. IX, 295 pages. 1995.

Vol. 932: J. Iivari, K. Lyytinen, M. Rossi (Eds.), Advanced Information Systems Engineering. Proceedings, 1995. XI, 388 pages. 1995.

Vol. 933: L. Pacholski, J. Tiuryn (Eds.), Computer Science Logic. Proceedings, 1994. IX, 543 pages. 1995.

Vol. 934: P. Barahona, M. Stefanelli, J. Wyatt (Eds.), Artificial Intelligence in Medicine. Proceedings, 1995. XI, 449 pages. 1995. (Subseries LNAI).

Vol. 935: G. De Michelis, M. Diaz (Eds.), Application and Theory of Petri Nets 1995. Proceedings, 1995. VIII, 511 pages. 1995.

Vol. 936: V.S. Alagar, M. Nivat (Eds.), Algebraic Methodology and Software Technology. Proceedings, 1995. XIV, 591 pages. 1995.

Vol. 937: Z. Galil, E. Ukkonen (Eds.), Combinatorial Pattern Matching. Proceedings, 1995. VIII, 409 pages. 1995.